AIR POLLUTION BY PHOTOCHEMICAL OXIDANTS

AIR QUALITY MONOGRAPHS

Vol. 1 Air Pollution by Photochemical Oxidants (I. Colbeck and A.R. MacKenzie)

AIR QUALITY MONOGRAPHS·VOL.1

AIR POLLUTION BY PHOTOCHEMICAL OXIDANTS

I. COLBECK

Institute for Environmental Research,
Department of Chemistry and Biological Chemistry,
University of Essex, Colchester, U.K.

A.R. MACKENZIE

Centre for Atmospheric Science,
Department of Chemistry, University of Cambridge,
Cambridge, U.K.

ELSEVIER
Amsterdam - London - New York - Tokyo 1994

ELSEVIER SCIENCE B.V.
Sara Burgerhartstraat 25
P.O. Box 211, 1000 AE Amsterdam, The Netherlands

ISBN: 0-444-88542-0

This book is printed on acid-free paper.

Printed in The Netherlands.

PREFACE[1]

It has become customary to perceive of Science in terms of *the limit*: the frontier, the cutting edge, the receding boundary. The images are of a well-disciplined army on the march, of an excelsior moving ever onward and upward. We all recall being told as wide-eyed undergraduates that PhD research entails the discovery of something utterly new, something never before understood. And, of course, being unique, each of us does indeed produce something utterly new. This kind of uniqueness is not necessarily what we thought we were being told about, however.

In many ways, the idea of *the liminal*, the threshold, fits atmospheric science much better than the idea of the limit. "Pure", laboratory, science, having set out the rules, the atmospheric scientists set out to see how well they perform in practice - interfacing with one another, and with the data from a properly open system. Much of the work of atmospheric scientists involves the subtle application of well-tried science to the particular case of the atmosphere. The need for new laboratory science may become apparent in the course of this application, but that only serves to emphasise atmospheric science's place as an applied science *par excellence*. Dr. Johnson appreciated the uncertainty attached to thresholds and acknowledged the fact by jumping over, rather than stepping upon, any he came across. Perhaps atmospheric scientists have more curiosity, or more foolhardiness, than the good doctor. At any rate, aspects of physics, chemistry, geology, geography, botany and statistics are all touched upon in the following text. The result, if we have done our job well enough, is more than the sum of the parts.

This emphasis on the applied-ness of atmospheric science is not to suggest that the work of the atmospheric scientist is in any way inferior to that of her/his colleague in laboratory science. Much of the work reported below is the result of profound insight into the bases of the sciences, assiduously applied to the atmosphere which, as Reynolds noted in the 1930s, society tends to take for granted all too often. We would like to pay tribute here to the scientists whose work we discuss and who have in many cases contributed much more to the understanding of photochemical oxidants than we can claim to have ourselves. The contribution below is an attempt to explore some of the thresholds which must be crossed if the impact of human activity on the behaviour of photochemical oxidants is to be fully understood. During this exploration we may occasionally appear to wander far from

---

[1] This preface is a pastiche of the analysis of Foucaldian political philosophy given by I.M. MacKenzie (PhD thesis, Glasgow University, 1994).

the subject at hand, but we would hope that the reader shares with us the opinion that, in a full-length exposition, it is only right to entertain a catholic definition of what is pertinent to the topic.

Our approach attempts to balance data and theory. We open with a brief historical note, setting the goals of the current research effort against those of a century ago. We then give some details of the physical and chemical properties of the compounds in which we are interested, and go on to describe their sources in the atmosphere. Descriptions follow, of the atmospheric chemistry, and of the measurement techniques, relevant to the study of photochemical oxidants. These descriptions bring to a close what is really, in the manner of Sir Walter Scott, an extended preamble to the matter at hand. Chapter 6 is that matter: concentrations of photochemical oxidants measured over several spatial and temporal scales. Chapter 7 outlines the case for concern over air pollution by photochemical oxidants, and the final chapter outlines some of the strategies which are being adopted in an effort to reduce acute pollution episodes.

We have adopted one or two strategies ourselves, to make our task a more tractable one. Most significantly, when dealing with data, we have adopted a "Mercator's Projection": almost like the map, our perspective is, by stages, Anglocentric, Eurocentric and then global. This is not meant to imply any political or scientific elitism, but is rather a framework for the reduction of the huge body of literature which now exists. We have not attempted a complete literature survey, but have sought to include a reasonably representative portion of the literature in the areas that we discuss. The literature reviewed here is reasonably representative of that published up to 1990: the more recent sources which are quoted reflect our personal prejudices and sources of information.

A note or two on terminology and style is in order. Throughout the book, VOC and NMHC are used, more or less interchangeably, to mean that fraction of the organic compounds emitted into the atmosphere which play a part in the production of secondary air pollutants such as ozone. We are aware that more precise definitions exist for these abbreviations, but we don't think any ambiguities will arise from our loosening of the lexicographical reins. Volume mixing ratios (ppbv = 1×10^{-9} v/v, and pptv = 1×10^{-12} v/v) are used throughout, except: in the very few instances where conversion from other units was not possible; when considering rates of chemical change; and when discussing the abundances of radicals. References are given as superscripts except where some confusion may arise, such as near chemical formulae and units of measure, and when the reference is used as an example of a larger body of literature. Chemical equations and mathematical equations have been numbered separately in each chapter.

Many people have helped us prepare this volume, and our heartfelt thanks is extended to them all. Linda Appleby, Marie Chan and Kate Fitzgerald helped in the arduous task of preparing camera-ready copy. Paul Brown, John Stringer and Deb Fish provided some of the figures. Oliver Wild calculated the photolysis rates in chapter 4. Paul Brown, Deb Fish, Rod Jones, Kathy Law, Marion Martin, Gary Mongan, Dave Nedwell, John Pyle, Claire Stirling, Bill Sturges, Ralf Toumi and Oliver Wild read various parts of the manuscript in draft, each providing comments which have much improved on our initial efforts. The responsibility for any errors or inaccuracies remaining is entirely our own, of course.

Ian Colbeck
Robert MacKenzie

December 1993

Acknowledgements

We would like to thank the authors and publishers who have given permission for the following illustrations to be used:

Figure 3.2 reproduced from *The Bacteria*, C.R. Woese & R.S. Wolfe (ed), 1985. With permission from Academic Press, New York.

Figure 3.4 reproduced from *Biology of Anaerobic Microorganisms*, A.J.B. Zehnder (ed), 1988; Figure 3.29 reproduced from *Fossil Fuel Combustion*, W. Bartok & A.F. Sorofim (eds), 1991. With permission from John Wiley and Sons, Chichester.

Figure 3.6 reprinted from *Journal of Geophysical Research* (1989,94,16405);
Figure 3.7 *op cit* (1987,92,2181); Figure 3.11 *op cit* (1991,96,1033); Figure 3.18 *op cit* (1988,93,3751); Figure 4.5 *op cit* (1993,98,18377); Figure 4.7 *op cit* (1989,94,13013); Figure 4.9 *op cit* (1990,95,22319); Figure 5.9 *op cit* (1990,95,10147); Figure 5.12 *op cit* (1985,90,12819); Figure 6.2 *op cit* (1986,91,5229). All copyright by the American Geophysical Union.

Figure 3.9 reproduced from *Mineral Nutrients in Tropical Forest and Savanna Ecosystems*, J. Proctor (ed), 1989. With permission from Blackwell Scientific, Oxford.

Figure 3.10 reprinted from *Soil Biology and Biochemistry* (1982,14,393). With kind permission from Elsevier Science Ltd, The Boulevard, Langford Lane, Kidlington, OX5 1GB, UK.

Figures 3.15 & 3.17 reprinted from *Forest Science* (1985,31,132). ©Society of American Foresters.

Figure 3.16 reprinted from *Plant Physiology* (1989,90,267). ©American Society of Plant Physiologists.

Figures 3.19 reprinted from *Atmospheric Environment* (1991,25A,251);
Figure 3.31 *op cit* (1990,24A,2105); Figure 3.35 *op cit* (1990,24A,43); Figures 5.3 & 5.4 *op cit* (1988,22,1335); Figure 5.5 *op cit* (1988,22,225); Figure 5.11 *op cit* (1990,24A,2839); Figure 6.7 *op cit* (1991,25A,1577); Figure 6.16 *op cit* (1989,23,2003); Figure 6.24 *op cit* (1992,26A,1609); Figure 6.32 *op cit* (1990,24A,2369); Figure 6.33 *op cit* (1991,25A,2039). All with kind permission from Pergamon Press Ltd, Headington Hill Hall, Oxford, OX3 0BW, UK.

Figure 3.20 reprinted with permission from *Nature*(1990,346,552);
Figure 3.22 *op cit* (1991,351,135); Figure 6.17 *op cit* (1986,319,655). All copyright Macmillan Magazines Limited.

Figure 3.21 reprinted from *Science* (1990,250,1669);
Figure 4.10 *op cit* (1989,243,745). All copyright by the American Association for the Advancement of Science.

Figure 3.24 reprinted from *Berichte der Bunsengesellschaft fuer Physikalische Chemie* (1983,87,1008). With permission from VCH Publishers.

Figures 3.25; 3.26; 3.27; 3.30 & Table 3.10 reproduced from *Progress in Energy and Combustion Science*, N. Chigier (ed), 1979. With permission from Pergamon Press Ltd, Headington Hill Hall, Oxford, OX3 0BW, UK.

Figure 3.28 reprinted from *Combustion Science and Technology* (1973,7,77). With permission from Gordon and Breach Science Publishers.

Figure 3.34 reprinted with permission from SAE Technical Paper 880313, ©1988 Society of Automotive Engineers, Inc.

Figure 3.36 reprinted from *Journal of the Air Pollution Control Association* (1990,40,747);
Figure 5.19 *op cit* (1987,37,244), Air & Waste Management Association.

Figure 3.37 reprinted from *Science of the Total Environment* (1993,134,263); Crown copyright, reproduced courtesy of Warren Spring Laboratory.

Figures 4.2 & 6.25 reproduced from *Tropospheric Ozone*, I.S.A. Isaken (ed), 1988. With permission from Kluwer Academic Publishers.

Figure 4.3 reprinted from *Chemical Physics Letters* (1990,167,205). Copyright permission from Elsevier Science Publishers.

Figure 4.4 reprinted from *Annual Reviews in Physical Chemistry* (1988,39,367). With permission from Annual Reviews Inc.

Figure 4.6 reprinted from *International Journal of Chemical Kinetics* (1989,21,593). Reprinted by permission of John Wiley and Sons, Inc.

Figure 5.6 reprinted from *Review of Scientific Instruments* (1992,63,1867). ©American Institute of Physics.

Figure 5.10 reprinted from *Environmental Science and Technology* (1986,20,1269). ©American Chemical Society.

Figures 5.13 & 5.14 reprinted with permission from *Tellus* (1988,40B,437).

Figure 5.15 reprinted with permission from *Journal of Atmospheric Chemistry* (1986,4,413);
Figure 5.16 *op cit* (1990,11,271); Figure 6.3 *op cit* (1988,7,73). Copyright permission from Kluwer Academic Publishers.

Figure 5.18 reprinted from *Analytical Chemistry* (1983,55,937);
Figure 5.24 *op cit* (1990,62,1055); Figure 5.26 *op cit* (1988,60,1074). All copyright by the American Chemical Society.

Figure 5.20 reprinted from *Journal of Chromatography* (1988,442,157). Copyright permission from Elsevier Science Publishers.

Figure 5.21 reprinted from *International Journal Environmental Analytical Chemistry* (1990,38,143);
Figure 5.22 *op cit* (1986,27,97); Figure 5.23 *op cit* (1983,13,129). With permission from Gordon and Breach Science Publishers.

Figure 5.25 reprinted from *Aerosol Science and Technology* (1990,12,64). Copyright permission from Elsevier Science Publishers.

Figure 6.4 reproduced from *Pollution: Causes, Effects and Control*, R.M. Harrison (ed), 1989. With permission from Royal Society of Chemistry.

Figures 6.9 & 6.11 reprinted from *Science of the Total Environment* (1990,93,245);
Figure 6.10 *op cit* (1986,59,181). Copyright permission from Elsevier Science Publishers

Figures 6.13 & 6.15 reproduced from LR941; Figures 6.22 & 6.23 reproduced from LR793. All courtesy of Warren Spring Laboratory. HMSO, London, Crown copyright

Figure 6.21 reproduced from *Advisory Group on the Medical Aspects of Air Pollution Episodes, 1st report, Ozone*. Crown copyright 1991. Reproduced with the permission of the Controller of Her Majesty's Stationery Office.

Contents

CHAPTER 1
OZONE IN THE NINETEENTH CENTURY

"To the Philosopher, the Physician, the Meteorologist, and the Chemist, there is perhaps no subject more attractive than that of Ozone" so wrote Fox[1] in his book on Ozone and Antozone in 1873. Andrews[2] in 1856 wrote *"Among many interesting bodies which the researchers of modern chemists have brought to light, few are more remarkable than the substance to which the name of ozone has been given"*. Alter the slightly outmoded grammar and both of these statements could have been made today.

The problem of atmospheric ozone has been studied since Schönbein[3] suggested in 1840 the existence of an atmospheric constituent having a particular odour. In a letter to Arago, he wrote *"Since I am fairly sure that the source of the odour must belong to the same category of substances as chlorine and bromine, that is the elementary halogens, I propose to call it ozone. ... I am convinced that this substance is always freed in noticeable amounts during thunderstorms"*. (translation).

Schönbein's opinion that ozone was a definite chemical substance was based on his finding that the oxygen produced in the electrolysis of water has a smell identical to the 'electrical odour'. Conclusive evidence that ozone was a form of oxygen was provided by de la Rive[4] and Marignac[5] in 1845. Ozone was chemically proven to exist in the troposphere, at ground level, by Houzeau[6] and the first clear spectroscopic detection, related to the atmosphere was made by Chappuis[7]. The following year, 1881, Hartley[8] detected, in the laboratory, the strong ultraviolet spectrum of ozone below 300 nm. As a result of Houzeau's conclusions, Hartley was able to propose that (1) ozone in the earth's atmosphere is sufficient to account for the limitation of the solar spectrum in the ultraviolet and (2) the concentration of ozone is greater in the upper atmosphere than near the ground. He erroneously concluded that the blue colour of the sky was due to ozone.

Following Schönbein's initial discovery, "ozone fever" spread rapidly in scientific circles. Schönbein gave instructions for making ozone test papers from a mixture of starch and potassium-iodide. Ozone in the air liberated iodine from the iodide and turned the starch blue. The papers were usually exposed for 12 hours and then compared with a scale of blue tints numbers from 0 to 10. Numerous other techniques were also suggested[1]. From the mid-1850s, more than 300 stations, commenced measurements to determine atmospheric ozone concentrations. However, only a few were observed continuously for longer than a few years. Unfortunately, most measurements were made using Schönbein test paper and

1

it has since been shown that such measurements give only semi-quantitative information due to poor standardization and the influence of humidity and wind speed on sensitivity[9]. In fact, doubts over the method had been raised by Fox[1] in 1873.

On the subject of fevers, medical doctors soon recognized the beneficial germicidal property of ozone and were recommending ozone rich rooms over the customary dark sickrooms of the day. The possible influence of ozone on health, for better or worse, was clearly one of the chief causes of the ozone fever[10].

Many of these pre 1900 measurements are now being re-evaluated to provide information on pre-industrial or background levels of air pollution. In an attempt to convert Schönbein data into ozone concentrations Linvill et al.[11] calibrated samples of test paper in a chamber with variable humidity where the ozone concentration was monitored by a uv analyzer. They established that the influence of humidity on coloration is nonlinear and at humidities of below 75% any correction must be applied with extreme caution. If one wishes to establish a correction for the influence of the relative humidity on Schönbein data using the Linvill et al. chart, it is very important not to use an average daily humidity. This results from maximum coloration of the test paper being reached during the first 1-2 hours of exposure[12]. Linvill et al examined ozone at several sites in Michigan between 1876 and 1880 and concluded that day-to-day ozone levels exhibited similar patterns to those of the day. Ozone levels peaked during the warm summer months and fell during the winter. However, it has been postulated[12] that the conversion chart of Linvill et al. overestimates the ozone concentrations. For a much larger data set in Michigan, Bojkov found that nineteenth century surface ozone partial pressures exhibited a pronounced annual cycle, a maximum in April-May, a minimum in October-December and an annual amplitude virtually identical to current midlatitude background sites.

A qualitative method for the determination of ozone was used from 1876 to 1910 at the Observatoire de Montsouris, located on the outskirts of Paris. The method utilized the iodide/iodine catalyzed oxidation of arsenite (AsO_3^{3-}) to arsenate (AsO_4^{3-}). The amount of arsenite transformed was established by titration with an iodine solution allowing the quantity of ozone to be determined. The method was known to be sensitive to other strong oxidants such as hydrogen peroxide, and subject to negative interference by reducing gases[13,14]. Levy[15] was aware that many places around the world were making observations with varieties of the Schönbein test paper. Hence he conducted simultaneous experiments at Montsouris using the test papers and the new quantitative method. He noted that data from both methods were "very variable and to establish a simple correction was difficult".

Bojkov[12] re-examined the results and established the following relation between surface ozone concentrations and the Schönbein ten-number scale: O_3 (ppbv) = 3.4 + 4.6 $N_{Schönbein}$. This includes a correction which makes the regression valid for 78% average relative humidity. Due to the humidity influence the above relationship should only be used for converting average monthly or average annual cycles only.

The conversion methods of Linvill et al.[11] and Bojkov[12] have recently been employed by various workers to analyze nineteenth century Schönbein data in Athens[16], Vienna[17], Zagreb[18], Moncalieri, northern Italy[19], Montevideo, Uruguay[20] and Cordoba, Argentina[20].

The Montsouris data are generally accepted to be more precise and reliable than that obtained from the Schönbein test papers. In fact Volz and Kley[21] rebuilt the Montsouris apparatus and compared its performance with a modern uv photometer. They found that the Montsouris method is accurate to within 3%. They also found that SO_2 interferences due to the vicinity of Paris rendered the data low by an average of 2-3 ppbv. However the Montsouris data were sufficiently detailed that allowances could be made for this. The results suggest that in clean air from a southwesterly direction, the yearly-average ozone concentration was approximately, 10 ppbv in the period 1876-86. Comparison with measurements at Arkona, Germany since 1956[22] indicate that the annual average ozone

Figure 1.1

Surface ozone at Montsouris, France 1876-1905 compared to Arkona, Germany 1956-1984 (After Volz and Kley[21]).

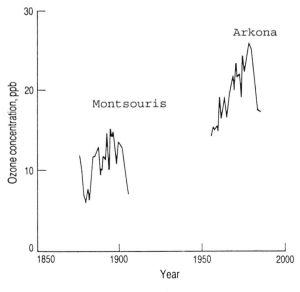

concentration has increased by approximately 100% since the turn of the century (Figure 1.1). These findings will be discussed in more detail in Chapter 6. One interesting point is that the seasonal variation observed in the last century is completely different from that observed now. A small spring maximum observed between February and May has been replaced by a much stronger maximum which peaks in May and extends through the summer. Large summer values are now associated with photochemical pollution.

It is fitting to end this section with two more quotes. Reynolds[23] in 1930 concludes *"If we paid for our air, like our water and milk, we should long ago have submitted it to routine tests; we get it and its impurities for nothing and we almost ignore its existence"*. Fox[1] lists seven subjects for investigation:

1. What are all the sources of atmospheric ozone?
2. How is it formed, and in what circumstances does it arise?
3. What is its precise action on animals and plants?
4. Has an excess or deficiency of ozone any effect on the public health?
5. If so, what is the nature of that influence?
6. What is the effect of the presence of epidemics on its amount, as calculated by the improved ozonometric method?
7. Does ozone oxidise one only, or all of the difficult kinds of organic matter found in the air?

With the exception of item 6, these same subjects are very much in the forefront of research today. The United Kingdom Photochemical Oxidants Review Group[24] made several recommendations for research. They highlighted the need for research into the biological effects of ozone. Additionally, they foresaw a period of intensifying research into the relationship between ozone and it precursors. An intensive research programme involving filed measurements, laboratory studies, atmospheric monitoring and computer modelling was strongly recommended. Such research is vital not only within the United Kingdom but globally. The US vice-president, Al Gore, has written a treatise[25] on global environmental management; the Earth Summit in Rio has at least signalled the willingness of governments to discuss environmental concerns; and the massive scientific effort to understand photochemical oxidants, of which this book attempts a far from exhaustive summary, continues apace.

1. Fox, C.B. (1873) *Ozone and Antozone*, J. and A. Churchill, London.

2. Andrews, T. (1856) On the constitution and properties of ozone, Phil. Trans. Roy. Soc., London, 6, 1-13.

3. Schönbein, C.F. (1840) Recherches sur la nature d l'odeur qui se manifeste dans certaines actions chimique, C.R. Acad. Sci., Paris, 10, 706-710.

4. de la Rive, A. (1845) Sur les movements vibratoires que determinent dans le corps, soit la transmission des courants electriques, soit leur influence extérieuse. C.R. Acad. Sci., Paris 20, 1287.

5. Marignac, J. C. de (1845) Sur la production et la nature de l'ozone. C.R. Acad. Sci., Paris, 20, 808-811.

6. Houzeau, A. (1858) Preuve de la présence dans l'atmosphère d'un nouveau principe gazeux, l'oxygène naissant, C.R. Acad. Sci., Paris, 46, 89-91.

7. Chappuis, J. (1880). Sur le spectre d'absorption de l'ozone, C.R. Acad. Sci., Paris, 91, 985-986.

8. Hartley, W.N. (1881) On the absorption spectrum of ozone, J. Chem. Soc., 39, 57-60.

9. Kley, D., Volz, A. and Mulheims, F. (1988) Ozone measurements in historic perspective, In: *Tropospheric Ozone*, I.S.A.Isaksen (Ed.), D. Reidel, Dordrecht, 63-72.

10. Walshaw, C. D. (1992) The Early History of Ozone, In: *Physicists Look Back*, J. Rocke (Ed.), Adam Hilger, Bristol.

11. Linvill, D.E., Hooker, W.J. and Olson, B. (1980) Ozone in Michigan's environment 1876-1880, Monthly Weather Review, 108, 1883-1891.

12. Bojkov, R. D. (1986) Surface ozone during the second half of the nineteenth century, J. Climate Appl. Met., 25, 343-352.

13. Levy, A., Henriet, H. and Bouyssy, M. (1905) L'ozone atmospherique, Ann. Observ. Municipal de Montsouris, 6, 18-22, 315-321.

14. Henriet, H. (1904) Sur la présence d l'aldehyde formique dans l'air atmospherique, Ann. Observ. Municipal de Montsouris, 5, 259-262.

15. Levy, A. (1878) Analyse chimique de l'air-Ozone, Annuaire de l'Observ. de Montsouris, 495-505.

16. Mariolopoulos, E.G., Repapis, C.C., Zerefos, C.S., Varotsos, C., Ziomass, I. and Bais, A. (1984) A neglected long-term series of ground level ozone, In: *Atmospheric Ozone*, C.S. Zerefos and A. Ghazi (Eds.), D. Reidel, Dordrecht, 788-790.

17. Lauscher, F. (1983) Aus der Frühzeit atmosphärischen ozonforschung, Wetter Leben, 35, 69-80.

18. Lisac, I. and Grubišić, V. (1991) An analysis of surface ozone data measured at the end of the 19th century in Zagreb, Yugoslavia, Atmos. Environ., 25A, 481-486.

19. Anfossi, D., Sandroni, S. and Viarengo, S. (1991) Tropospheric ozone in the nineteenth century : the Moncalieri series, J. Geophys. Res., 96, 17349-17352.

20. Sandroni, S., Anfossi, D. and Viarengo, S. (1992) Surface ozone levels at the end of the nineteenth century in south America, J. Geophys. Res., 97, 2535-2539.

21. Volz, A. and Kley, D. (1988) Evaluation of the Montsouris series of ozone measurements made in the nineteenth century, Nature, 332, 240-242.

22. Feister, U. and Warmbt, W. (1987) Long-term measurements of surface ozone in the German Democratic Republic, J. Atmos. Chem., 5, 1-21.

23. Reynolds, W.C. (1930) Notes on London and suburban air, J. Soc. Chem. Ind., 49, 168-172.

24. Photochemical oxidant Review Group (1987) *Ozone in the United Kingdom*, Department of The

Environment.

25. Gore, A. (1992) *Earth in the Balance: Forging a Common Purpose*, Earthscan, London.

CHAPTER 2
THE NATURE AND PROPERTIES OF OXIDANTS AND THEIR PRECURSORS

Ozone and other photochemical oxidants are almost exclusively secondary pollutants that are formed in the troposphere by chemical reactions between primary pollutants emitted into the ambient air from a variety of sources. The term photochemical is applied to many secondary pollutants when the main force driving these reactions is sunlight. Photochemical oxidants are primarily ozone but include other species (eg NO_2, H_2O_2, PAN), capable of oxidising aqueous solutions of iodide ions. Observations have shown that these other oxidants and other intermediate oxidation products like aldehydes are less abundant in the atmosphere than ozone. Their formation is strongly linked to ozone chemistry, and is most pronounced in strongly polluted areas where free radical formation is efficient.

The primary pollutants, often referred to as precursors, that are involved in the formation of ozone and other photochemical oxidants are volatile organic compounds (VOC) and nitrogen oxides. Methane is only negligibly reactive in ambient air, and hence the VOCs are sometimes referred to as non-methane hydrocarbons (NMHC). Aromatic and olefinic hydrocarbons contribute significantly to ozone formation.

In this chapter we present an overview of the physico-chemical properties of oxidants and their precursors.

2.1   OZONE

Ozone (trioxygen) is a dark blue explosive liquid or pale blue gas. The chemical reactions of ozone (as well as H_2O_2 and PAN) are electrophilic as a result of the presence of only six electrons on one of the oxygen atoms. Hence it is characterised by its ability to remove electrons from or to share electrons with other molecules or ions. The redox potential of ozone, ie. the capability of a chemical species for oxidising is 2.07 volts in aqueous systems.

Ozone has a characteristic odour and may induce severe irritation of the respiratory tract and eyes at concentrations in excess of 200 ppbv. At high concentrations it is extremely poisonous. It also produces oxidative degradation in certain non-biological materials, especially elastomers, textile fibres and dyes and certain types of paint. By virtue of its oxidising powers, ozone is used as a disinfectant for air and water, for bleaching waxes, textiles and oils and in organic syntheses. The physical properties of ozone are summarised in Table 2.1.

Table 2.1: Physical properties of ozone

Molecular weight	48
Melting point	-192.7°C
Boiling point	-111.9°C
Gas density (0°C, 760 mm Hg)	2.144 g l^{-1}
Heat capacity (0°C)	9.1 cal/g mol °C
Liquid density (-195.4°C, 760 mm Hg)	1.614 g ml^{-1}
OSHA PEL : TWA	0.2 mg m^{-3}
ACGIH TLV : TWA/CL	0.1 ppm
Conversion factors at 0°C, 760 mm Hg	1 ppm = 2141 μg m^{-3}
	1 μg m^{-3} = 4.670 x 10^{-4} ppm

TLV - Threshold limit value; PEL - Permissible exposure limit

TWA - Time weighted average; CL - Ceiling concentration

OSHA - Occupational Safety and Health Administration

ACGIH - American Conference of Governmental Industrial Hygienists

Data for all Tables in this chapter from references 1 and 2

2.2   HYDROGEN PEROXIDE

Hydrogen peroxide is a colourless, rather unstable liquid with a bitter taste. It will decompose violently when contacted with impurities, especially organic matter or traces of heavy metals. The standard potential for the redox system H_2O_2/H_2O is 1.776 volts. Undiluted it can cause burns of the skin and eyes and is extremely irritating to the respiratory system. In aqueous form it has numerous applications ranging from a dough conditioner, maturing and bleaching agent in food to rocket propulsion.

2.3   PEROXYACYL NITRATES

The compound $CH_3C(O)OONO_2$ was given the name peroxyacetyl nitrate about 30 years ago. According to IUPAC rules, PAN should be named as ethane peroxoic nitric anhydride. As a compromise, so as to relate to the acronym PAN, the name peroxyacetic nitric anhydride has been proposed. The class name should be the peroxy carboxylic nitric anhydrides, however this class of compounds are generally referred to as PANs. The most abundant and best studied of these species is peroxyacetyl nitrate (PAN). Peroxyproprionyl

8

nitrate (PPN), peroxybenzoyl nitrate (PBzN), peroxymethacrylyl nitrate (MPAN) and peroxy(hydroxy)acetyl nitrate (HPAN) have been detected in the atmosphere[3,4].

There is considerable physical and chemical evidence that PAN exists in two basic conformations: a linear one and a cyclic one. The predicted cyclic structure has extended bond-length for the peroxy bond of 1.4 Å, when compared to the linear form which is short, 1.278 Å. Pure PAN is a colourless explosive liquid at room temperature. Its thermal instability necessitates that precautions be taken in handling, synthesising and storing PAN. Studies have indicated that PAN is not very soluble in water but extremely soluble in non-polar organic solvents. A fact has been used to advantage in the aqueous synthetic chemical production of PANs. No standard potential is available in the literature for PAN. It has, however, been observed that on a molecule per molecule basis PAN is a more active toxic agent than ozone.

Table 2.2: Physical properties of hydrogen peroxide

Molecular weight	34.01
Melting point	-0.41°C
Boiling point	150.2°C
Density (25°C, 760 mm Hg)	1.4067 g l^{-1}
OSHA PEL : TWA	1 ppm
ACGIH TLV : TWA	1 ppm
Conversion factors at 0°C	1 ppm = 1520 μg m^{-3}
	1 μg m^{-3} = 6.594 x 10^{-4} ppm

Table 2.3: Physical properties of PAN

Molecular weight	121.06
Boiling point	106°C
Triple point	-49.3°C
Vapour pressure curve	ln p = 4586/T + 18.78
	(p in torr, T in degrees Kelvin)
Conversion factors at 0°C, 760 mm Hg	1 ppm = 5398 μg m^{-3}
	1 μg m^{-3} = 2.022 x 10^{-4} ppm

2.4 NITROGEN DIOXIDE

Nitrogen dioxide is a reddish brown gas with an irritating odour. The colourless dimer N_2O_4 exists in equilibrium with pure NO_2. The commercial brown liquid under pressure is called nitrogen tetroxide.

Nitrogen dioxide is used in the nitration of organic compounds and explosives and in the manufacture of oxidised cellulose compounds. It is one of the most insidious gases. Inflammation of lungs may cause only slight pain, but the resulting edema several days later may cause death. 100 ppm is dangerous even for a short exposure and 200 ppm may be fatal.

Table 2.4: Physical properties of nitrogen dioxide

Molecular weight	46.01
Melting point	-11.2°C
Boiling point	21.2°C
OSHA PEL : CL	5 ppm
ACGIH TLV : TWA	3 ppm
Conversion factors at 0°C, 760 mm Hg	1 ppm = 2053 μg m^{-3}
	1 μg m^{-3} = 4.872 x 10^{-4} ppm

Table 2.5: Physical properties of nitric oxide

Molecular weight	30.1
Melting point	-163.6°C
Boiling point	-151.8°C
OSHA PEL : CL	25 ppm
ACGIH TLV : TWA	25 ppm
Conversion factors at 0°C, 760 mm Hg	1 ppm = 1339 μg m^{-3}
	1 μg m^{-3} = 7.468 x 10^{-4} ppm

2.5 NITRIC OXIDE

Nitric oxide is a colourless gas, deep blue when liquid and a bluish white snow when solid. NO is the direct precursor of NO_2. It is paramagnetic and contains an odd number of electrons. NO is used in large quantities in the manufacture of nitric acid and also in the bleaching of rayon. Since NO is converted quickly to NO_2, suitable precautions must be taken even when handling small amounts in the laboratory.

2.6 HYDROCARBONS

Hydrocarbons are compounds consisting only of hydrogen and carbon. At ordinary temperatures, hydrocarbons with a carbon number of one to four are gaseous, while those with a higher carbon number are liquid or solid in pure state. Gas-phase mixtures in ambient air normally include compounds that are liquid in their pure form. Hydrocarbons with a carbon number greater than about 12 are in small quantities in the atmosphere, while those with a carbon number below 8 are abundant.

Hydrocarbons containing no double or triple bonds are saturated. Alkanes, or paraffins, are straight- or branched-chain compounds having the general formula C_nH_{2n+2}, $n = 1,2,3, ...$, they are a subset of acyclic hydrocarbons known as aliphatic hydrocarbons. The simplest saturated acyclic hydrocarbon is methane, CH_4. Hydrocarbons that contain one or more double bonds are unsaturated, such as alkenes or olefins. The general formula for alkenes with one double bond is C_2H_{2n} (eg ethene C_2H_4, propene C_3H_6). Like alkanes, alkenes are aliphatic straight- or branched-chain hydrocarbons. Alkenes are the most photochemically reactive hydrocarbons.

Alkynes are open-chain hydrocarbons that contain one or more triple bonds. The general formula is C_nH_{2n-2}, and for each additional triple bond in the molecule four hydrogen atoms must be removed from the general formula. The simplest member of this class is acetylene, C_2H_2, which is relatively unreactive in the atmosphere. Since acetylene is predominantly emitted from mobile sources, it is often used as an indicator of car exhaust emissions.

Aromatic hydrocarbons generally are considered those which have the characteristic chemical properties of benzene (C_6H_6). The carbon and hydrogen atoms are arranged in a six-membered ring, linked by three conjugated double bonds. The double bonds in aromatics are not nearly as chemically active as those in alkenes due to resonance stabilisation.

Terpenes, having the formula $C_{10}H_{16}$, are a naturally occurring (emitted from

Table 2.6: Physical properties of hydrocarbons

	Alkanes			Alkenes			
	Ethane	Propane	n-Butane	Ethene	Propene	trans-2-Butene	Acetylene
Molecular weight	30.07	44.11	58.12	28.05	42.03	56.12	26.02
Melting point °C	-183.3	-189.7	-138.4	-169.0	-185.2	-105.5	-80.8
Boiling point °C	-88.6	-42.1	-0.5	-103.7	-47.4	0.9	-84.0
Properties	colourless odourless flammable asphyxiant luminous flame	odourless explosive limits % volume in air 2.37 - 9.5	flammable asphyxiant	colourless gas burns with luminous flame asphyxiant	flammable gas burns with yellow sooty flame asphyxiant		burns with sooty flame, explosive limits : % volume in air, 3-65%
Uses	refrigerant fuel gas	refrigerant fuel gas	synthetic rubber and motor fuel manufacturing	fruit ripening plastic manufacturing oxyethylene welding	manufacture of acetone		welding, illuminant fuel

	Aromatics			Methane
	Benzene	Toluene	o-Xylene	
Molecular weight	78.12	92.15	106.17	16.04
Melting point °C	5.5	-95.0	-25.2	-182.6
Boiling point °C	80.1	106.2	144.4	-161.4
Properties	clear colourless flammable liquid carcinogen	flammable refractive liquid	colourless liquid	colourless, odourless flammable gas, faintly luminous flame
Uses	manufacture of medicinal chemicals, dyes, organic compounds	petrol additive solvent for paint lacquers	solvent, manufacture of dyes, organic compounds	illuminant organic synthesis manufacture

vegetation) subset of alkenes. α and β pinene have been identified in ambient air. Isoprene, C_5H_8, also occurs naturally.

2.7 ALDEHYDES

Aldehydes, formula $C_nH_{2n}O$, contain at least one oxygen atom and are not true hydrocarbons. The trivial names of aldehydes derive from the fatty acid which an aldehyde will yield upon oxidation (eg formaldehyde from formic acid) or are named after the alcohol from which it is derived. All aldehydes, with the exception of gaseous formaldehyde, up to C_{11} are neutral, mobile volatile liquids. Above C_{11} they are solid under ambient conditions. The solubility of aldehydes decreases with formula weight: the high carbon aldehydes are essentially insoluble in water.

The presence of the double bond (carbonyl group C=O) determines the chemical properties. The hydrogen atom connected directly to the carbonyl group is not easily displaced. The carbonyl group forms the basis for one analytical method to measure aldehydes. The main atmospheric source is emissions from vehicles due to incomplete combustion in the engine and photochemical formation via reactions involving hydrocarbons.

Table 2.7: Physical properties of aldehydes

	Acetaldehyde	Formaldehyde
Molecular weight	44.05	30.03
Melting point °C	-121	-92
Boiling point °C	20.8	-19.5
	Flammable liquid pungent odour	Flammable colourless gas

2.8 KETONES

Ketones are similar to aldehydes and have the same formula. A ketone is isomeric with the aldehyde that contains the same number of carbon atoms. Structurally, ketones consist of a carbonyl group between two radicals. The chemical and physical properties are

13

similar to those described for aldehydes.

1. The Merck Index, 11th edition, Merck & Co, New Jersey, 1989.

2. Sax, N.I. and Lewis,R.J. (1989) *Dangerous Properties of Industrial Materials*, Van Nostrand Reinhold, New York.

3. Heuss, J.M. and Glasson, W.A. (1968) Hydrocarbon reactivity and eye irritation, Environ. Sci. Technol., 2, 1109-1116.

4. Singh, H.B. (1987) Reactive nitrogen in the troposphere, Environ. Sci. Technol., 21, 320-327.

CHAPTER 3
SOURCES AND EMISSIONS OF OXIDANT PRECURSORS.

The wide variety of sources - their disparate distribution and intensities - represents a substantial obstacle to our understanding of photochemical oxidant chemistry. Acute photochemical oxidant pollution episodes are clearly linked to our increased use of cars in urban areas but recent concern about regional- and global-scale pollution has stimulated interest in sources which, although not usually intense enough to produce acute pollution episodes, are of sufficient magnitude to contribute to global oxidant production. This change of emphasis has, ironically, led to the discovery that some "global" sources, most notably the biogenic production of non-methane hydrocarbons (NMHCs), can be important in local-scale oxidant production[1]. The widening perspective has also, perhaps more alarmingly, brought into focus the fact that "acute" pollution can achieve almost global proportions[2].

Photochemical oxidant production occurs as an intermediate process in the atmospheric oxidation of carbonaceous material at ambient temperatures. We discuss the details of the chemistry in the next chapter; here we note only that efficient photochemical oxidant production requires a plentiful supply of carbonaceous fuel and nitrogen oxide catalyst. Carbonaceous fuel emissions into the atmosphere can be divided into two groups: those, biogenic emissions, which are the products of the ordinary biochemistry of particular organisms (and so, in this narrow sense at least, natural); and those resulting from the incomplete combustion of carbonaceous material in flames. The nitrogen oxide catalyst also has biogenic and combustion sources. Were this a book on geophysiology[3, 4, 5], it might be fruitful to speculate on the Gaian significance of the fact that fuel and catalyst are both on hand without help from man. As it is, we note instead that both biogenic and combustion sources are highly sensitive to human manipulation of the environment and so both may contribute to "pollution" in this sense.

Having outlined the biogenic sources of fuel and catalyst, we turn to their production in flames. Biomass burning provides a bridge between the discussion of biogenic and anthropogenic sources, so it is described before fossil fuel combustion. The final section considers the scaling-up of emission meaurements into emission inventories. It is not possible, or appropriate, to give details of all the existing inventories, so we adopt a "Mercator's Projection": examples from the U.K. are used before working outwards to Europe and the rest of the World.

It has also proved impossible, for reasons of space, to include a full discussion of deposition processes. A very brief description of some of the work carried out to date is given as an appendix to this chapter.

3.1 MICROBIAL PRODUCTION OF METHANE

The prokaryotes are unicellular micro-organisms which do not contain cell nuclei. Their phylogeny, that is, their position in evolutionary development, is a subject of some controversy at present[6]. Prokaryotes are now believed to fall into two groups: the eubacteria and the archaebacteria. The archaebacteria include extreme halophiles, thermophiles, and the methanogens which are of atmospheric importance because they are the only significant biogenic source of methane[7]. Most of the archaebacteria are confined to harsh environments, such as hot springs and saturated brines, but the methanogens are found over a wide range of habitats, as shown in Figure 3.1[8]. Broadly speaking, the habitats fall into three categories: aquatic or water-logged ecosystems; animal digestive tracts; and waste disposal sites. The habitats are all oxygen-free: methanogens are strict anerobes. Nitrate and sulphate also tend to be scarce where methanogens are present because nitrate- and sulphate-reducing bacteria can out-compete methanogens for any available hydrogen in such systems. Hydrogen is required for methane production and only when nitrate and sulphate are not available can the less thermodynamically feasible CO_2 reduction take place. Methanogens and sulphate-reducing bacteria can co-exist, however, if non-competitive substrates, such as dimethyl sulphide, are available[9].

Methane production is the only catabolic process of the methanogens, and the number of compounds which can be utilised, is limited. Most species can reduce CO_2 with H_2:

$$CO_2 + 4 \ H_2 \ \rightarrow \ CH_4 + 2 \ H_2O \qquad\qquad [3.1]$$

Other, methylotrophic, species can also use small organic compounds such as methanol, methylamines, dimethyl sulphide and acetate. For example[9,10]:

$$4 \ CH_3OH \ \rightarrow \ 3 \ CH_4 + CO_2 + 2 \ H_2O \qquad\qquad [3.2]$$

$$CH_3OH + H_2 \ \rightarrow \ CH_4 + H_2O \qquad\qquad [3.3]$$

$$4 \ CH_3NH_3Cl + 2 \ H_2O \ \rightarrow \ 3 \ CH_4 + CO_2 + 4 \ NH_4Cl \qquad\qquad [3.4]$$

$$CH_3COOH \ \rightarrow \ CH_4 + CO_2 \qquad\qquad [3.5]$$

Formic acid can also be used as a feedstock and is effectively equivalent to H_2/CO_2, the formate being converted into hydrogen and carbon dioxide prior to the formation of methane[10]:

Figure 3.1

A Phylogeny of methanogens, with examples of habitat. After Whitman[8].

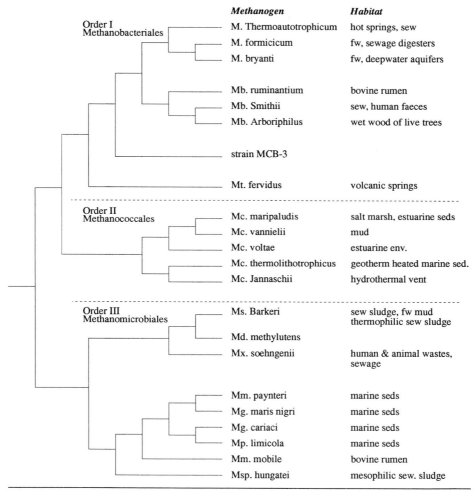

	Methanogen	Habitat
Order I Methanobacteriales	M. Thermoautotrophicum	hot springs, sew
	M. formicicum	fw, sewage digesters
	M. bryanti	fw, deepwater aquifers
	Mb. ruminantium	bovine rumen
	Mb. Smithii	sew, human faeces
	Mb. Arboriphilus	wet wood of live trees
	strain MCB-3	
	Mt. fervidus	volcanic springs
Order II Methanococcales	Mc. maripaludis	salt marsh, estuarine seds
	Mc. vannielii	mud
	Mc. voltae	estuarine env.
	Mc. thermolithotrophicus	geotherm heated marine sed.
	Mc. Jannaschii	hydrothermal vent
Order III Methanomicrobiales	Ms. Barkeri	sew sludge, fw mud thermophilic sew sludge
	Md. methylutens	
	Mx. soehngenii	human & animal wastes, sewage
	Mm. paynteri	marine seds
	Mg. maris nigri	marine seds
	Mg. cariaci	marine seds
	Mp. limicola	marine seds
	Mm. mobile	bovine rumen
	Msp. hungatei	mesophilic sew. sludge

Legend. M = methanobacterium; Mb = methanobrevibacter; Mt = methanothermus; Me = methanocoecus; Ms = methanosarcina; md = methanococcoides; Mx = methanothrix; Mm = methanomicrobrium; Mg = methanogenium; Mp = methanoplanus; Msp = methanospirillum.

$$4 \ HCOOH \ \rightarrow \ CH_4 + 3 \ CO_2 + 2 \ H_2O \qquad\qquad [3.6]$$

A few species, *methanobrevibaxter ruminantium* for example, are obligately methylotrophic and so must be provided with a methylated substrate (acetate in this case) rather than CO_2, for growth. The energy given out in forming methane by reduction of CO_2 is sufficient to synthesise adenosine triphosphate (ATP)[a, 11], which is the molecular energy store in a cell, and so fulfill all the methanogens' energy requirements. The acetoclastic reaction 5, however, is not formally a reduction and does not provide enough energy for direct phosphorylation of ADP to ATP on a molecule-for-molecule basis. A more circuitous process must then link methanogenesis and ATP formation[8].

An example of the biochemical production of CH_4 from CO_2 is given in Figure 3.2.

Figure 3.2

The sequential reduction of CO_2 to CH_4. From Whitman[8]. MFR is methanofuran (carbon dioxide reducing factor), MPT is tetra hydromethanopherin (formaldehyde activating factor), HS-CoM is co-enzyme M. The reduction of CH_3-S-CoM to CH_4 (step 4) is coupled to CO_2 activation but the mechanism is not yet known. MFR-CO_2^- has not been identified as an intermediate in the process.

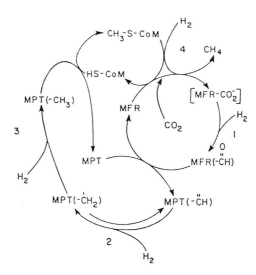

Initial CO_2 activation is coupled to the formation of CH_4 at the other end of the biochemical pathway. This enables the methanogen to overcome the thermodynamically unfavourable steps in the activation of CO_2 (Ref. 8). In a four-step process the carbon dioxide is reduced

---

[a] Descriptions of the general biochemistry of ATP, NADH, enzymes, and co-enzymes, can be found in any good biochemistry textbook - Stryer[11], for example.

to methane via aldehydic, alkynyl, alkenyl, and alkyl intermediates. Immediately prior to release, the methyl group is attached to the co-enzyme M (-S-$(CH_2)_2$-SO_3^-) which is unique to methanogens.

As can be seen in Figure 3.2, methane production requires a plentiful supply of molecular hydrogen. This is provided by other organisms which live in close proximity to,

Figure 3.3

Ruminococcus albus fermentation in the absence and presence of a methanogen[10]. In monoculture ethanol and H_2 are formed; when a methanogen is present acetate and CH_4 are formed.

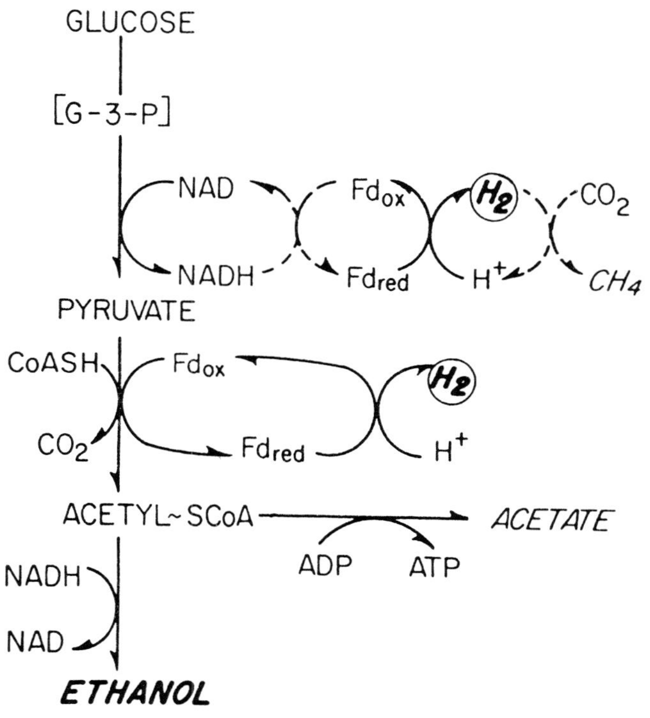

or in some cases symbiotically with, the methanogens. The effect of such interspecies hydrogen transfer is demonstrated in Figure 3.3. Fermentation by the bacteria *ruminococcus albus* produces ethanol in pure culture, but acetate in the presence of a methanogen. This is because, in pure culture, the hydrogen evolved from the breakdown of pyruvate to acetyl co-A, hydrogen and CO_2, inhibits the release of hydrogen from NADH (Ref. 10), which is a biochemical sink for the hydride ion. Instead the NADH is used to convert acetyl co-A into ethanol. When a methanogen is present, the partial pressure of hydrogen is maintained at low levels, hydrogen release from NADH is not inhibited, and acetyl co-A is oxidised to acetate rather than reduced to ethanol. Electron-sink (reduction) reactions may also be inhibited in other species when they are grown in co-culture with methanogens. Hence, bacteria that produce succinate ($^-OOC(CH_2)_2COO^-$), lactate ($CH_3CH(OH)COO^-$) or propionate ($CH_3CH_2COO^-$) may all convert to the production of acetate in the presence of the methanogen.

Some species grow poorly, or cannot grow at all, if interspecies hydrogen transfer is not available[10]. This syntrophic effect can be explained by the fact that the free energy for the conversion of the substrates used by the non-methanogenic syntroph is large, particularly at high partial pressures of hydrogen. Co-culture with methanogens reduces the partial pressure of hydrogen in the system, and so renders the bioconversion process of the syntroph more energetically feasible.

Interspecies hydrogen transfer, methylotrophy and syntrophy are phenomena which indicate the complex nature of methanogenesis in anaerobic ecosystems. Rather than a

Figure 3.4

The anaerobic food web involved in complementary and competitive microbial interactions associated with methanogenesis[12]. Methanogens and sulphate-reducing bacteria compete for H_2, CO_2 and acetate.

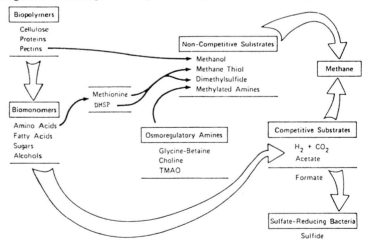

simple food chain, an intricate *food web*, such as that shown in Figure 3.4, is established[12, 9]. The overall effect is the decomposition of biopolymers to small organic molecules and CO_2, with methane and sulphide tending to act as electron sinks. Although the sulphate-reducing bacteria will tend to win the competition for available hydrogen, the methanogens can utilise other, methylated, substrates for which there is no competition.

The extent to which substrates are converted into carbon dioxide and methane varies for different anaerobic systems[10]. In sewage digesters and aquatic sediments, for example, bioconversion is almost complete:

$$57.5 \text{ Hexose } \rightarrow 172.5 \text{ CH}_4 + 172.5 \text{ CO}_2 \qquad [3.7]$$

whereas rumen fermentation typically produces a range of organic acid salts along with the gaseous products:

$$57.5 \text{ Hexose } \rightarrow 65 \text{ Acetate} + 20 \text{ Propionate} + 15 \text{ butyrate}$$
$$+ 35 \text{ CH}_4 + 60 \text{ CO}_2 + 25 \text{ H}_2\text{O} \qquad [3.8]$$

The principal difference between the two types of environment is their turnover time. Material remains in the rumen for one or two days; in other anaerobic systems the residence times can be much longer. The short residence of substrate in the rumen prevents, fortunately perhaps, the build-up of methanogens which are capable of utilising the acid salts. It should also be noted at around this point that non-ruminants, including humans, carry methanogens in their digestive tracts. Methane production is not as intense, however, because only ruminants possess gastrointestinal tracts that are adapted for the efficient breakdown of plant matter[13].

Even without complete bioconversion, the capacity for methane production by anaerobic ecosystems is enormous. However, the flux to the atmosphere from soils and water-logged ecosystems is reduced by *in situ* oxidation of methane back to carbon dioxide by methanotrophic eubacteria (a small amount of the methane produced is used by the methanogens themselves). Bacterial methane oxidation may occur under aerobic or anaerobic conditions[14, 9]. When oxygen is present it can act as an electron acceptor (Figure 3.5). Methane is first hydroxylated to form methanol using NADH. A methanol dehydrogenase enzyme then carries out the further oxidation to formaldehyde which is finally converted to CO_2 using NAD^+. The figure has been drawn to emphasise the coupling between NADH oxidation and NAD^+ reduction at either end of the biochemical pathway. The structure of the prosthetic group in methanol dehydrogenase, that is, the non-protein part of the active site of the enzyme, along with its proposed function, is also shown in Figure 3.5.

Anaerobic oxidation of methane appears to be coupled with sulphate reduction but, as yet, the biochemical pathway has not been established[9]. The existence of methanotrophic bacteria in soils implies, of course, that soil surfaces can act as sinks for atmospheric

Figure 3.5

The biochemical oxidation of CH_4 to CO_2. After Anthony[14]. a) the oxidation pathway used by aerobic methanotrophs. b) the structure and function of the prosthetic group, PQQ, of methanol dehydrogenase during the oxidation of methanol to formaldehyde.

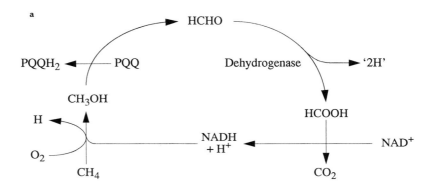

methane[15, 16]. Seiler *et al.*[17] compared methane production from archaebacteria in termite guts with destruction of methane at the soil surface and concluded that, in tropical and sub-tropical savannah, the net effect is methane loss. Desert surfaces also constitute a surface for methane loss[18].

To begin to quantify the contribution of methanogenesis and microbial oxidation to the global budget of methane, estimates of local fluxes per unit area or individual animal,

Table 3.1: Rates of methane production from ruminants and other animals. After Crutzen *et al.*[13].

Animal	Methane Yield, %	Energy Intake MJ d^{-1}	CH$_4$ Production kg head^{-1} d^{-1}
Cattle			
Industrialised Countries	5-7	200-300	55
S. Africa, Argentina, Aus.	7.5	110	54
Industrialising Countries	9	60.3	35
Other Ruminants			
Buffalo	9	85	50
Sheep (Industialised)	6	20	8
Sheep (Industrialising)	6	13	5
Goats	6	14	5
Camels	9	100	58
Other Domestic Animals			
Pigs (Industrialised)	0.6	38	1.5
Pigs (Industrialising)	0.6	25	1.0
Horses	2.5	110	18
Donkeys, mules, etc.	2.5	-	10
Humans	-	-	0.06
Wild Ruminants			
Moose, elk	9	53	31
Deer	9	26	15
Roe Deer	9	5	3
Wildebeest	9	22	13
Buffalo	9	57	34
Thompson's Gazelle	9	4	2
Other Wild Animals			
Zebra	2.5	31	5
Elephant	2.5	157	26
Insects	< 1.5	-	-

are required. Crutzen *et al.*[13] have compiled the most comprehensive survey of animal emissions to date and Table 3.1 gives a summary of their findings. When combined with population statistics, the annual emissions per head provide estimates for the global-scale contribution of each species. The amount of methane emitted by any individual animal will depend mainly on the species it belongs to and the type of fodder on which it feeds. These factors are combined to derive a *methane yield* parameter which is defined as the fraction of gross energy (food) intake lost in the generation of methane[13]. This energy loss is then converted in to a mass emission rate using the conversion factor 55.65 $1kg \equiv CH_4$ MJ.

The dominance of methane production in ruminants is immediately apparent from Table 3.1. When feeding on a low quality diet, high in fibre, ruminants can have methane yields of about 9 % : this might be the case for cattle on the range and undomesticated animals. Improving the quality of the feedstock reduces the methane yield but also, of course, increases the energy intake. Hence the cattle population of the industrialised countries, which have large numbers of intensively-farmed dairy herds, has a lower average methane yield than the stock of other nations, but a higher overall emission per head. Non-ruminant herbivores have, in general, small emission rates per head and so can be discounted from the global budget unless present in very large numbers. The same is true of humans: the ratio of the per-capita human emission to that of cattle in industrialised countries is about 1/1000, whereas the ratio of human and cattle populations is about 8. Clearly, therefore, humans are a negligible direct source of methane and this is unlikely to change even with the continuing increase in world population (and the increasing popularity of vegetarianism in the West).

The only class of animal which could rival the contribution of the ruminants is the insect. Few measurements have been made of the production of methane by insects other than termites. Moreover, it is not certain what proportion of any ecosystem's net primary productivity (NPP) is consumed by invertebrates. It would be preferable to have more definite quantifications of the input from insects but, if emission from termites is the major contributor, then recent inventories suggest that this is a small (though non-negligible) fraction of the total methane source.

Some results from soil and wetland flux studies are shown in Table 3.2[19, 20, 21, 22, 23, 24, 25, 26, 27, 28, 29]. We have classified the data into broad land types and give average fluxes for the season or measurement time indicated. Note, however, that substantial variations can be found within any one land type and during any one season. Plant growth itself has been shown to be an important factor in methane production[30, 31,]

[32]. Growing plants reduce the residence time of methane in the soil by transporting methane through their stems and out into the atmosphere. This reduces soil oxidation of methane. Plants also exude nutrients which feed the methanogens, so increasing methane production.

A substantial capacity for methane production exists in each of the broad land classes shown in Table 3.2. As would be expected from our discussion of microbial methanogenesis, the most productive environments, in terms of methane emission, are those which are water-logged and hence completely anaerobic. This need not imply, of course, that dessicated soils are the most effective sink for methane; Keller et al.[15] have reported that moist, but still presumably aerobic, soils are more effective. The dependence of CH_4 emission on water-logging suggests a correlation with mean precipitation rates and dry/wet seasonality. When compiling a global inventory, Fung et al.[33] used this correlation to produce a seasonal cycle of wetland methane emissions. The wet season is defined as those months when monthly precipitation exceeds potential evaporation. This system predicts that tropical sources are active for 5 - 12 months.

An effect of temperature on methane fluxes is not evident from Table 3.2 at first glance: unit fluxes during the emission season are more strongly controlled by the ecology and nutrient supply than by climate. An interesting example of this is given by Nisbet[34]. The population of beavers in North America has increased dramatically over the last century, and with it the area of water-logged land. Increasing the amount of water-logging increases the total regional CH_4 flux, of course, but beavers also increase the unit area flux by providing nutrients to the ponds in the shape of harvested wood. The effect on the global methane budget of the increase in the beaver population has not, as yet, been quantified. For any one system, however, there are seasonal variations that can be attributed to, or correlated with, soil temperature. Whalen and Reeburgh[35] report that over 80% of the total mean annual flux from tussock and low shrub tundra occurs during the months when the tundra is thawed. Moore et al.[36] observed an exponential dependence of methane flux to 0.1 m soil temperature. These two results have been incorporated into the database of Fung et al.[33] by considering that a monthly mean soil surface temperature T_s of > 5 °C indicates thawed conditions and that, in the season of active efflux, methane emissions are related to T_s by

$$k(T_s) = k(T_o).Q_{10}^{(T_s-T_o)/10} \qquad (3.1)$$

Table 3.2: Methane fluxes from soils, wetlands and bodies of water. After Bartlett et al.[38].

Surface	Flux mg CH_4 m^{-2} h^{-1}	Season, or measurement time	Ref
Arctic/Boreal			
Subarctic mire, Sweden	6.0	June-Sept	19
Nonforested bog, Alaska	5.0	August	20
Arctic Tundra, Alaska	0.6	*thaw*	35
Boreal Fen, Quebec	0.7	*May-Sept*	36
Forested bog, Minnesota	4.2	May-Aug	21
Non-forested bog, Minnesota	10.8	May-Aug	21
Temperate/Subtropical			
Temperate swamp, Virginia	6.5	*annual*	22
Cypress swamp, SE USA	2.0	June	23
Shrub swamp, Georgia	6.2	*annual*	38
Forested swamp/alluvial, Georgia	4.8	*annual*	38
Non-forested swamp, Florida	4.5	*annual*	24
Non-forested swamp, Florida	2.5	*annual*	25
Forested swamp, Florida	2.5	*annual*	25
Tropical			
River floodplain, Brazil	7.7	*wet season*	37
River floodplain, Brazil	2.8	*dry season*	37
Non-forested swamp, Brazil	6.2	*wet season*	38
Non-forested swamp, Brazil	5.0	*dry season*	38
Forest soil, Brazil	-0.015,-0.055	*annual*	29
Forest soil, Panama	-0.03	*annual*	15
Desert, USA	-0.03	*annual*	18
Agricultural			
Rice paddy, worldwide	4.58	*grow. seas.*	Table 3.3
Agricultural Soil, Panama	-0.008	*annual*	15
Pasture, Brazil	-0.002,0.01	-	29
Aquatic systems			
Ocean surface	0.002,0.005	*annual*	52
Freshwater lake, USA	≤ 50	*annual*	26
Meromictic lakes, USA	≤ 3	*annual*	26
Tidal freshwater, USA	5.8	*May-Nov*	28

where $Q_{10}=2$ and $k=0.05$ g CH_4 m^{-2} d^{-1} at $T_0=10$ °C. The dependence of methane emission to soil temperature will produce both diurnal and seasonal cycles[35].

Tropical regions, having relatively constant temperatures, do not display such a large seasonal variation[37, 38], except in those regions where floating meadows are common. Whereas emissions from pools and flooded forests do not show appreciable seasonality, the seasonal cycle of organic plant matter production on floating meadows could influence methanogenesis by providing an anaerobic environment in the dense root mat of the growing raft[39, 37]. Methane release in tropical floodplains is dominated by ebullition[40], whilst that occurring in more temperate wetlands is dominated by transport through plants[30, 32].

By far the largest fluxes are claimed for rice paddy fields. A more detailed

---

Table 3.3: Average methane fluxes from rice paddies during the growing season.

Site	Flux mg CH_4 m^{-2} h^{-1}	References
Sezchuan, China	58	44
Hangzhou, China	40	43
Po valley, Italy	12	42
California, USA	12	41
Uterra, Spain	4	17
Texas, USA	1.2	46

---

breakdown of measurements which have been made is given in Table 3.3. First attempts to estimate the global flux from this source used measurements made in Italy and the United States where unit area fluxes are of the same magnitude as strong natural sources[41, 42]. Recent experiments[43, 44] indicate, however, that Chinese rice paddies produce very much more methane per unit area than European or American equivalents. It will be necessary, therefore, to carry out a more extensive survey than that done to date if the global methane flux from rice paddies is to be more firmly established.

The fluxes measured from rice paddies so far exhibit a great deal of temporal, as well as geographical, variability. Diurnal cycles of emission correlate well with those of soil temperature[42] and a broadly linear dependence of fluxes on soil temperatures is seen throughout the growing season[44]. This seasonal dependence on temperature should not be taken as a causal link, however, but rather as the compounded effect of all seasonally

Figure 3.6

Seasonal variation of daily average CH_4 emission rates from unfertilized rice paddies[42]. The vertical bars denote diurnal variations. a) 1984; c) 1985; d) 1986. b) shows the emission rate from an unplanted and unfertilized field in 1984. Note the three peaks in emission rates which are evident in 1985 and 1986.

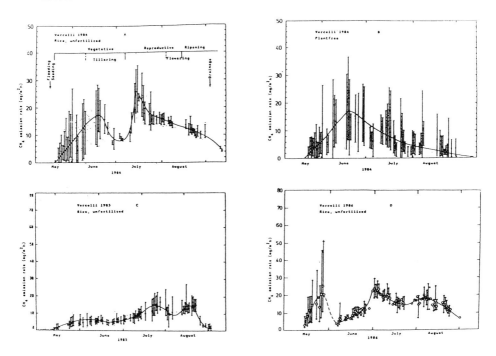

varying factors. Fine structure is also evident in some measurements during the growing season[45, 42, 46, 44]. Three peaks in the flux can be discerned (Figure 3.6):

(i) a peak early in the season which is characterised by large diurnal variations. This is attributed to the mineralisation of the organic remains of previous crops, producing substrates on which methanogens can feed. Consequently, this peak may not be observed in fields that have been left fallow for some time[46];

(ii) a second peak in the middle of the season is probably due to the build-up of root exudates in the soil[42]. Organic acids percolate away from plant roots and are converted into the methanogenic substrates H_2, CO_2, and CH_3COO^-. Sass et al.[46] have demonstrated the control of paddy soil methane emission by redox conditions and acetate supply.

(iii) the third, and generally largest, peak in emissions comes at the end of the growing season when root systems are extensive and deterioration of root matter can act as a source of substrates[42]. This maximum flux may continue up to field drainage after harvest, or may tail off prior to this[46].

Agricultural practices are also observed to exert a large effect on the production of methane from rice fields. Fluxes were observed to increase with the supply of organic fertiliser (rice straw) up to a maximum at a fertiliser loading of 12 tonnes ha^{-1} (Refs. 42 and 47). No further increase in emissions accompanied heavier fertiliser doses, possibly due to the production of phytotoxic compounds in the soil. Addition of nitrogen fertilisers ($CaCN_2$, urea, $(NH_4)_2SO_4$) gives a more complicated response which is related to the time, rate, and mode of application. Initial studies[48] found that the application of ammonium sulphate resulted in increased methane production as might be expected given the intimate connection between CH_4 flux and crop growth. However, Schutz et al.[42] have reported that mineral fertiliser can inhibit methanogenesis by up to about 60%, particularly when dug into the soil rather than spread on its surface. It is supposed that this effect is due to sulphate-reducing bacteria feeding on the ammonium sulphate and out-competing the methanogens for available hydrogen. Sass et al.[49] have shown that the timing and duration of flooding can greatly influence methane emissions from rice paddies without affecting rice yields. Emissions are lowest when the paddy is allowed to dry out several times during the growing season; emissions are highest when the paddy is flooded late in the growing season. Obviously the effects of agricultural practice will be important when trying to scale up local flux measurements but it is a factor that may prove singularly difficult to quantify.

Microbial methanogensis can also occur in the anaerobic interior of municipal waste landfill sites[50]. Much of the municipal solid waste (MSW) generated by human activity is composed of degradable organic carbon (DOC), and a large fraction of this can be processed into bio-gas (methane and CO_2) by microbial degradation (Figure 3.7). The exact percentage of MSW that is made up of DOC will depend largely on economic factors: industrialised countries produce more DOC per ton of waste (as well as producing more waste per capita) due, primarily, to disposal of wastepaper[51]. Similarly, the amount of MSW which is dumped in concentrated sites, which can develope anaerobic environments, is also a function of industrialisation, and hence geography. For example, much of the urban waste of industrialising countries is dumped crudely, without compaction, and is turned over by scavenging. This may prevent the onset of anoxic conditions and hence prevent methane emission. The relative amounts of processed DOC that are released as bio-gas and retained to construct microbial cell material, varies with the temperature inside the landfill where C_{oe} is that part of the total processed organic carbon C_o, that is released as bio-gas

Figure 3.7

The processing of municipal solid waste (MSW) in landfill sites. After Bingemer an Crutzen[51]. DOC is degradable organic material.

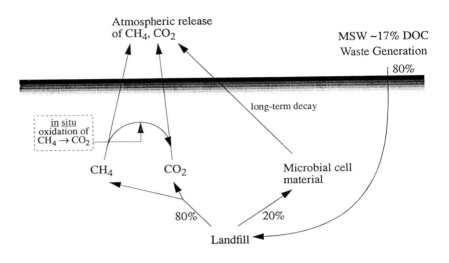

$$\frac{C_{oe}}{C_o} = 0.014\,T_i + 0.28 \qquad (3.2)$$

and T_i is the temperature of the landfill's interior[51].

One must also consider the time lag between waste dumping and methane production. This delay can be as long as 20 years for paper products, and up to 100 years for the non-lignin, decomposable, fraction of wood. Inventory calculations for the present should, therefore, use the MSW dumping rates of, say, 20 years ago. Furthermore, because dumping rates increased, at least in the United States, during the previous decades we can expect the output from landfill sites to likewise change. Lastly, the *in situ* oxidation of methane to carbon dioxide by aerobic bacteria must also be taken into account. Given some values for all the above factors, estimates of the total contribution from landfills to the global budget of methane can be made but, as yet, these may only be regarded as crude.

It has been known for some time that open waters, such as the oceans and large freshwater lakes, are supersaturated with respect to methane, so that an outward flux is expected from them (Ref. 52 and references therein). This flux can be calculated using a gas-exchange model, or measured directly. From the work reported in Table 3.2 it would appear that freshwater lakes can be as productive as swamps and ponds; meromictic (alkaline saline) lakes and the oceans are less productive per unit area but the large surface coverage of the oceans in particular will make them significant contributors to the global budget.

3.2 MICROBIAL PRODUCTION OF NITROGEN OXIDES

That part of the biogeochemical nitrogen cycle, to which soil micro-organisms contribute, involves *fixation* of molecular nitrogen to ammonia, *nitrification* of ammonia to nitrite and thence to nitrate, and *denitrification* of nitrate back to molecular nitrogen[53]. This is the original meaning of denitrification; the term is also commonly applied now to

Figure 3.8

The "leaky vessel" analogy to nitrogen oxide emissions from nitrification and denitrification.

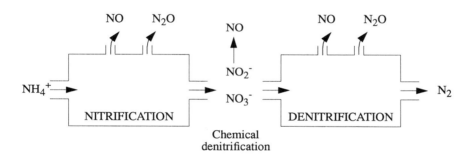

the removal of non-N_2O nitrogen oxides (NO_y) from the wintertime polar lower stratosphere by physical processes; we do not discuss stratospheric denitrification here. In the course of biogeochemical nitrogen cycling, *immobilisation* of inorganic nitrogen as organic matter, the reverse process of *mineralisation* can occur, as can the release of volatile nitrogen intermediates to the environment. This is shown schematically in Figure 3.8 by analogy to a series of "leaky" reaction vessels. The release of volatile nitrogen oxides, and hence the upper limit to the source strength of the soil, is governed primarily by the rates of nitrification and denitrification (*i.e.* the flow through the reaction vessels) and secondly by the percentage of intermediate evolved per mole of product formed, that is, the leakiness of the vessels[54]. Before a realistic soil emission rate can be calculated, however, we must also predict the physico-chemical behaviour of the evolved compounds as they percolate through the soil to be released at the surface. Whilst some, increasingly detailed, statements can be made about various parts of the mechanism, a properly *a priori* estimate of release rates is still some way off[55].

As shown in Figure 3.8, nitrogen oxides are released during nitrification and denitrification in the form of NO and N_2O. Since N_2O is stable towards tropospheric oxidation it will not be discussed further here (it is, of course, an important source of reactive nitrogen compounds in the stratosphere). NO, on the other hand, is central to tropospheric oxidant chemistry and the strength of any biogenic sources must be well established if the effects of industrial emissions are to be gauged, and if the "natural" biogenic signal is to be distinguished from any enhancement due to human activity.

Except in extreme conditions of low pH and high temperature, soil nitrification is dominated by the chemoautotrophic nitrifiers[53]. Heterotrophic nitrification does occur in some organisms, but is probably not an important source of soil NO emissions (see below). Chemoautotrophes require only inorganic substrates and synthesize all their organic cell material from CO_2. They are strict aerobes, requiring oxygen to act as an electron sink (oxidant). In soil, the genera *nitrospira*, *nitrosomas*, and *nitrosolobus* make up the ammonium/nitrite oxidiser population; nitrite/nitrate oxidation appears to be carried out exclusively by *nitrobacter*. The route of ammonium/nitrite oxidation is

$$NH_4^+ \rightarrow NH_2OH \rightarrow [HNO] \rightarrow NO_2^- \rightarrow NO \text{ or } N_2O \qquad [3.9]$$

The two electrons from the oxidation of ammonium to hydroxylamine are believed to provide energy for the synthesis of ATP which can be used to fuel CO_2 reduction. Nitric oxide is thought to be formed by the reduction of the nitrite product. Similarly, nitrite oxidation in *nitrobacter* releases electrons to cytochromes which can implement ATP synthesis:

$$NO_2^- + 0.5\ O_2 \rightarrow NO_3^- \qquad [3.10]$$

Nitrification rates will depend, therefore, on the availability of NH_4^+ and O_2 as well as other factors which might influence the well-being of the nitrifier population[54, 194].

The hierarchy of these effects is shown in Figure 3.9. Environmental factors, which effect nitrification in a broad manner, are termed *distal*; *proximal* factors are those which act at the cellular level[56]. In the example shown, nitrification is controlled predominately by the supply of NH_4^+. This supply is in itself a function of the rate of diffusional transport, the ratio of mineralisation to immobilisation, uptake of nitrogen by plants and cation exchange which, in their turn, depend upon more distal factors such as rainfall, temperature

Figure 3.9

Hierarchy of the major factors regulating nitrification in soils. From Robertson[56]. This example is for rainforest soils and shows how general factors such as climate and soil type influence more specific factors such as oxygen and NH_4^+ abundance.

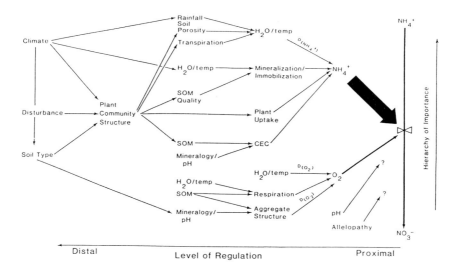

and soil pH. Inhibition of nitrification by high and low temperatures[57], a characteristic of biological processes, is demonstrated in Figure 3.10. The temperature dependence of NO emissions from soils does not appear to follow this pattern, however (see Figure 3.11), emphasising the complex interplay of biology and physico-chemistry of which it is the result.

Nitrification rates are also affected non-linearly by NH_4^+ supply and soil moisture. Rates generally increase with the availability of NH_4^+, but at high loadings and at high pH, toxic levels of ammonia can occur. Increasing soil moisture increases the rate of nitrification until, when water-logged, the onset of anaerobic conditions inhibits NH_4^+/NO_2^- oxidation[53]. Soil emissions of NO exhibit a similar pattern although it has not been shown to be quantitatively related. Of course, quantities such as soil moisture and temperature are themselves correlated so that individual relationships between factors are not easily distinguished in the field. Some normalisation procedure must be applied if individual effects are to become apparent[58].

In the biochemical pathway of denitrification, both nitrous and nitric oxides appear

Figure 3.10

The effect of soil temperature on the normalized nitrification rates of three soils from different climates. From Malhi and McGill[57]. The mean annual temperatures of the regions are: Northern Australia, 25°C, Iowa 10°C and Alberta 2.5°C.

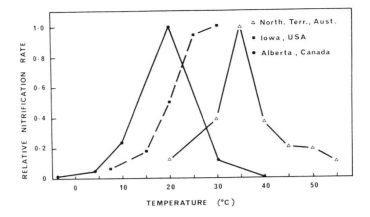

Figure 3.12

The role of nitrogen oxides in denitrification, and their release into the environment. After Firestone and Davidson[54, 194]. Note that the releases of NO and N_2O are reversible. X is an unknown intermediate.

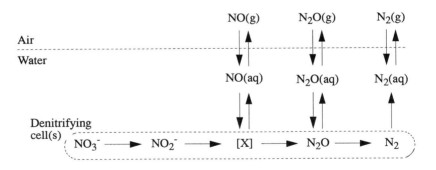

as exchangeable intermediates (Figure 3.12). The process can act, then, as either a source or a sink of these gases depending on the relative rates of reduction and intermediate release[54, 194]. Furthermore, it is likely that the ratio of consumption to emission will be a function of environmental parameters. For example, increasing soil moisture will increase the rate of NO production into the aqueous phase surrounding the producer, but will also

Figure 3.11

The flux of NO as a function of temperature for a variety of soils[58]. Vertical bars are the standard deviation of fluxes measured over the temperature range given by the horizontal bars.

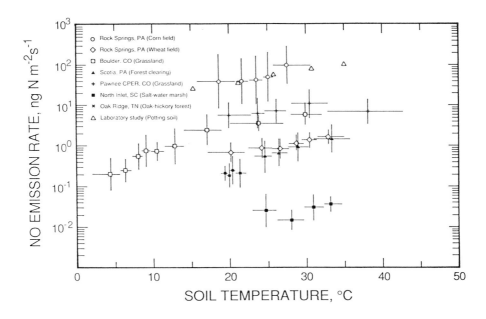

slow the rate of transfer through the soil by decreasing the effective diffusion rate. Factors such as that above which act to retain NO in the soil will enhance its re-utilisation and so decrease the ultimate soil emission rate[55].

In addition to the biological pathways outlined above, NO_x may be produced in soils by purely chemical means[55]. The major processes all involve the nitrite product which results from the ammonium/nitrite nitrification, and are summarised in Figure 3.13. the equilibrium

$$HONO(g) \; \rightleftharpoons \; H^+(aq) + NO_2^-(aq) \hspace{3cm} [3.11]$$

has $HK_a = 0.025 \; M^2 \; atm^{-1}$, where HK_a is the combined Henry's Law and dissociation constant (chapter 4). Appreciable equilibrium partial pressures of HONO(g) can, therefore, exist above acid/nitrite-rich soils. Even in more typical soils gas phase concentrations of around 1 ppbv could be sustained by this mechanism. HONO(g) is highly reactive in the troposphere where it is photolysed to produce hydroxyl radicals and NO (chapter 4).

In anaerobic acid conditions, aqueous nitrous acid may disproportionate to give nitric

Figure 3.13

The chemical production of NO and HONO in soils. Once in the gas phase HONO will liberate NO very quickly as a result of photolysis (chapter 4).

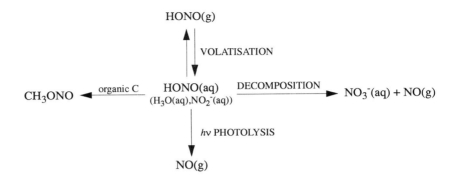

oxide and nitric acid

$$3\ HNO_2(aq)\ \rightarrow\ 2\ NO(g) + HNO_3(aq) + H_2O \hspace{2cm} [3.12]$$

This process has been observed to occur in soils of relatively high pH suggesting, perhaps, that disproportionation occurs at colloidal surfaces which have a local pH lower than that of the bulk[53]. Equilibrium values of NO from disproportionation are estimated to be substantial, but consideration of the reaction kinetics indicates that the chemical process will have a much slower rate than the microbial pathway[55].

The addition of nitrous acid to moist, acid, soils results in a small emission of methyl nitrate[59]. The lifetime of this compound in the soil is short, however, due to its breakdown into aqueous nitrite or nitrate. No field measurements have yet been made to ascertain the magnitude of any flux to the atmosphere.

Aqueous nitrite is photolysed at an appreciable rate when exposed to actinic fluxes that are typical of surface daylight radiation[55]:

$$NO_2^-(aq) + H_2O + h\nu\ \rightarrow\ NO(aq) + OH + OH^- \hspace{2cm} [3.13]$$

Most of the NO produced, however, reacts in aqueous solution and does not escape to the atmosphere. This mechanism is unlikely, therefore, to contribute significantly to the emission of NO_x from soils.

Tortoso and Hutchinson[60] quantified the contribution of each of the above mechanisms, chemical and biochemical, to the overall flux from a mildly acidic soil containing low organics. A control emission rate of 4 μg N kg^{-1} h^{-1}, a typical value for this type of soil[61] could be extinguished by the addition of nitropyrin to the soil, or by sterilisation. Emissions initially increased on the addition of glucose, but subsequently decreased. Addition of sodium chlorate led to an increase in the emission flux. Nitropyrin is believed to be an inhibitor of NH_4^+ oxidising nitrifiers so that the disappearance of the flux on addition of this compound indicates the importance of these bacteria as a source of NO. The lack of emission from sterilised soil discounts chemical NO_x production. Sodium chlorate is presumed to inhibit the nitrite oxidisers specifically: the increase in NO emission and accummulation of soil nitrite for this experiment provides evidence that these organisms are not the principal NO emitters in this case. To gauge the contribution from heterotrophic nitrifiers glucose is added to act as a foodstuff. Although NO emissions were seen to rise initially, the flux decreased as heterotrophic growth began to rise exponentially. This is probably due to the immobilisation of NH_4^+ in microbial biomass. Taking all the above results into consideration, the authors conclude that NO production in the soil studied is due primarily to the action of ammonium-oxidising chemoautotrophic nitrifiers. Their estimated global flux of 6 Tg N a^{-1} lies within current estimates[52]. Previous work[62] had also concluded that the autotrophic nitrifiers were an important source of NO in Californian chaparral soil. When acetylene, an inhibitor of nitrification, was added to the system, NO emissions fell below the limit of detection for dry soils, and became negative for wet soils. This suggests that, under moist conditions and in the absence of nitrification, soils can act as a sink for NO, probably as a result of heterotrophic denitrification.

Other recent flux measurements are presented in Table 3.4[63, 64, 65, 66, 67, 68, 69, 70]. Emission rates are given per unit area for a variety of soil types and geographical locations[58]. With the exception of the measurements taken on Venezuelan savanna during rainy season, the average fluxes from grasslands are reasonably close to one another. Since the measuring sites span four continents and a wide range of soil/climatic conditions, the apparent agreement suggests that some reasonably simple parameterisation may be suitable in global budget estimates: Williams and Fehsenfeld[58] suggest an exponential function of soil temperature. Of course, the problem of quantifying dry/rainy season differences remains. Table 3.4 also lists some flux measurements for temperate forest soils, and one measurement of the flux at an estuarine site. The average fluxes from the forest sites agree well between the four sites and over two continents. Forest soils and estuaries are unlikely to be major contributors to the global NO source.

Biomass burning, which is an important direct source of NO_x emissions to the troposphere (see § 3.2, below), can also significantly enhance soil emissions of NO from

Table 3.4: NO emissions from a variety of ecosystems. After Williams and Fehsenfeld[58].

Site Description and Location	Mean Range (NO) Emission, ng N m^{-2} s^{-1}	Reference
Grasslands		
Ungrazed pasture Aspendale, Australia	1.6 (0.6 to 2.6)	63
Meadow (bare soil) Finthen, Germany	2.2 (-6 to 14.2)	70
Grassland Boulder, Colorado, USA	3.0 (0.028 to 65)	64
Shortgrass Prairie Nunn, Colorado, USA	10.0 (1.15 to 52.9)	58
Crested wheat grass Erie, Colorado, USA	(-20 to 30)	65
Savanna (dry season) Guarico State, Venezuela	8 (3 to 15)	66
Savanna (rainy season)	(9.5 to 117)	69
Chaparral (unburned) Southern California, USA	10 (0 to 35)	62
Temperate Forests		
Coniferous forest Sorentorp, Sweden	0.35 (0.10 to 0.76)	67
Jadraas, Sweden	0.23 (0.10 to 0.56)	67
Coniferous/deciduous forest Scotia, Pennsylvania, USA	1 (0.2 to 4.1)	68
Deciduous forest Oak Ridge, Tennessee, USA	0.3 (0.06 to 1.12)	58
Coastal Marine		
Salt Marsh/estuarine North Inlet, SC, USA	0.03 (N.D. to 0.09)	58

nitrification[62]. Biomass burning serves to increase soil pH, mineralise NH_4^+ and remove plants (which are the microbes' major competitors for ammonium). The population of nitrifiers recovers quickly after the burn and utilises the NH_4^+ released from the thermal degradation of proteins and from the mineralisation of dead microbial matter. This results in increased emissions of NO from the soil, particularly on wetting. Fluxes can remain above the level measured over unburnt soil for many months.

3.3 THE PRODUCTION OF BIOGENIC NON-METHANE HYDROCARBONS

The emission of unsaturated organic compounds by plants, and the subsequent participation of these compounds in atmospheric chemistry, has been a subject of research since the mid-1950's at least[71], yet a fully, or even sufficiently, detailed account of the biochemical and micrometeorological processes involved is not so far available. The *isoprene unit* has long been established as a basic building block in plant and insect biosynthesis[72, 73] although the exact biological form of this unit took some time to establish. Dimerisation and polymerisation of the isoprene unit leads to the production of terpenes, the lighter of which are moderately volatile. On volatilisation, from leaf surfaces and chemical reaction in the atmosphere, the characteristic 'blue haze' aerosol forms over forested areas[74, 75, b]. Isoprene itself may be evolved[76] and in some species, such as deciduous trees, it is the primary component of the emission. All terpenes are reactive towards tropospheric oxidants, particularly the hemiterpene, isoprene[77] (chapter 4). Recent estimates of the global production of biogenic non-methane hydrocarbons (BNMHC) from vegetation suggest that they are by far the largest source of reactive non-methane hydrocarbons in the troposphere. As such they will be important in determining the spatial distribution, and chemical speciation, of tropospheric oxidants.

The biochemical equivalent of isoprene in the natural synthesis of terpenes, the isoprene unit, is 3-isopentyl pyrophoshate (3-ipp) which is produced from acetate subunits (Figure 3.14). Monoterpenes result from the dimerisation of the isoprene unit to give a ten-carbon compound, geranyl pyrophoshate, which can cyclise in a variety of ways. The exact pathway to isoprene synthesis has not been reported, but it seems reasonable to assume that it is via the dehydration of 3-ipp. Note that the cellular fuel, ATP, is consumed in the decarboxylation of mevalonic acid to produce 3-ipp. Subsequent conversion to isoprene is compartmentalised primarily in the chloroplasts[78]; monoterpene biosynthesis occurs in the epithelial cells at resin ducts. One might expect that this difference in manufacturing site results in phenomenological differences in isoprene and monoterpene emissions, and this is indeed the case.

Table 3.5 ranks a variety of plant species by isoprene emission rate measured under controlled conditions[79]. The highest rates are found for trees and shrubs, the lowest for herb and crop species. Dicotyledonous trees are generally strong emitters, the ferns are

---

[b] A simple demonstration of aerosol formation by the reaction of ozone and terpenes can be given by squeezing the peel of an orange in a stream of ozone[75].

Figure 3.14

The biosynthesis of isoprene and terpenes from acetate. The exact method of isoprene release from Δ^3-isopentyl pyrophosphate is not known. The conversion of ATP to ADP provides energy for the decarboxylation and dehydration of mevalonic acid.

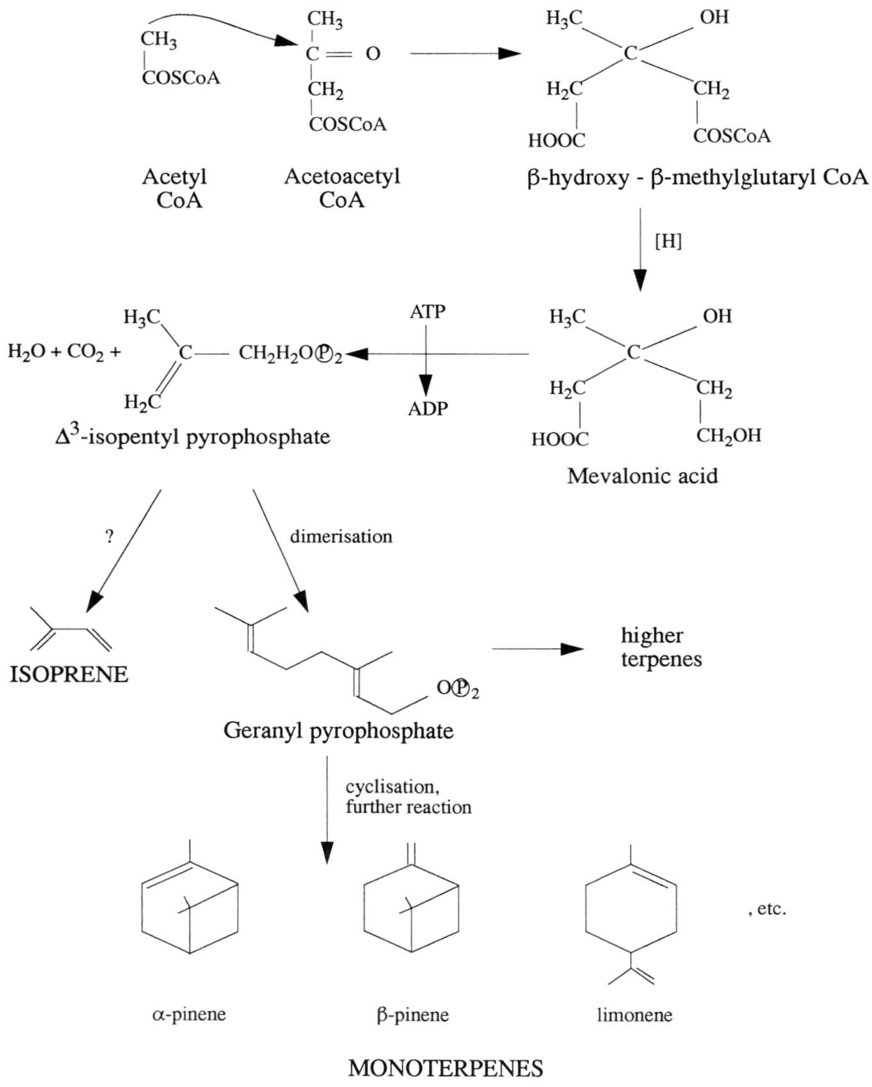

MONOTERPENES

Table 3.5: Plant species ranking based on isoprene screening[79].

Class and Family	High	Medium	Low	Nondetectable
I. Pteridophyta:				
Cyatheaceae		Dicksonia antarctica		
Polypodiaceae		Thelypertis decursive-pinnata, Thelypertis dentata		
II. Coniferophyta:				
Pinaceae	Picea engelmannii	Picea sitchensis		Pinus monticola, Pinus contorta, Pinus elliottii
III. Anthophyta:				
Dicotyledoneae:				
Aceraceae				Acer rubrum, Acer saccharum
Asclepiadaceae				Asclepias speciosa
Berberidaceae	Berberis aquafolium			
Betulaceae			Betula papyrifera, Betula lutea	
Chenopodiaceae				Beta vulgaris
Cruciferae				Brassca oleracea
Fagaceae		Quercus borealis, Quercus virginiana		Lithocarpus densiflora
Hamamelidaceae		Liquidabar styraciflua		
Labiatae				Salvia mellifera
Leguminosae		Cystisus multiflorus, Pueraria lobata, Robina pseudoacacia, Arachis glabrata	Phaseolus vulgaris, Pisum sativum, Vigna unguiculata unguiculata, Cercis canadenis, Gleditsia triacanthos, Glycine max, Lathyrus latifollius, Lupinus albicanlis, Medicago sativa, Trifolium pratense, Lotus pedunculatus, Lotus corniculatus var. arvensis, Arachis hypogaea	Vicia faba, Vicia pannonica
Malvaceae				Gossympium hirsutum
Myrtaceae	Eucalyptus globulus			
Papaveraceae	Eschscholzia californica			
Platanaceae		Platanus occidentalis		
Rhamnaceae		Rhamnus californica		Ceanothus maritimus
Salicaceae	Populus deltoides, Populus tremuloides	Salix nigra		
Solanaceae				Solanum tuberosum, Nicotiana tabacum
Monocotyledoneae:				
Gramineae			Zea mays, Triticum aestivum	Sorghum vulgare

Isoprene Emission Ranking[a]

[a] Ranking was based on the total isoprene emissions per gram leaf dry weight for three replicates of each plant species. The ranges were: high, 10-30µg C g⁻¹; medium 1-10µg C g⁻¹; low 0.01-1.1µg C g⁻¹; and non-detectable isoprene emissions, < 0.01µg C g⁻¹.

middle-ranking, and leguminous plants emit weakly. However, there are also wide variations in the emission rates of species in the same family, and from specimen to specimen within any one species[79]. There are, moreover, additional environmental effects[80, 81, 82].

It has been known for some time that isoprene production and emission is a light-dependent process occurring at the site of photosynthesis, the chloroplasts[78, 83]. As with other biological processes, emission is temperature-dependent with a maximum around 40 °C (Refs. 83 and 84). An example of this behaviour with respect to light and

Figure 3.15

The effect of light and temperature on isoprene emissions from Engelmann spruce (A) and Sitka spruce (B) seedlings[80]. The surfaces can be represented by:

$ln(E_{Englemann})$ $= -2.548 - 3.575 \times 10^{-4}I + 0.2233T + 4.412 \times 10^{-5}IT - 2.835 \times 10^{-3}T^2$

$ln(E_{Sitka})$ $= -3.384 - 7.458 \times 10^{-4}I + 0.2359T + 3.446 \times 10^{-5}IT - 3.197 \times 10^{-3}T^2$

where E = isoprene emissions (μ g^{-1} h^{-1}), I = light flux, μEm^{-2}s^{-1}, T = temperature, K.

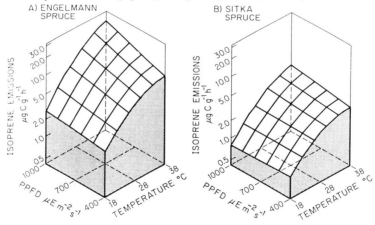

temperature is given in Figure 3.15 for two species of spruce[80]. The interspecific difference in emission strength is immediately apparent (if we take these samples as being typical of the species), the isoprene release rate of the Engelmann spruce being approximately five times that of the Sikta spruce. Sensitivity to light intensity is relatively high for the Engelmann, but less so for the Sikta. It is argued that this is due to the adaptation of the Sikta to shady conditions so that it becomes light-saturated at lower photon fluxes. This then results in saturation of the isoprene emission. Monson and Fall[82] have demonstrated that isoprene emission rates follow photosynthesis assimilation rates as a function of light intensity.

The form of the dependence on temperature is similar for both species and follows the pattern reported earlier for other species[83, 84]. Emission rates increase by a factor of 2-4 for a 10 °C increase in temperature (*ie.* $Q_{10} \approx 2\text{-}4$) until levelling off and decreasing sharply at high temperature. The dependence can be modelled over the entire range of ambient temperatures using an expression of the form given by Tingey[85] for live oak:

$$\log_{10}(\varepsilon_i^H) = \left(\frac{3.162}{1 + \exp[-0.236(T-22.39)]} \right) - 1.097 \qquad (3.3)$$

where ε_i^H, µg C dm^{-2} (leaf) h^{-1}, is the emission rate of a high ranking isoprene emitter at temperature T °C.

Ambient concentrations of oxygen and carbon dioxide can also affect isoprene emissions significantly[81, 82]. Figure 3.16 shows the rates of photosynthetic carbon

Figure 3.16

Isoprene emissions (I_s) as a function of CO_2 and O_2 partial pressures (O_2 is given as a volume percentage). The photosynthetic assimilation rate (A) is also shown (as open bars)[82].

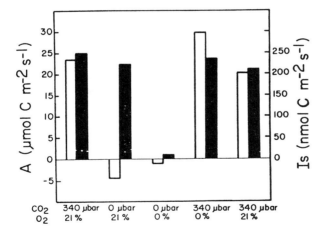

assimilation and isoprene emission for a variety of test atmospheres. Only if either CO_2 or O_2 is present does emission occur. It does not seem to matter much which of the two gases is, in fact, present since the emission rate in an oxygen-free environment is not significantly different from that in a carbon dioxide-free one. This suggests that isoprene can be synthesised from carbon that has been incorporated into the plant via either photosynthesis or photorespiration [82]. Earlier work tended to emphasise one or other of these routes, some workers stating that isoprene is synthesised from photosynthetic carbon [86], others concentrating on the link with photorespiration [87].

When photosynthesis becomes insentive to changes in ambient oxygen (*i.e.*, becomes phosphate-limited), isoprene emissions display a different dependence on oxygen concentration than during the normal oxygen-sensitive process [82]. If O_2-insensitive photosynthesis is induced by the addition of mannose, isoprene emissions fall dramatically. Possible reasons for this are reduced carbon exportation from the chloroplasts, or a reduction of available ATP for 3-ipp synthesis. Increases in ambient CO_2 levels also cause a reduction in isoprene emissions presumably due to either induced phosphate imbalances in the chloroplasts and cytoplasm, or increased competition from photosynthesis reducing the available ATP. This effect of increased CO_2 may have to be considered when building a global budget of BNMHC emissions given the observed trend in CO_2 levels [88]. In sum, the effects above, whilst still uncertain in some points, reveal the close connection between isoprene emission and biosynthetic pathways. They indicate that no substantial reservoir of isoprene exists which might cause the response to changing environmental conditions to differ from that of the biosynthetic pathways themselves.

A strong dependence of isoprene emissions on ambient humidity has been observed by Monson and Fall [82] but this does not appear to be caused by stomatal closure at low humidity. This has been demonstrated by using abscisic acid to alter stomatal conductance at constant humidity: a relatively small drop in emissions was recorded as conductance decreased. These responses can be explained if a large fraction of the evolved isoprene is emitted by the leaf cuticle, and decreasing ambient humidity causes a reduction in cuticular conductance. Conclusive proof of the cuticular diffusion of isoprene, and its dependence on humidity, has yet to appear although earlier work does support the claim that emissions are not strongly correlated with water stress [81].

In addition to isoprene, monoterpenes may be emitted from plants. Sometimes monoterpenes are the only terpenoid compounds observed in emissions (Table 3.6). When isoprene *is* released it always comprises the largest fraction of the total hydrocarbon source [79]. Of the monoterpenes, α-pinene is usually the most abundant [76, 89], but not always. For example, in Table 3.6, the terpene 1,8-cineole is the largest component in the emission from *eucalyptus globulus*, and sabinene dominates terpene emission from *acer saccharum*. Arey *et al.* [90] recently reported that the open-chain monoterpene linalool makes up the major portion of the emission from Valencia orange blossoms (*citrus sinensis*).

Table 3.6: Emission Rate Estimates for Selected Plant Species[79]. Leaf Temperature of 28 °C, Light Intensity of 10^3 μE m^{-2} s^{-1}.

Plant Species	No.	Isoprene (μg C/gh)	α-Pinene	Camphene	β-Pinene	Myrcene	Ocimene	β-Phell-andreen	1,8 Cineole	Limonene	Sabinene	Total (μg C/gh)	% isoprene Detected	Ranking
Eucalyptus globulus	8	38.5 (1.10)	1.14 (1.11)	...	0.58 (1.25)	...	0.84 (1.24)	...	3.53 (1.24)	0.69 (1.34)	...	45.28	85%	High
Populus tremuloides	8	33.89 (1.08)	...	...	...	...	...	...	...	...	...	33.89	100%	High
Populus deltoides	8	24.95 (1.09)	...	...	...	...	...	...	...	...	...	24.95	100%	High
Rhamnus californica	5	19.80 (1.07)	...	...	...	...	...	...	...	...	...	19.80	100%	(Medium)[a]
Platanus occidentalis	7	18.55 (1.29)	...	...	...	...	...	...	...	...	...	18.55	100%	(Medium)
Salix nigra	12	17.00 (1.11)	...	...	...	...	...	...	...	...	...	17.00	100%	(Medium)
Thelypteris decursive-pinnata	8	16.50 (1.05)	...	...	...	...	...	...	...	...	...	16.50	100%	(Medium)
Quercus borealis	7	13.33 (1.16)	...	...	...	...	...	...	...	...	...	13.33	100%	(Medium)
Liquidambar styraciflua	12	12.00 (1.10)	1.20 (1.17)	...	0.35 (2.33)	...	...	...	...	0.57 (1.98)	...	14.12	85%	(Medium)
Picea engelmannii	12	11.00 (1.12)	1.40 (1.26)	0.17 (2.14)	0.65 (1.32)	0.10 (1.97)	...	0.19 (2.14)	...	...	...	13.51	81%	High
Pueraria lobata	8	6.45 (1.36)	...	...	...	...	...	...	...	...	...	6.45	100%	Medium
Picea sitchensis	12	2.70 (1.28)	0.21 (1.31)	0.13 (1.14)	0.18 (1.31)	0.32 (1.13)	...	0.00[b] (2.44)	...	...	...	3.54	76%	Medium
Glycine max	9	0.018 (1.14)	...	...	...	...	...	...	...	...	...	0.018	100%	Low
Cercis canadensis	4	0.013 (1.19)	...	...	...	...	...	...	...	...	...	0.013	100%	Low
Pinus elliottii	5	...	2.79 (1.19)	...	2.00 (1.17)	0.19 (1.19)	...	0.14 (1.21)	...	0.01 (1.15)	...	5.13	0%	Nondetectable
Acer saccharum	8	...	0.45 (1.31)	0.21 (1.14)	... (1.16)	0.38	...	...	...	...	0.58 (1.44)	1.62	0%	Nondetectable
Citrus sinensis[c]	1	...	linalool 13.0	...	...	2.6	...	...	...	0.06	0.13	15.79	0%	Nondetectable

NOTE - Three dots indicates that these compounds were not detected in the emission profiles by the methods used in this study. The values in parentheses are the standard geometric errors.
[a] Preliminary ranking resulting from screening encapsulaion when different from final ranking.
[b] This compound was detected in the foliar emissions of only one of the 12 plants tested.
[c] Data from Arey et al.[90]

45

Monoterpene emission rates are similar for species that emit isoprene and those which do not, so that it seems safe to assume that monoterpene release is not influenced by the biosynthesis of isoprene.

Monoterpene emissions are a strong function of temperature[80, 81, 91]. Figure 3.17

Figure 3.17

Influence of temperature on monoterpene emissions from Englemann (A) an Sitka (B) spruce seedlings under controlled conditions. From Evans *et al.*[80]. Expressions for the temperature dependence of each compound are given in the figure (t = temp., °C).

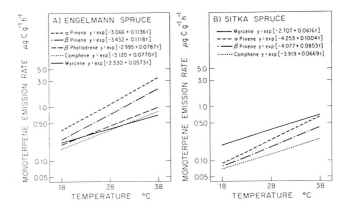

gives the dependence on temperature of several monoterpenes emitted from two species of spruce. The general log-linear relationship is common to all the compounds shown, although the gradient varies. Such a correlation is consistent with an emission mechanism controlled by phase equilibria between monoterpene reservoirs within the plant leaf, and the ambient air, rather than a mechanism controlled by any biosynthetic process. Myrcene shows the smallest increase with temperature, α- and β-pinene the largest[80]. Tingey[85] gives the following equation for α-pinene emission of this type from slash pine

$$ ln\left(\varepsilon_p^H\right) = 0.067\,T - 0.85 \tag{3.4} $$

where ε_p^H μg C g^{-1} (leaf) h^{-1} is the α-pinene emission from a high ranking source species at temperature, T °C. As is clear from Figure 3.17, however, applying this formula to all monoterpene emissions will result in over-estimation of their rate of change with temperature.

Monoterpene emissions are generally regarded as independent of light absorption by

Figure 3.18

The variation of isoprene and α-pinene mixing ratios in an agricultural area of Japan[94]. Isoprene (open squares) and α-pinene (closed squares) concentrations are both given in ppbv C. Note the anticorrelation between the species' concentrations due, probably, to the different biochemistries of the two compounds (see text).

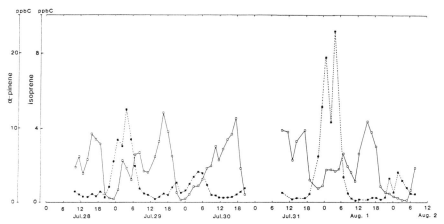

the plant[80, 85], although Yokouchi and Ambe[92] have reported an effect of solar radiation on plants grown in controlled environment gas chambers. The lack of correlation between light intensity and emission strength, if this is indeed the case, supports the hypothesis that monoterpene emissions are controlled by the physical process of evaporation from leaf reservoirs. The abundance of any single compound in the vapour will then be a function of its concentration in the oleoresin, and its volatility. α-pinene, β-pinene and camphene are the most volatile of the commonly emitted monoterpenes. As shown in Table 3.6, these compounds often comprise a large part of the total monoterpene emission. If this effect of volatility is taken into account, it appears that the composition of the emitted vapours generally follows that of the oleoresin. Figure 3.18 gives an example of the diurnal variation of monoterpene concentrations in open-chambers due, primarily to the temperature dependence of the flux. More usually, in the open, maximum concentrations are observed at night[92, 93, 94], when concentrations of O_3 and OH are low, and turbulent dispersion is reduced.

Although diurnal and day-to-day variations appear to be due to the dependence of evaporation on the ambient temperature, there may also be changes in the longer term associated with a seasonal cycle in the monoterpene pool size. Arey *et al.*[90] measured emissions from an Olinda Valencia orange tree (*citrus sinsensis*) over several months and

Figure 3.19

Monoterpene emissions from Olinda Valencia orange trees as a function of time[90]. Temperatures were usually $25 \pm 2°C$ when the measurements were taken, or were normalized to 25°C using previously computed temperature correlations. The highest emission rates occur in March and April when the plant is in bloom.

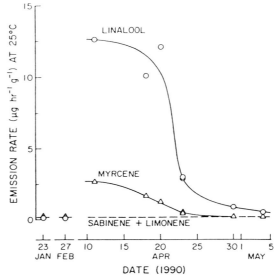

found a distinct seasonality associated with flowering (Figure 3.19). The seasonal variation shown in this figure, in which all emission rates have been normalised to 25 °C, must represent the influence of some biological process on the availability of monoterpenes within the plant itself. This biologically-induced seasonality makes monoterpene fluxes rather more akin to methane emission than to the biosynthetically rate-limited isoprene flux.

The effect of air pollutant stresses must also be considered for both isoprene and monoterpene release, particularly since plant damage is one of the major dangers associated with air pollution by photochemical oxidants. Plants subject to ozone and other photo-oxidants showed that emissions of ethene increased by several orders of magnitude. The effect on terpenoid emissions is not so clear, however. Original studies indicated that the reaction of ozone with unsaturated compounds within the plant tissue led to the formation of hydroperoxides and to a decrease in the potential for isoprene emission[95]. Current research, however, appears to indicate that emissions are not substantially changed by ozone stress (C.N. Hewitt, personal comm., 1991).

Subject to the provision of data about leaf area indexes, forest speciation and land use, the emission rates calculated from individual specimens can be scaled-up to provide

Table 3.7: Biogenic non-methane hydrocarbon fluxes from a variety of ecosystems.

Climate	Comments	Flux ngC m^{-2} s^{-1}	References
(i) isoprene			
Temperate			
	SW France, twilight, over groves and crops, July	28	99
	11 °C	120	
	28 °C, WA, USA, isolated white oak forest	22060	96
Tropical	30 °C, PA, USA, hickory forest	1960	97
	0800-1600 LT, Amazonia, Jul-Aug.	760	100
	1000-1600 LT, tropical Australia	1373	101
(ii) α-pinene			
Temperate		324	97
	max. value, Douglas fir, WA, USA, Oct-Nov.	343	
Tropical	15 °C	637	98
	30 °C, WA, USA, Loblolly pine	22	
	night	29	100
	day, Amazonia, Jul-Aug		

estimates of regional- and global-scale source strengths. Such an approach is subject to large uncertainties, however, due to the variability associated with each of the ecological/geographical scaling factors, so that some independent verification is required. Source strength estimates of this second, corroborating, kind are presented in Table 3.7[96, 97, 98]. Differences in the ecosystem and temperature produce large ranges in the reported areal emission rates of both isoprene and α-pinene, but there is insufficient data to detect any seasonality in the measurements.

It is clear from Table 3.7 that isoprene has strong sources in both temperate and tropical climes (note that the temperate measurements are for intense and isolated sources whereas the tropical values are more general). Even over open countryside the flux of isoprene can contribute up to 10 % of the atmospheric NMHC loading[99]. In the tropics, remote from urbanisation, conditions are such that isoprene can be the major sink for OH at the surface[100, 101]. Sources of α-pinene and, by extension, other monoterpenes appear to be predominantly temperate. This reflects the importance of coniferous trees as monoterpene sources. It is also clear that, even for temperate regions, monoterpenes make up a minor fraction of the total biogenic non-methane carbon emitted in the areas studied. A little caution should be exercised when drawing conclusions from such a limited dataset, however. There is no reason, other than simple optimism perhaps, for believing that the results from these few ecosystems can be extrapolated meaningfully to produce a global inventory, or regional inventories for areas not yet studied. Clearly, further field studies are needed.

Considering the emissions of photochemically active compounds in general, the following points can be made:
(i) there is unanimous agreement that biogenic compounds are important in tropospheric photo-oxidant chemistry;
(ii) it is desirable to have an *a priori* mechanism for emissions that takes into account biosynthetic and physico-chemical processes;
(iii) several links in this mechanistic chain are missing for all the compounds discussed above; and
(iv) to fill this gap a rather arbitrary combination of laboratory data and field/ecosystem measurements is currently being used.

The upshot of our lack of sufficient data is a large uncertainty associated with each source term, and a concomitant margin of error in estimated budgets. By combining expertise, a more holistic view of the agronomic, microbiological, chemical and physical factors affecting processes such as biogenic source gas emission is, however, already becoming possible[102].

3.4 OXIDANT PRECURSORS FROM BIOMASS BURNING

The combustion of plant matter is an integral part of land cultivation in many parts of the world. There are many purposes for biomass burning: forest clearing, pest control, disposal of plant litter, nutrient mobilisation (mineralisation), charcoal production, energy production, game hunting and even religious ceremonial[103]. With the current rapid increase in human population, some, if not all, of these activities are increasing and evidence of significant land-use changes is already coming to light[104, 88]. During the dry season, when it most commonly occurs, biomass burning is sufficiently widespread and intense to be visible from space[2, 105, 106]. It has been reported that emissions from fires of this sort can substantially enhance the concentration of tropospheric ozone[2, 106, 107]. They are also predicted to be a major source of nitric acid in the tropics[108].

Dry plant matter approximates to the empirical formula CH_2O and has the following minor constituents: 0.3-3.8 %w/w nitrogen; 0.1-0.9 % sulphur; 0.01-0.3 % phosphorous; and 0.5-3.4 % potassium[103]. Hegg *et al.*[109] use a slightly different carbon mass fraction of 0.497. From the point of view of photo-chemical oxidant chemistry, we are interested in the volatilisation of partially oxidised carbon (oxidation number less than +4) and of oxides of nitrogen (NO_x); organic nitrates may also be evolved in some quantity but, as yet, no measurements have been made. The ratio of hydrocarbon to NO_x emissions from the fire will then determine the rate of production, or destruction, of secondary pollutants and oxidising radicals[110].

Laboratory measurements (*e.g.* Ref. 111) or field studies (*e.g.* Ref. 109) can both be used to ascertain the composition of smoke from biomass fires. Apparatus for the

Figure 3.20

An experimental apparatus for the analysis of the gaseous exhaust from biomass burning[112].

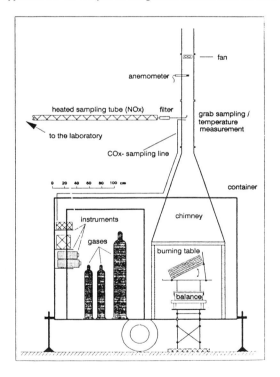

laboratory measurement of plume concentrations is shown in Figure 3.20. The use of laboratory-scale simulations of bushfires is obviously convenient in that unambiguous measurements of variables such as fuel composition, fire temperature and plume composition are possible (and reproducible). There is, however, always the chance that conditions within the experimental chamber do not accurately represent those under which real fires occur, so that *in situ* observations serve to provide important corroborating evidence.

The results from a laboratory test fire are shown in Figure 3.21. Hydrogen, methyl chloride, molecular nitrogen, carbonyl sulphide, particulates, and various other minor constituents are emitted[103, 111, 112]. Two phases have been identified in the burn: an early, flaming stage in which temperatures are high and bright flames can be seen; and a later, smouldering phase when the fire cools and flames are not present. As is obvious from Figure 3.21, flaming stage emissions are characterised by high oxidation states (*e.g.* CO_2, SO_2, NO_x) whereas the smouldering stage emits elements in less oxidised forms (*e.g.* CH_4, and N_2O). The exact duration of each stage will depend on the conditions at the time of the fire. This could substantially complicate the estimation of emission factors but, fortunately, it appears that the ratio of CO to CO_2 is a good indicator of the degree of flaming combustion which correlates linearly with the other emissions thus enabling a simpler parameterisation of trace gas release[111]. Figure 3.22 demonstrates this: in part (a) there is a good linear relationship between the fraction of nitrogen in the fuel released as N_2 and CO/CO_2 (smouldering/flaming ratio). In this case emissions are favoured by flaming combustion (see below). The good correlation between N_2 and CO_2 in the smoke, ratifies this conclusion (Figure 3.22(b)). In part (c) of the figure, however, it can be seen that CH_4 emissions are favoured by smouldering combustion.

Integrating over the entire burn, the carbon and nitrogen emissions can be speciated. Mass balance measurements together with elemental analysis of the ash will give the fraction of total carbon and nitrogen emitted. These values can then be scaled-up to give global- and regional-scale estimates of the source strength (Table 3.8). This procedure is fraught with uncertainties[113]. The exact formula used will depend on the geographical and ecological data available, and may have to be further adjusted to take account of other variables. For example, the amount of wood used per annum, and the amount of land cleared, may both be functions of family size.

Lobert *et al.*[112] found that $\approx$90 %w/w of the nitrogen and $\approx$95 %w/w of the carbon present in the biomass, volatilised. About 13 %w/w of the volatilised N is released as NO, about 11 %w/w of the C as CO, and around 1 %w/w as CH_4 (Ref. 103). Hence we can expect something like 0.5 g of N as NO_x, 42 g of C as CO, and 4 g of C as CH_4, to be released from the burning of 1 kg (dry weight) of biomass. The enormous quantities of biomass burnt each year make these emissions substantial contributors to the global flux of each compound except, perhaps, for methane. It should perhaps be noted in passing here

Figure 3.21

The concentration of trace gas emissions during the controlled combustion of *Trachypogon* grass from Venezuela as a function of time. The stack gas temperature is also shown, and the flaming and smouldering stages of combustion identified. Note that the oxidized gases (CO_2, NO_x, SO_2) and N_2O are emitted mainly during flaming combustion. The less oxidized gases (CO, methane, NMHC) are emitted mainly during smouldering combustion.

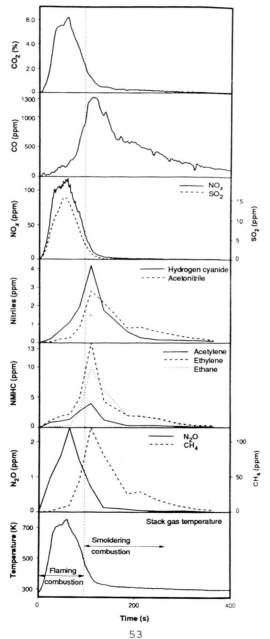

Figure 3.22

The correlation between the molar CO/CO_2 ratio and other variables in a biomass burning experiment. From Kuhlbusch *et al.*[111]. (a) The correlation with the ratio of nitrogen emitted as N_2 to the fuel nitrogen content; (b) the correlation with the ratio of carbon emitted as methane to fuel carbon content. The opposite slopes of the lines in (a) and (b) show that N_2 is a product of the relatively high temperature, flaming, combustion whereas CH_4 is a product of lower temperature, smouldering combustion.

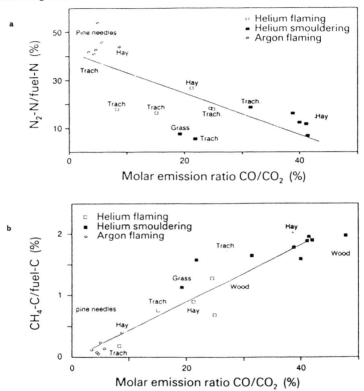

that, whilst the somewhat surprisingly low emission ratio for NO_x undoubtedly lessens the potential for photochemical ozone formation in biomass fire plumes, the loss of the biomass N as molecular nitrogen (so-called *pyrodenitrification*) represents a major loss of fixed nitrogen from the biosphere[111].

Unlike fossil-fuel combustion, temperatures in a bushfire do not reach those required for the oxidation of atmospheric nitrogen by the Zeldovich mechanism (see below). The NO_x emitted can therefore come only from the nitrogen in the fuel itself (it may also come from "prompt" NO_x production in the flame, but this is not discussed here) and, in general, there is a good correlation between fuel-N and NO_x release (Figure 3.23) for a given fire type[114]. Dignon *et al.*[108] have converted this into the linear formula

Table 3.8: Some formulae for the estimation of global trace gas emissions from biomass burning. After Robinson[113].

```
(a) M, the annual mass of biomass burnt

(i)    area burnt x    biomass x    biomass above ground x   above ground biomass burnt
       year            unit area    unit biomass             above ground biomass

(ii)   area burnt  x   biomass  x   biomass burnt
       year            unit area    unit biomass

(iii)  wood used  x    wood burnt
       year            unit wood used

(iv)   total area   x  area burnt   x        biomass   x      biomass burnt
                       total area x year      unit area       unit biomass

(v)    total area  x   no. of fires x  biomass x   biomass burnt
                       year            unit area   unit biomass

(vi)   total area  x   biomass burnt
                       total area x year

(vii)  population   x  wood used   x   wood burnt
                       capita x year   wood used

(viii) population clearing land  x  area cleared  x  biomass  x  biomass burnt
                                    capita x year     unit area   unit biomass

(ix)  crop yield  x    crop waste  x   waste exposed to fire x     waste burnt
                       usable crop     total waste               waste exposed to fire


(b) E_i, the trace gas emission rate per year

(i) M  x    mass of element,χ  x   mass of χ evolved  x   mass of χ evolved as i
            unit biomass           unit mass of χ         unit mass of χ evolved

(ii) M  x  nutrient loading x volatilisation efficiency x emission ratio
```

$$E_F \ (NO_x) \ = \ 3.9 n_f \ - \ 1.5 \tag{3.5}$$

where $E_F(NO_x)$ is the emission rate of NO_x in g N kg^{-1} (dry matter) and n_f is the percentage of nitrogen in the dry biomass. The nitrogen content of fuel is about 0.5 - 2 % of the carbon content, which is itself between 40 and 50 % of the dry biomass[103]. Substituting the average NO_x emission factor of 3.9 g(NO_x) kg^{-1} (~1.4 g N kg^{-1} if $NO:NO_2$ is 1:1) from Hegg *et al.*[109] into equation (3.5) gives a fuel-N of 0.75 %, which is in reasonable agreement with the estimate of Crutzen and Andreae[103].

Non-methane hydrocarbons are also released by biomass burning and some results of field measurements are given in Table 3.9. It is these compounds, rather than CO and CH_4, that will be important in local-scale photochemical oxidant pollution. Each hydrocarbon will contribute to ozone formation in a unique way depending on its structure and reactivity (see chapter 4), but we can make some initial approximation to the behaviour of the system by using the NMHC/NO_x ratio. From the mean fluxes, or "emission factors", given in Table 3.9 we find a NMHC/NO_x flux ratio (expressed as moles C kg^{-1}/moles N kg^{-1}) of around unity. This does not, of course tell us anything about the quality of the air

Figure 3.23

The emission of NO_x as a function of fuel-N[108]. Data used is from Clements and McMahon[114].

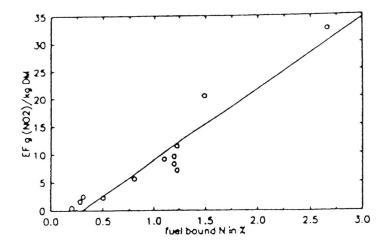

into which the compounds are emitted, but it is at least an indication of the potential for ozone formation due solely to the fire. An examination of typical ozone isopleth diagrams (eg. Refs. 115 and 116 and Fig. 4.10) reveals that $NMHC/NO_x$ ratios of unity are far into the "hydrocarbon-limited" region, so that ozone production from the plume emissions alone is unlikely. Some plumes act as temporary sinks for ozone by titrating O_3 with NO and shifting the photostationary state (chapter 4). However, plumes are released into air which is likely to be very NO_x-poor and enriched in biogenic hydrocarbons such as isoprene, so that ozone production can occur and, indeed, is often observed[2, 109, 107]. Elevated concentrations of CO and CH_4 are available for oxidation too, of course, but even at the high temperatures of the smoke near to the fire their lifetimes against attack by OH are too long for any effect to be seen.

Biomass combustion is becoming, on the laboratory scale at least, reasonably well understood, particularly now that the nitrogen emissions have been speciated[111, 112]. Measurements made *in situ* (eg. Ref. 109) are also yielding important results and may help to make the scale-up to regional and global inventories more accurate. This will also depend on the quality of geographical, ecological and other data, however, and on the method employed to compile the inventory[113].

Table 3.9: Emission factors (g kg^{-1} dry biomass) measured over real fires. From Hegg *et al.*[109].

Fire	Vegetation	CO	CH_4	C_2H_4	C_2H_6	C_3H_4	C_3H_2	n-C_4	NO_x
Lodi I	Chaparral, Chamise	74±16	2.4±0.15	0.58±0.05	0.35±0.12	0.21±0.12	0.32±0.05	0.11±0.07	8.9±3.5
Lodi II	Chaparral, Chamise	75±14	3.6±0.25	0.46±0.03	0.55±0.15	0.32±0.12	0.21±0.03	0.10±0.05	3.3±0.8
Myrtle/ Fall Creek	Pine, brush, Douglas Fir	106±20	3.0±0.8	0.7 ±0.04	0.60±0.13	0.25±0.05	0.22±0.04	0.02±0.04	2.54±0.7
Silver	Douglas Fir, True Fir, Hemlock	89±50	2.6±1.6	0.08±0.01	0.56±0.33	0.42±0.13	0.19±0.09	0.02±0.1	0.81±0.69
Hardiman	Jack Pine, Aspen, Birch	82±36	1.9±0.5	0.58±0.09	0.45±0.26	0.18±0.13	0.31±0.35	0.02±0.04	3.3±2.3
Eagle	Black Sage, Sumae, Chamise	34± 6	0.9±0.2	0.25±0.06	0.18±0.05	0.05±0.02	0.08±0.02	0.02±0.08	7.2±3.8
Battersby	Jack Pine, White & Black Spruce	175±91	5.6±1.7	0.9 ±0.15	0.57±0.45	0.27±0.12	0.33±0.06	0.07±0.06	1.05±1.33
Overall average	-	91±21	2.9±0.66	0.51±0.14	0.47±0.11	0.24±0.06	0.24±0.04	0.10±0.05	3.9±1.6

n-C_4 is a straight chain paraffin with carbon number 4.

3.5 COMBUSTION OF FOSSIL FUELS

The concerted utilization of fossil fuels marks industrialised societies out from other societies which have existed - and which remain, albeit under pressure from those dominant and energy intensive industrialised societies. It is not difficult to understand the importance of fossil fuel combustion to photochemical pollution since it is the high temperature analogue of atmospheric oxidation. Retrieving fuel from geological reservoirs means that there is additional oxidation to that which would occur naturally. The important factors which will determine the significance of fossil fuel combustion exhaust to photochemical oxidant production will be: the degree to which hydrocarbons have been oxidised in the flame; the amount of catalytic NO_x produced; and, of course, the magnitude of the emission.

The energy released by fossil fuel combustion is used as a direct source of heat, as a source of the heat required to vaporise water and so drive steam turbines in the power industry, and as a source of mechanical power in the internal combustion engine. In these two latter applications the fuel must be atomised for fast, efficient combustion. This is achieved by pre-vaporisation or spraying in the case of liquids, and pulverisation in the case of coal. Table 3.10 shows methods for introducing liquid fuel into a flame, and their applications[117]. By careful control of conditions in this way the molar ratio of fuel to oxygen (the equivalence ratio, ϕ) can be maintained at a nearly uniform level. It is ϕ which determines the relative importance of competing chemical processes, and hence, the exhaust products of combustion.

The details of fossil fuel combustion will differ, therefore, according to the phase of the fuel and the design of the combustor, and ϕ, but some general points can be made. "Hot" combustion, of the kind seen in power generation and internal combustion, involves the production of O and OH radicals:

$$H + O_2 \rightarrow HO + O \hspace{4cm} [3.14]$$
$$H_2 + OH \rightarrow H_2O + H \hspace{3.5cm} [3.15]$$
$$H_2 + O \rightarrow H + OH \hspace{3.7cm} [3.16]$$
$$H_2O + O \rightarrow 2OH \hspace{4cm} [3.17]$$

and subsequent attack at a C-H bond in the fuel

$$RH + OH \rightarrow R + H_2O \hspace{3.5cm} [3.18]$$
$$RH + O \rightarrow R + OH \hspace{3.7cm} [3.19]$$
$$RH + H \rightarrow R + H_2 \hspace{3.8cm} [3.20]$$

The reactions above might also be addition reactions, if the fuel contains unsaturated

Table 3.10: Systems used for the combustion of liquid fuels as sprays[117]. The designation "premixed" or "diffusion flame" is a general indication only of the dominant behaviour.

Application	Configuration	Independent variables*	Structure
Prevaporizing system afterburners lean combustors carburetor ramjets		Z	Steady nonburning
Liquid rocket engines		Z	Steady, more or less premixed
Gas turbine combustors		R, Z	Steady, diffusion flame
Industrial furnaces		X, Y, Z	Steady, diffusion flame
Diesel engines		t, X, Y, Z	Transient diffusion flame, ignition characteristics needed

* Simplest realistic approximation, all systems are axisymmetric or three dimensional near the injector.

carbon-carbon bonds[118]. Initiation of the free-radical chain process is via

$$RH + O_2 + M \rightarrow R + HO_2 + M \qquad [3.21]$$

although some high temperature thermolysis can also occur

$$RH + M \rightarrow R + H + M \qquad [3.22]$$

$$RH + M \rightarrow R' + R'' + M \qquad [3.23]$$

Further breakdown of the hydrocarbon moieties is by H-atom abstraction, addition and rearrangement, in a manner similar to atmospheric degradation (See Chapter 4). Figure 3.24

Figure 3.24

Reaction pathways in the stoichiometric combustion of methane with air at 1000 mbar. Thickness of the arrows is proportional to calculated reaction rates integrated over the whole flame front[120].

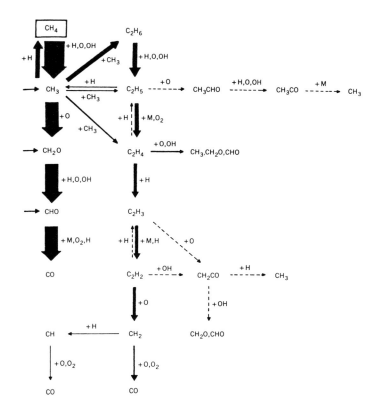

shows one such degradative pathway[119, 120]. Note that radical-radical recombination i.e.,

$$CH_3 + CH_3 \rightarrow C_2H_6 \qquad [3.24]$$

is significant, particularly in "rich" flames ($\phi > 1$). Further carbon-chain extension can occur, but the products quickly revert to methyl and ethyl species. Conversely, the combustion of longer hydrocarbons is rate-limited by the decomposition of methyl and ethyl species, so that all alkanes and alkenes have similar flame properties. Addition and recombination reactions also result in the formation of polycyclic aromatic hydrocarbons (PAH) which, although not important in photochemical oxidant production, pose a direct threat to health due to their carcinogenicity.

The evolution of chemical species during combustion is shown in Figure 3.25 for

Figure 3.25

The concentration of gases, and distributions of temperature and flow velocity, downstream of a flame from an air-blast injector[117]. The diagram shows results from a propane flame (a) and from a kerosene spray flame (b). The results for both flames are broadly similar. The hydrocarbon (HC) and droplet concentration distributions in (b) suggest that droplet evaporation occurs prior to oxidation.

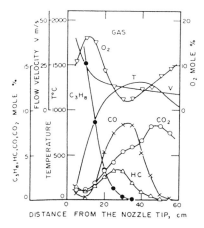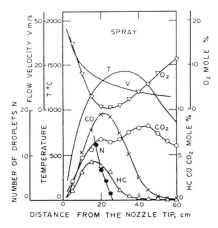

a spray of kerosene droplets, and for propane gas, injected into the system in the same manner[117]. Temperatures reach 1200-1500 K and oxygen is rapidly consumed. Lower hydrocarbons and CO are formed by decomposition of the fuel and subsequently decay as they themselves are oxidised. When the droplet flame is viewed in the radial plane (Figure 3.26), it can be seen that droplet evaporation occurs before significant oxidation takes place. Model prediction of the axial distribution of chemical species in the combustion of coal is given in Figure 3.27[121].

Due to the high temperatures occurring in a flame, combustion of volatile non-methane hydrocarbons (NMHC) to CO usually proceeds to completion. Release of NMHC from combustors, particularly internal combustion engines, is, therefore, something of a surprise, and one which is not easy to explain quantitatively[119]. Early attempts to explain NMHC release as an artefact of the boundary layer between flame and wall (e.g. Ref. 122) require unrealistic layer thicknesses or diffusion coefficients, and it now appears that unburnt fuel seeping into and out of crevices in the cylinder and local inhomogeneities in the equivalence ratio may be the best explanations of fuel surviving passage through the combustor.

Figure 3.26

The radial distribution of gases, temperature and flow velocity for the spray flame in Figure 3.25(b)[117]. The section is taken upstream of the minimum in oxygen concentration.

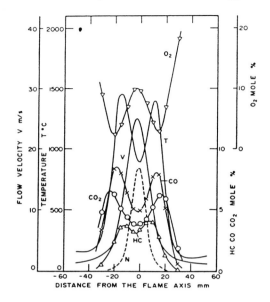

NO produced in flames is of three kinds: *Thermal* NO, *prompt* NO, and *fuel* NO. The first of these is due to O-atom catalysed oxidation of N_2, via the Zeldovich mechanism (e.g. Ref. 123):

$$O + N_2 \rightarrow NO + N \qquad\qquad [3.25]$$
$$N + O_2 \rightarrow NO + O \qquad\qquad [3.26]$$

Net: $\quad N_2 + O_2 \rightarrow 2NO$

Reaction [3.25] has an activation energy of 76.5 kJmol^{-1} and so is very sensitive to temperature. Cooling the flame is therefore a very efficient method of reducing the contribution of thermal NO to overall NO emissions. This can be achieved by lean burn, water injection, stratified charge (dilution downstream of the spark), and exhaust recirculation[123, 124, 125].

Measurements of NO very close to the flame front found that there is a rapid source of NO, particularly in "rich" flames ($\phi > 1$) which could not be accounted for by the Zeldovich mechanism. It has been shown that such "prompt" NO formation is due to the production of alkyl radicals by thermolysis followed by attack on N_2:

Figure 3.27

The predicted structure of a laminar coal-air flame for 0.3 kg m^{-3} of 30 μm bitumous Pittsburgh coal particles in air. From Smoot and Horton[121].

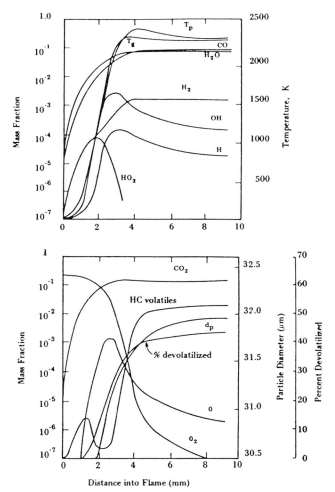

$$RH \rightarrow CH, CH_2 \text{ etc.} \tag{3.27}$$

$$CH + N_2 \rightarrow HCN + N \tag{3.28}$$

$$CH_2 + N_2 \rightarrow HCN + NH \tag{3.29}$$

$$NH + H \rightarrow N + H_2 \tag{3.30}$$

$$N + OH \rightarrow NO + H \tag{3.31}$$

$$N + O_2 \rightarrow NO + O \tag{3.26}$$

$$NH + O \rightarrow NO + H \tag{3.32}$$

$$\text{NH} + \text{O} \rightarrow \text{N} + \text{OH} \qquad\qquad\qquad [3.33]$$

$$\text{HCN} + \text{OH} \rightarrow \text{CN} + \text{H}_2\text{O} \qquad\qquad\qquad [3.34]$$

Figure 3.28

Prompt NO production as a function of the equivalence ratio, ϕ, for a variety of hydrocarbon fuels[126, 123]. The dotted lines in the propane graph show the uncertainties in the determination; similar curves were obtained for the other fuels. Notice the general tendency for prompt NO to peak on the rich side of stoichiometric.

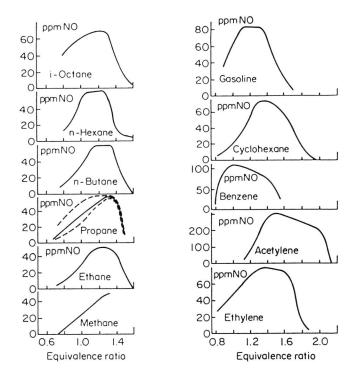

Figure 3.28 shows the production of prompt NO as a function of equivalence ratio for several hydrocarbon fuels[126]. Prompt NO production falls off very rapidly at very rich equivalence ratios ($\phi \geq 1.5$) because O and OH radical atom concentrations become too small to convert HCN to NO. (Bachmeier et al, 1973). In moderately rich flames prompt NO is proportional to the molar concentration of carbon atoms, and so is independent of the fuel used[123].

Production of NO from nitrogen contained in the fuel has already been discussed for biomass burning. Thermolysis produces NH_i ($i = 1, 2, 3$) radicals and HCN which then go on to form NO by the same route as that for prompt NO and by

$$\text{NH}_2 + \text{O} \rightarrow \text{HNO} + \text{H} \qquad\qquad\qquad [3.35]$$

$$NH_2 + O_2 \rightarrow HNO + OH \qquad\qquad [3.36]$$

$$HNO + X \rightarrow NO + XH \qquad\qquad [3.37]$$

$$HNO + M \rightarrow NO + H + M \qquad\qquad [3.38]$$

NO production from fuel is therefore a very rapid process, like prompt NO production, and is only slightly temperature dependent, so it cannot be avoided in the same way as thermal NO production. Fuel NO production as a fraction of N-content is greatest for "lean" flames ($\phi > 1$) where O atom concentrations are relatively high. In rich flames, reaction of fuel-

Figure 3.29

Schematic diagram of the principal reaction paths in the fuel nitrogen conversion process in flames. After Bowman[125]. Notice the path to N_2 which has been shown to be important in biomass burning.

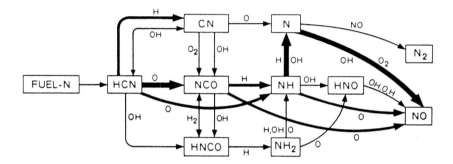

nitrogen products with other nitrogen radicals becomes competitive with NO formation, and N_2 production results[127] (see Figure 3.29):

$$NH_2 + NO \rightarrow N_2 + H_2O \qquad\qquad [3.39]$$

$$N + NCO \rightarrow N_2 + CO \qquad\qquad [3.40]$$

$$N + CN \rightarrow N_2 + C \qquad\qquad [3.41]$$

Flames produced by the burning of biomass are very fuel-rich which explains the N_2 production which has been observed in test burns[111]. This process is known as pyrodenitrification because fixed nitrogen from plants is converted back into N_2.

NO$_2$ production has been observed in flames although, at flame temperatures, the equilibrium

$$NO + 0.5 \, O_2 \rightleftharpoons NO_2 \qquad\qquad [3.42]$$

should lie very far to the left. It has been suggested that NO_2 production occurs via

$$NO + HO_2 \rightarrow NO_2 + OH \qquad\qquad [3.43]$$

in the low temperature region of the flame where HO_2 is at a maximum. NO_2 can be destroyed by reaction with O or H in the high temperature region

$$NO_2 + O \rightarrow NO + O_2 \qquad\qquad [3.44]$$

$$NO_2 + H \rightarrow NO + OH \qquad\qquad [3.45]$$

so that it is presumed that the presence of NO_2 in exhaust gas is indicative of inhomogeneities in the fluid as it passes through the combustor[125].

The characterization of individual flames is important in understanding the chemical processes occurring during combustion but, in order to scale up exhaust emissions to regional and global scales, some information regarding the variability of combustor performance is required. This is a relatively straightforward task for the few, large, power generating combustors but for small, mobile sources like cars the best strategy is to measure the actual exhaust composition of a characteristic sample under characteristic conditions of use. Standard test cycles are used for laboratory studies and development of regulations[128] and are designed to simulate different driving conditions such as urban, motorway, etc. Emissions of total hydrocarbons (THC), carbon monoxide (CO) and oxides of nitrogen (NO_x) are regulated and hence come under the closest scrutiny. However, in order to understand the effect of vehicle emissions on photo-oxidant production, the speciation of hydrocarbon emissions is also important. The primary variables which require investigation are then engine design, fuel composition, timing, ambient temperature, and the equivalence ratio, ϕ.

A schematic of the general behaviour of internal combustion engine exhaust emissions with ϕ is given in Figure 3.30[122]. NO emissions peak slightly to the lean side of a stoichiometric burn, whereas THC and CO emissions increase rapidly on the rich side. The lean peak is a result of the kinetics of NO formation which is favoured by higher O atom concentrations and inhibited by the presence of alkyl radicals. Exhaust gas NO levels are much higher than predicted from equilibrium values at the exhaust gas temperature implying that the products from combustion at higher temperatures have been "frozen" into the composition of the exhaust gas. The increase of THC and CO emissions on the rich side of stoichiometric is obviously the result of incomplete combustion in the engine cylinder.

The temperature at which standard tests are performed can have a significant impact on the resultant emissions[129, 130]. In general, THC and CO emissions increase markedly with decreasing temperature. NO_x emissions also increase, but much less markedly (Figure 3.31). Figure 3.31 also shows the sensitivity of emissions to fuel type and test cycle: Fuel with the higher olefinic and aromatic content typical of winter gasoline (WG)

Figure 3.30

Schematic diagram of the variation of NO, CO and THC emissions with equivalence ratio, ϕ, in a conventional spark-ignition engine[122].

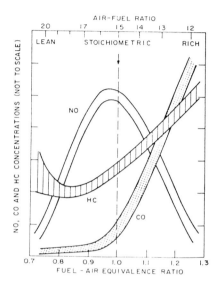

gave slightly higher CO and THC emissions; testing which included idling at the start of the cycle to allow the engine to warm up, resulted in lower emissions of THC and CO. For both these tests NO levels remained effectively unchanged. These tests reveal the fortuitous result that exhaust emissions of THC and CO are lower in the high temperature conditions typical of photo-oxidant pollution episodes.

More radical changes in fuel composition have also been studied, including the use of alternative fuels such as methanol and other alcohols[131, 132, 133, 134]. Reformulating gasoline to reduce its vapour pressure leads to moderate reductions in THC and CO of 5-18% and 11-21%, respectively, with no change in NO_x emissions[131]. The effects of blending commercial gasoline with ethanol and of blending refiner's base stock with methyl *tertiary* butyl ether (MTBE) are shown in Table 3.11 and Table 3.12 as a function of temperature on an urban test cycle[130]. At 40 °F (4 °C), addition of 8.1 %v/v ethanol to commercial summer grade gasoline produced an increase in THC and CO emissions; addition of 16.2 %v/v MTBE to refiner's base stock substantially reduced THC and CO emissions. At higher temperatures the effect of both blends becomes negligible. There is little effect on NO_x emissions throughout, with the possible exception of an increase of some 30 % for the

Figure 3.31

Effects of temperature, fuel type and test cycle on emissions of THC, CO and NO$_x$. FTP = Federal Test Procedures (UDDS); MFTP = modified FTP (5 minute idle at start of test). SG = summer gasoline blend; WG = winter gasoline blend[130].

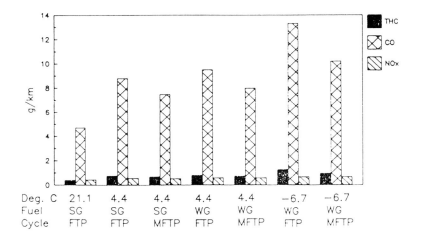

addition of MTBE to refiner's base stock at 40 °F (no indication of the statistical significance of the result is given in Stump et al.[134]). The emissions of individual hydrocarbon species are also shown in Table 3.11 and Table 3.12. Differences generally reflect the chemical composition of the fuel: The ethanol blend shows increased emissions of ethanol itself and other oxygenates (acetaldehyde and other aldehydes); the MTBE blend emits MTBE itself and increased amounts of paraffins.

The effect of φ on the emissions from blended fuel is shown in Figure 3.32. Note that in this figure the air-fuel equivalence ratio is used: This is the reciprocal of our φ, so that lean (fuel-lean) mixes occur to the right of unity, and rich (fuel- rich) mixes to the left. The general picture is as would be expected from Figure 3.30 with CO and THC (here expressed as the mass of carbon and hydrogen emitted) increasing for rich mixes. Blending with alcohols decreases CO emissions in rich combustion. This effect increases as the molecular wt of the alcohol decreases and is due to the fact that volumetric addition of alcohols effectively makes the combustion leaner. In lean combustion the "equivalent hydrocarbon" emission increases with blending and with decreasing molecular wt of the blending alcohol (Figure 3.32b). Blending decreases NO$_x$ emissions (Figure 3.32c) by

Table 3.11: Effects of ambient temperature and fuel blend on emissions from a 1988 GM Corsica with current control technology during an Urban Dynamometer Driving Cycle (UDDS)[134].

Vehicle Test Conditions

Test temp. deg. F (°C) Test fuel	40(4) Base A	40(4) Blend A	75(21.5) Base A	75(21.5) Blend A	90(29) Base A	90(29) Blend A
THC, g/km	0.35	0.49	0.47	0.46	0.2	0.29
CO, g/km	6.46	6.89	6.46	6.54	2.74	2.62
NOx, g/km	0.3	0.3	0.47	2.90	0.24	0.23
Ethanol, mg/km	0.00	16.43	0.00	10.60	0.00	14.38
Formaldehyde, mg/km	1.7	1.78	4.14	1.78	1.49	1.75
Acetaldehyde, mg/km	1.33	3.31	1.69	1.80	0.62	1.77
Aldehydes, mg/km	4.22	6.16	7.86	4.54	2.68	4.14
Benzene, mg/km	13.76	17.3	17.83	9.49	4.22	10.60
1,3-Butadiene, mg/km	0.5	0.68	0.78	0.25	0.19	0.31
Paraffins, wt%						
Methane	12.94	12.68	10.90	12.76	12.32	11.78
Iso-butane	0.32	0.30	0.35	1.21	1.21	1.64
N-butane	2.50	1.97	2.61	7.27	7.27	8.06
Iso-pentane	3.74	3.15	4.37	6.34	6.34	5.05
N-pentane	1.62	1.46	1.64	2.31	2.31	1.86
Total paraffins	46.05	44.48	44.08	53.29	53.29	51.14
Olefins, wt%						
Ethylene	7.74	6.75	6.28	6.42	6.42	4.06
Propylene	2.51	2.08	2.95	3.22	3.22	2.59
1-Butene	0.68	0.59	0.71	0.64	0.64	0.58
Iso-butylene	0.88	0.68	0.83	1.01	1.01	1.22
Total Olefins	19.69	16.86	18.03	18.39	18.39	16.20
Aromatics, wt%						
Benzene	4.01	3.55	3.85	2.26	2.26	3.61
Toluene	3.58	3.53	3.13	3.32	3.32	2.83
p/m-xylene	4.16	4.20	3.16	3.06	3.06	3.11
o-xylene	1.30	1.47	1.10	0.96	0.96	1.13
Total aromatics	32.70	34.10	34.76	24.99	24.99	25.85
Acetylene, wt%	0.57	0.63	0.75	0.40	0.40	0.82
Unknowns, wt%	1.23	0.47	1.71	2.58	2.58	0.98

Base A - Unleaded regular summer grade gasoline.
Blend A - 8.1% Ethanol blended with Base A.

Table 3.12: As for Table 3.11, but with Refiners base blend (Base B) and a MTBE blend (Blend B)[134].

Vehicle Test Conditions

Test temp. deg F (°C) Test fuel	40(4) Base B	40(4) Blend B	75(21.5) Base B	75(21.5) Blend B	90(29) Base B	90(29) Blend B
THC, g/km	0.31	0.21	0.18	0.16	0.13	0.19
CO, g/km	6.56	4.15	2.39	1.96	2.16	2.17
NOX, g/km	0.21	0.27	0.22	0.22	0.22	0.17
MTBE, mg/km	0.00	2.17	0.00	0.93	0.00	3.91
Formaldehyde, mg/km	1.49	1.66	1.37	1.97	1.35	1.62
Adldehydes, mg/km	3.50	2.85	2.81	2.96	2.71	2.5
Benzene, mg/km	11.72	3.72	6.57	2.60	5.52	3.22
1,3-Butadiene, mg/km	0.19	0.12	0.19	0.12	0.12	0.12
Paraffins, wt%						
Methane	14.01	9.97	12.40	15.54	13.86	11.07
Iso-butane	1.10	1.17	1.08	2.06	2.40	8.36
N-butane	2.28	2.50	2.00	3.32	3.23	11.28
Iso-pentane	1.26	2.03	1.22	1.85	1.21	2.74
N-pentane	0.29	0.41	0.30	0.30	0.23	0.40
Total paraffins	44.46	51.11	40.87	52.15	44.53	61.90
Olefins, wt%						
Ethylene	8.07	5.83	5.33	6.02	6.42	4.96
Propylene	2.51	3.07	2.92	3.63	3.77	3.84
1-Butene	0.85	0.52	0.90	0.64	0.63	0.91
Iso-butylene	1.37	5.27	1.08	4.82	1.83	5.05
Total Olefins	19.62	21.49	18.70	20.97	17.62	21.70
Aromatics, wt%						
Benzene	3.75	1.77	3.60	1.62	4.18	1.70
Toluene	21.76	5.76	17.63	5.97	18.44	4.47
p/m-xylene	1.63	3.34	1.88	2.00	1.35	1.51
o-xylene	0.65	0.88	0.54	0.56	0.46	0.56
Total aromatics	34.06	23.62	36.91	24.48	34.65	12.61
Acetylene, wt%	1.27	1.70	1.14	1.33	1.70	1.26
Unknowns, wt%	0.52	0.42	1.97	1.31	1.36	0.29

Base B - Refiners blend stock fuel.
Blend B - 16.2% Methyl tertiary butyl ether in Base B.

Figure 3.32

Variation of CO (a), THC (b) and NO$_x$ (c) emissions as a function of the inverse equivalence ratio, $1/\phi$, for a variety of fuels[133]. Hydrocarbon emissions are given as the amount of carbon and hydrogen emitted - the organic material "hydrocarbon equivalent". The engine was stationary and running at 1500 rpm.

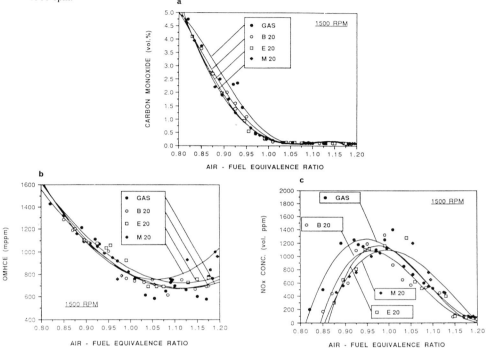

lowering the flame temperature in the engine. Switching to high methanol blends (85 %v/v methanol in gasoline) produces a decrease in the THC and NO$_x$ emissions, a small increase in CO emissions, and a large increase in formaldehyde emissions[132, 135, 136]. Weighing up this change in emission speciation, a change from gasoline to high-methanol fuel would probably result in an overall decrease in the potential for ozone formation in urban air[137], but the cost effectiveness of such a measure remains controversial[138].

Diesel engines generally produce less THC and CO, but more NO$_x$ (Ref. 139). The exhaust component from diesel engines of most concern is aerosol because it contains mutagenic and carcinogenic substances; it does not, however, play any part in photochemical oxidant production (unless it can provide a surface for heterogeneous chemistry). Exhaust emissions from heavy-duty vehicles form a substantial part of the total automotive exhaust inventory and these vehicles generally run on diesel. Emissions from

Figure 3.33

Emission of NO$_x$, CO and methanol and total aldehydes from buses running on diesel or methanol. From Eberhard *et al.*[142]. Tests were conducted at Chevron and SWRI; with two-way catalyst in place on the methanol fuelled vehicle, and removed.

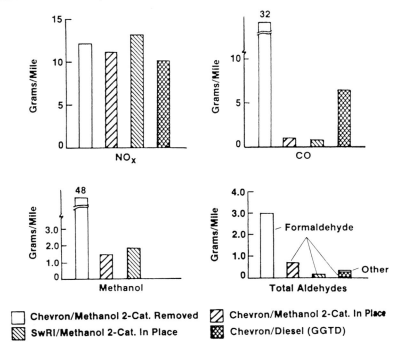

heavy duty vehicles are generally a complex function of engine speed, engine load, the rate at which these change, and temperature[140, 141]. Figure 3.33 shows the emissions from buses running diesel or methanol[142]. As for the light-duty gasoline vehicles, conversion to methanol results in a decrease in CO emissions and an increase in methanol and formaldehyde emissions, but only if an effective catalyst is in place. NO$_x$ emissions increase slightly on switching to methanol.

Whilst quantifying the emissions of THC, CO and NO$_x$ integrated over a standard test cycle is essential for the establishment of standards which vehicles must meet, accurate estimation of the contribution of vehicles to regional pollution requires knowledge of emission characteristics under non-standard conditions. This can be done by breaking down test cycles into segments of uniform operation (e.g. Ref. 143), or by recording the emissions of vehicles in use[144, 145, 146]. Figure 3.34 shows some on-the-road emissions

Figure 3.34

Emissions of CO (a), THC (b) and NO$_x$ (c) measured on the road using a portable exhaust analyser. The figure shows the envelope of emissions for the 20 cars surveyed: these were selected as being typical of the 1987 U.K. fleet. After Potter *et al.*[144]. Note the non-linear dependence of emissions on vehicle speed, and the increase in emissions at the low speeds typical of urban driving.

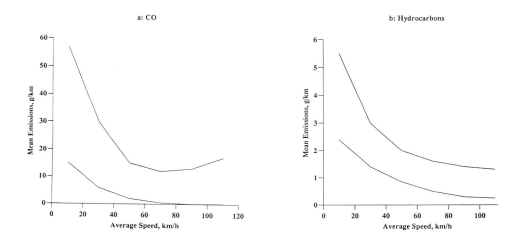

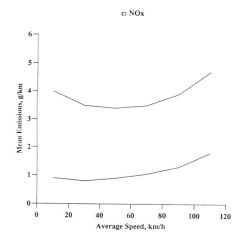

Figure 3.35

Variation of THC composition with vehicle speed for a survey similar to that used to compile
Figure 3.34[145]. Gasoline components tend to decrease in abundance as vehicle speed increases. For
combustion products the trend is opposite.

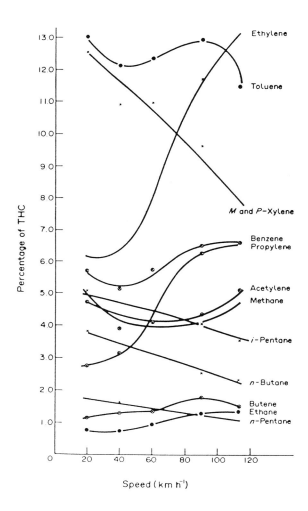

as a function of vehicle speed. NO_x emissions vary only weakly with speed, but THC and CO decrease markedly as speed increases. The speciation of hydrocarbons also changes with vehicle speed (Figure 3.35), reflecting a change from direct gasoline emissions (saturated alkanes and akyl benzenes) to combustion product emissions (ethylene and propylene). This is equivalent to an increase in overall reactivity of the emissions with speed (see Chapter 4) although, as can be seen from Figure 3.34, the total amount of HC emission is decreasing.

Figure 3.36

Contribution of various phases of the test cycle to total emissions in the FTP urban test cycle, as a function of fuel composition[132]. There are large contributions from both "hot soak" and "diurnal" evaporative emissions. "Cold", "Stabilized" and "Hot" refer to the beginning, middle and end of the test cycle, respectively.

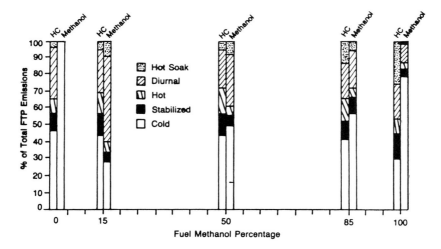

The impact of motor vehicle use on hydrocarbon emissions must also take account of evaporative emissions, both during refuelling and during operation. Three situations are important in evaporative emissions from the mototr vehicle itself: "Diurnal" emissions arise from the exchange of air in the vehicle as ambient temperatures vary throughout the day; "hot soak" emissions arise from the evaporation of fuel from the hot engine immediately after use; "running losses" arise from the expulsion of vapour from the fuel tank during operation. A standard formula converts these emissions to the same units as exhaust emissions:

$$E(g\ mi^{-1}) = \frac{3.05\,g_{hs} + g_d}{31.1} \qquad (3.6)$$

where E is the evaporation rate in g mi^{-1} assuming usage of 31.1 mi d^{-1} made up of 3.05 trips d^{-1}, g$_{hs}$ is the hot soak emission rate (g trip^{-1}), and g$_d$ is the diurnal emission rate (g d^{-1}). Evaporative emissions are calculated using the U.S. Federal Test Procedure standard Sealed-House Evaporative Determination (SHED). An example of the contribution of evaporative emissions to the total emission is given in Figure 3.36. Diurnal emissions generally form a larger fraction of the total than hot soak emissions: Together, up to 40% of the total can be made up of evaporative emissions[132]. The mass of evaporative emissions will, of course, be a direct reflection of the volatility of the fuel, so that high-methanol fuel (85% v/v, Reid vapour pressure, RVP = 8 psi) produces less than gasoline[134, 136] (RVP = 10.2 psi). There may also be a long-term trend of increasing RVP in conventional gasoline, at least as sold in the U.S.A.[147, 148] which would lead to increased evaporative and refuelling emissions.

Even given relatively well-characterised test-cycle and individual vehicle emission measurements, it is not a trivial task to scale these up into area emission rates. As well as emission data for all vehicle types (as a function of age, condition, fuel type, etc.), factors describing the distribution of vehicle types, the distribution of technology (catalysts, etc.), vehicle miles travelled per day, number of journeys, and average speed per journey, are

Figure 3.37

Speciated hydrocarbon emission rates (mg C km^{-1}) for urban, rural and motorway driving conditions. From Bailey and Eggleston[150]. Urban emissions per km are generally largest of the three categories, but distances travelled (per journey) will usually be less.

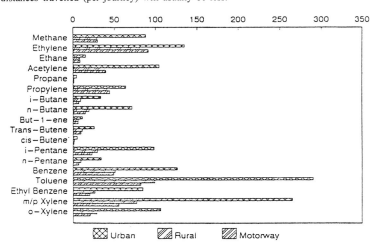

Table 3.13: 1985 Motor Vehicle Fleet Average Emission Factors[151], g km^{-1}. VKT = vehicle kilometres travelled per year.

	LDGV	LDGT	HDGV	LDDV	LDDT	HDDV	MC	ALL
Veh. speed km h^{-1}	32.3	32.2	32.3	32.2	32.2	32.2	32.2	-
*fraction of VKT	0.652	0.218	0.040	0.023	0.008	0.054	0.007	-
Exhaust NMHC	1.06	2.08	3.20	0.24	0.38	2.80	2.22	1.45
Exhaust HC	1.14	2.19	3.44	0.25	0.39	2.88	2.37	1.54
Evap NMHC	0.86	1.33	4.51	-	-	-	2.11	1.05
Refuel HC	0.21	0.27	0.44	-	-	-	-	0.21
THC	2.21	3.79	8.37	0.25	0.39	2.88	4.48	2.79
CO	11.95	20.96	86.74	0.80	0.96	7.76	12.40	16.33
NO$_x$	1.24	2.09	3.46	0.86	1.02	12.84	0.52	2.13

* Proportion of total VKT due to each category

LD = light duty,   HD = heavy duty

GV = gasoline vehicle,   DV = Diesel vehicle

GT = gasoline truck,   DT = Diesel truck

MC = motorcycle

required. It is then usually easiest to calculate emission rates as a function of time and season for each link in a road network[149, 150]. A typical calculation is given by Black[151], and is shown in Table 3.13. Light-duty gasoline vehicles make up the major fractions of THC, CO and NO_x emission, and emissions due to evaporation together with refuelling contribute as much to the THC emissions as do those due to exhaust. For use as input to photochemical models the hydrocarbon emission data should be speciated and, with the onset of new measurement technology, this is now becoming possible[150] (Figure 3.37). This shows that, for the UK large amounts of vehicle hydrocarbon emissions occur as reactive aromatic and olefinic species. Figure 3.37 also shows the breakdown of emissions by road type. Here it is obvious that urban emissions dominate over rural and motorway emissions. This is due to the very rapid increase in hydrocarbon emissions with decreasing speed (Figure 3.34).

However, as with the scale up of other distributed sources, errors are attached to each of the geographical factors used in the calculation and so final estimates are quite uncertain. A comparison of scaled up emissions against measured emissions for one particular location showed observed-to-calculated ratios of 2.7±1.8, 4.0±1.8 and about unity for CO, THC and NO_x, respectively[152]. This suggests that emission inventories may be underestimating $NMHC/NO_x$ ratios substantially and hence underestimating the potential for ozone formation from these emissions[138]. In order to improve the agreement between calculation and measurement, the following improvements to current US emission calculation algorithms have been suggested:

(i) improve the treatment of evaporative losses;

(ii) output error estimates alongside results;

(iii) improve the algorithm's flexibility so that specific scenarios can be explored;

(iv) include additional algorithms to describe the failure of specific vehicle components;

(v) output the results on model grids for easy insertion into photo-chemical models; and

(vi) include a factor describing the reactivity of the fuel.

In addition to these refinements to the algorithm, the following improvements to input data were also identified as being of importance:

(i) accurate representation of the status of the fleet, with particular reference to high emitting vehicles;

(ii) a more accurate representation of actual driving patterns in vehicle testing;

(iii) accurate emission rates for heavy duty vehicles;

(iv) data on alternative fuel penetration into the market; and

(v) re-evaluated evaporative emissions over the fleet.

With such an impressive list of refinements to algorithms and data being required, it may be some time before true benchmark estimates of mobile emissions are available.

3.6 PRECURSOR EMISSION INVENTORIES

Compiling regional and global compendia of photochemical oxidant precursor emissions is, as will have become obvious during the foregoing discussion, a labour intensive and, currently at least, subjective task. Yet, recognized standards in emission inputs are at least as important as standard physical parameterizations and numerical procedures if prognostic models of the atmosphere are to be used as the basis for legislation. It is not possible, therefore, to wait until the uncertainties associated with many source processes are greatly reduced (or even quantified) before proceeding to compile areal compendia. Moreover, as we shall see, the requirement for mass continuity in global budgets can help to constrain emissions which, in terms of the uncertainties associated with individual emission measurements, appear very ill-constrained indeed.

An excellent compilation of inventories and the primary databases from which they are derived has been published recently[153]. In this, the central formula used to compute areal emissions is given - i.e.

$$E_{i,\Delta t, A(x,y)} = \sum_t \sum_{A(x,y)} \sum_s (N_{s,A(x,y)} \ U_{s,A(x,y),\Delta t} \ \Phi_{s,i,t})$$ (3.7)

along with a bibliography of the data needed in the computation. In equation (3.7), E is the emission rate of species i over time interval, Δt, and over the area A(x,y); N is the number of sources of type s in area A(x,y); U is the time-dependent usage of source s in area A(x,y); and Φ is the flux, or "emission factor", of species i from source s. The dependence of N, U, and of Φ on their variables will be very different for biogenic sources and anthropogenic sources, but so too will the dependence for individual species within either category. Sample data sets which can be used to derive N and U are given in Table 3.14 and Table 3.15. Climatic data sets[154, 155, 156, 157, 158, 159, 160] such as surface air temperature and precipitation are required for the calculation of biogenic inputs - for example, biogenic NO emissions are sensitive to temperature and soil water content (Figure 3.9). Agricultural and geographical data sets[161, 162, 163, 164, 165, 166, 167, 168], such as livestock populations[169] and rice cultivation[170], are obviously of prime importance for the calculation of methane emissions (Table 3.1, Table 3.2). The industrial

Figure 3.38

A global inventory of NO$_x$ emissions (tonnes N) from biomass burning (a) October - March, (b) April - September. This source is confined almost exclusively to the tropics. After Dignon et al.[108].

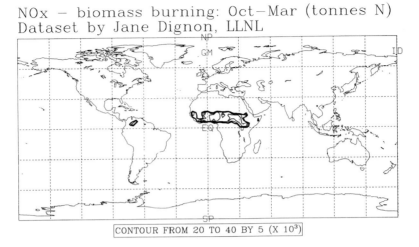

NOx — biomass burning: Oct—Mar (tonnes N)
Dataset by Jane Dignon, LLNL

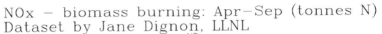

CONTOUR FROM 20 TO 40 BY 5 (X 10³)

NOx — biomass burning: Apr—Sep (tonnes N)
Dataset by Jane Dignon, LLNL

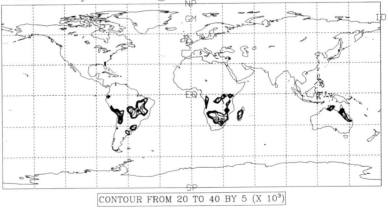

CONTOUR FROM 20 TO 40 BY 5 (X 10³)

data[171, 172] listed in Table 3.15 will establish anthropogenic emissions from stationary sources but some further information is required if mobile sources are to be calculated. Two examples of the kind of detailed emissions inventory which can be produced are given in Figure 3.38 and Figure 3.39. The first of these shows NO$_x$ production from biomass burning (see the beginning of § 3.3) averaged over the periods October to March and April to September. Emissions are concentrated in the tropics and are synchronised with the

Figure 3.39

A global inventory of NO_x emissions (tonnes N yr^{-1}) from fossil fuel combustion. The contribution from the Industrialised nations dominates. After Dignon *et al.*[108].

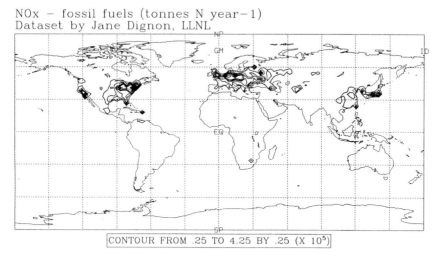

NOx – fossil fuels (tonnes N year−1)
Dataset by Jane Dignon, LLNL

CONTOUR FROM .25 TO 4.25 BY .25 (X 10^5)

occurrence of the dry season in each hemisphere. A similar inventory for fossil-fuel NO_x emissions (Figure 3.39) shows a very different spatial distribution, their being concentrated over the industrialised regions of the northern hemisphere. Notice also the difference in magnitude of the biomass-burning and fossil-fuel sources: annual emissions from industrialised countries are about ten times greater than those from the countries in which widespread biomass burning occurs.

Aggregate emissions are useful in the formulation of control strategies, and examples on the U.K. and European scales are given in Table 3.16 and Table 3.17[173, 174, 175, 176, 177]. The National Atmospheric Emissions Inventory for the U.K. in 1991, shows that whilst NO_x and NMHC emissions are of the same order (by mass), the contribution of mobile and stationary sources to each emission differs widely. This suggests, for example, that motor vehicle emissions will emit pollution at low NMCH/NO_x ratios into background air having a much higher NMHC/NO_x ratio and may, therefore, reduce the potential for ozone production locally[178] (see § 4.3).

Mass emissions of NO_x and NMHC are also about equal for the whole of Europe (Table 3.17). Notice that the biogenic source of NMHC is estimated to be much smaller than the anthropogenic source. A very rough estimate of the effect of these emissions in a regional sense can be found by making the crude approximation that all the NMHC is

Table 3.14: A selection of geophysical data sets which are useful in compiling global emissions inventories. After Graedel et al.[153].

Parameter	Spatial Resolution	Temporal Resolution	Effective Year[a]	Reference
Atmosphere				
Specific humidity	2.5°lat.x5°long. x11 levels	Monthly	(1958-73)	154
Surface air temp.	2°x2.5°	Monthly	(1958-73)	155
ΔT surface	4°x5°	Monthly	1880-1990	156
ΔT surface	hemisphere	Monthly	1861-1990	157
precipitation	2°x2.5°	Monthly	(1958-79)	155
Δ Precipitation	"global"	Seasonal	1880-1990	158
Clouds	280 km x 280 km	3-hourly	1983-90	160
Clouds	5°x5°	Monthly	1952-81	159
Tropical temperature	280km x 280km x 5 levels	Monthly	1983-90	160
Tropical water vapour	280 km x 280 km x 5 levels	Monthly	1983-90	160
Land				
Coal Resources	Country	None	Present	161
Vegetation	1°x1°	None	Present	162
Vegetation	0.5°x0.5°	None	Present	163
Land cover	1°x1°	None	Present	164
Land use	1°x1°	None	Present	162
Soils	1°x1°	None	Present	165
Soils	1°x1°	None	Present	164
Wetlands	1°x1°	None	Present	166
Wetlands	2°x2.5°	None	Present	167
Vegetation index	1°x1°	Monthly	1982-90	168

[a] Years in parentheses indicate the period of observation, or the period for which the data are typical. Years not in parentheses are the years for which data is actually available.

present as butane, which shows that the molar NMHC (as C)/NO_x ratio is about 3. This puts Europe as a whole on the "hydrocarbon-limited" side of the ozone production feedback (Fig. 4.11), so that uncertainties in hydrocarbon inventories will have a large effect on predictions of ozone formation. And, in fact, hydrocarbon emission inventories are not known to high accuracy. For example U.K. emissions have been revised upward recently by a factor of 1.4 (Ref. 174). Hydrocarbon-limitation implies two further points: that the reactivity of the "cocktail" of hydrocarbons being emitted must be known with some accuracy; and that reduction of hydrocarbons will be an effective control strategy. This latter point is dealt with in Chapter 8; here we discuss some of the efforts being made to tackle the problem of speciation. Table 3.18 shows a speciated hydrocarbon inventory for

Table 3.15: A selection of geographical data sets which are useful in compiling global emissions inventories. After Graedel *et al.*[153].

Parameter	Spatial Resolution	Effective Year	References
Agriculture			
Cultivation	1°x1°	1980s	162
Rice cultivation	1°x1°	1984	170
Fetiliser Application	1°x1°	1984	153
Human Population	1°x1°	1985	153
Cattle population	1°x1°	1984	169
Dairy cattle population	1°x1°	1984	169
Water buffalo population	1°x1°	1984	169
Sheep population	1°x1°	1984	169
Goat population	1°x1°	1984	169
Camel population	1°x1°	1984	169
Horse population	1°x1°	1984	169
Pig population	1°x1°	1984	169
Caribou population	1°x1°	1984	169
Industrial			
Coal mines	1°x1°	1980	172
coal-fired power stations	point sources	1985	172
oil wells	1°x1°	1980	172
oil refineries	1°x1°	1980	172
Natural gas wells	1°x1°	1980	172

the U.K.[174] The scale of the problem of speciation is immediately obvious: 87 compounds are identified and no single compound comprises more than 8 % of the mass emitted. The major compounds and classes contributing to the inventory are butane (12%), pentane (6%), ethylene (4%), toluene (8%), xylenes (10%), aldehydes/ketones (5%) and alcohols/esters (8%). Almost 7% of the total source is unidentified, and the total itself is only some 60% of the recent update[174]. Multivariate regression fitting of hydrocarbon abundances in atmospheric samples, accomplished using "fingerprint" emission compositions from a variety of source types, is proving to be one way in which emission inventories and ambient concentrations can be reconciled[179].

Globally, of course, sources and sinks must conserve mass. This is particularly useful for long-lived compounds, such as methane, which can be assumed to be well-mixed throughout the troposphere. In-situ chemical reaction and deposition to the Earth's surface constitute the total global sink for atmospheric trace gases although for ease of accounting, and because in many ways the two regions are physically and chemically distinct,

Table 3.16: Photochemical oxidant precursor emissions in the U.K., 1991 (Ref. 173).

Source	Pollutant (%) - 1991 NO_x	NMHC	CO
Power Stations	26	< 1	1
Industry (excluding Power)	11	56	1
Road Transport	51	36	89
Domestic	3	2	5
Other	9	< 6	4
Total Emissions (kT)	2750[a]	3245[b]	6800

[a] NO_x emission expressed in kT NO_2.
[b] Revised estimate given in Ref. 174.

troposphere-only budgets are calculated by including transport into the stratosphere as an additional sink. Quantifying the global tropospheric sink due to chemistry is generally a case of quantifying the global tropospheric hydroxyl radical concentration since this is the primary agent of tropospheric oxidation, particularly of long-lived compounds. This quantification is achieved using a "source $\rightarrow$ sink" algorithm: ambient concentrations of a trace gas which has an accurately known, usually industrial, source strength are used to imply the global concentration of OH (e.g. Ref. 180). If deposition and stratospheric transport terms are known, the algorithm can then be turned on its head to produce values for the source strength of gases having poorly quantified emissions. For methane, the resulting budget is: 420 ± 80 Tg a^{-1} reacted with OH; 30 ± 15 Tg a^{-1} deposited on the surface; 10 ± 5 Tg a^{-1} transported to, and destroyed in, the stratosphere; and 45 ± 5 Tg a^{-1} increase in the annual atmospheric loading. Total methane emissions are therefore 505 Tg a^{-1} (Ref. 181).

For more reactive gases such as CO, NO_y and ozone itself, chemical reaction is sensitive to local conditions and so budget calculations must be carried out within chemistry-and-transport models. The results of one such study are summarised in Table 3.19[182, 183]. The calculations of budgets within models is highly sensitive to the formulation of the model so that these budgets for reactive gases, whilst an important improvement in our understanding of global oxidant behaviour, are less useful for

Table 3.17: The regional variation of NO_x, anthropogenic hydrocarbon (NMHC), and biogenic hydrocarbon (BNMHC) emissions in Europe[175, 176].

Source	West	East	South
NO_x			
Power plants	1906	2290	775
Industry/domestic	1909	923	804
Gasoline	3408	414	1040
Diesel	2320	878	1283
Total[a]	9543	4505	3902
NMHC			
Total[b]	9094	3234	3460
BNMHC			
Total[b]	254	218	941

West = Austria, Belgium, Denmark, Finland, France, Germany (FRG as was), Ireland, Luxembourg, The Netherlands, Norway, Sweden, Switzerland and U.K.

East = Bulgaria, Czechoslovakia (as was), Germany (GDR as was), Hungary, Poland and Romania. European CIS states are not included.

South = Albania, Greece, Italy, Portugal, Spain, Turkey and Yugoslavia (as was).

[a] kT NO_x as NO_2 for 1985.
[b] kT for 1989.

constraining emissions than the budgets for long-lived gases like methane. For example, there are large discrepancies in the reported chemical source of ozone in the troposphere[182]. The apparent net trend reported in Table 3.20 is negative because the model has not been run out to steady state. The trend is not, therefore, comparable to the trend reported from long-term measurement programs (Chapter 6).

Table 3.18: UK hydrocarbon emissions by compound for 1991[174]. The total emission is only 60 % of that given in Table 3.16.

Emissions (T)	Petrol Exhaust	Diesel Exhaust	Petrol Evaporation	Stationary Combustion	Solvents	Proceses	Petrol Refn/Distn	Gas Leaks	Ind/Resid Waste	Species Sub-total	%
ethane	6,074	875	0	6,516	0	0	0	22,049	1,408	32,922	1.86
propane	1,374	0	0	3,506	0	0	0	5,512	1,408	11,800	0.59
n-butane	22,988	7,000	28,586	1,422	34,522	0	45,600	1,587	0	141,666	7.14
i-butane	10,517	3,500	14,986	3,200	34,522	0	25,125	1,279	0	93,129	4.69
n-pentane	10,704	4,900	10,325	5,260	1,561	0	11,685	573	0	44,953	2.27
i-pentane	31,382	7,350	14,515	852	1,561	0	18,318	706	0	74,682	3.76
n-hexane	10,226	1,750	2,889	349	12,534	0	1,662	278	0	29,686	1.50
2-methylpentane	12,377	3,675	4,319	349	1,561	0	3,095	186	0	25,561	1.29
3-methylpentane	8,611	1,925	2,874	349	1,561	0	1,930	115	0	17,363	0.88
2,2-dimethylbutane	1,615	700	858	349	1,561	0	777		0	5,859	0.30
2,3-dimethylbutane	3,230	700	1,085	349	1,561	0	804	62	0	7,763	0.39
n-heptane	4,407	0	1,005	171	1,561	0	224	145	0	7,513	0.38
2-methylhexane	8,265	0	1,321	171	1,561	0	402	68	0	11,788	0.59
3-methylhexane	6,614	0	1,333	171	1,561	0	382	76	0	10,136	0.51
n-octane	4,037	0	386	171	1,561	0	28	68	0	6,251	0.32
methylheptanes	32,298	0	964	171	1,561	0	99	75	0	35,167	1.77
n-nonane	438	0	0	171	22,807	0	0	43	0	23,459	1.18
methyloctanes	3,704	0	0	168	22,807	0	0	12	0	26,692	1.35
n-decane	869	0	0	0	22,807	0	0	17	0	23,694	1.19
methylnonanes	1,960	0	0	0	22,807	0	0	2	0	24,769	1.25
n-undecane	3,920	0	0	0	22,807	0	0	0	0	26,727	1.35
n-duodecane	3,920	0	0	0	3,121	0	0	0	0	7,041	0.35
ethylene	51,010	19,250	0	8,666		0	0	0	0	78,925	3.98
propylene	25,395	5,950	0	4,526		0	0	0	0	35,871	1.81
1-butene	4,268	875	2,131	20		0	3,484	0	0	10,778	0.54
2-butene	10,989	0	3,647	20		0	5,789	0	0	20,445	1.03
2-pentene	5,449	0	3,361	109		0	4,583	0	0	13,501	0.68
1-pentene	4,087	1,225	701	109		0	858	0	0	6,979	0.35
2-methyl-1-butene	1,060	0	958	109		0	603	0	0	2,731	0.14
3-methyl-1-butene	1,060	875	958	109		0		0	0	3,606	0.18
2-methyl-2-butene	1,060	875	1,873	109		0	603	0	0	5,941	0.30
2-methyl-1-propene	4,937	0	0	20		0	2,023	0	0	4,958	0.25
acetylene	30,711	5,600	3,346	531		0	0	0	0	36,841	1.86
benzene	39,251	4,500	8,094	943	0	0	1,367	377	0	49,835	2.51
toluene	88,015	1,400	2,274	152	52,797	0	1,166	105	0	152,729	7.65
o-xylene	30,584	1,400	2,688	38	23,235	0	80	9	0	57,619	2.90
m-xylene	38,927	1,400	2,668	38	23,235	0	255	9	0	66,552	3.35
p-xylene	38,926	1,400	1,888	38	23,235	0	121	9	0	66,417	3.35
ethylbenzene	25,493	1,400	586	38	6,394	0	121	1	0	28,940	1.46
n-propylbenzene	3,002	875	172	38	6,394	0	11	1	0	10,906	0.55
i-propylbenzene	1,498	875	443	38	6,394	0	3		0	8,980	0.45
1,2,3-trimethylbenzene	4,019	875	2,288	38	6,394	0	4		0	11,773	0.59
1,2,4-trimethylbenzene	17,141	875	588	38	6,394	0	27		0	26,767	1.35
1,3,5-trimethylbenzene	5,356	875	529	38	6,394	0	8		0	13,229	0.67
o-ethyltoluene	4,506	875	915	38	6,394	0	8		0	12,350	0.62
methyltoluene	7,507	875	915	38	6,394	0	16		0	15,745	0.79
p-ethyltoluene	7,507	875		38	6,394	0	16		0	15,475	0.79
formaldehyde	5,590	10,325		984	0	0	0	0	92	16,992	0.86
acetaldehyde	1,245	1,750		217	0	0	0	0	0	3,212	0.16
propionaldehyde	1,633	1,750		217	0	0	0	0	0	3,600	0.18
butylraldehyde	265	1,750		217	0	0	0	0	0	2,232	0.11
i-butraldehyde	0	1,750		217	0	0	0	0	0	1,967	0.10
valderaldehyde	136	0		217	0	0	0	0	0	352	0.02
benzaldehyde	1,344			217	0	0	0	0	0	1,560	0.08
acetone	499	3,500		217	14,966	0	0	0	0	19,182	0.97
methylethylketone	0	0		217	0	0	0	0	0	217	0.01
methyl-i-butylketone	0	0		217	7,488	0	0	0	0	7,665	0.39

Table 3.18(contd.): UK hydrocarbon emissions by compound for 1991[174]. The total emission is only 60 % of that given in Table 3.16.

Emissions(T)	Petrol Exhaust	Diesel Exhaust	Petrol Evaporation	Stationary Combustion	Solvents	Proceses	Petrol Refn/Distn	Gas Leaks	Ind/Resid Waste	Species Sub-total	%
methanol	0	0	0	217	0	0	0	0	0	217	0.01
ethanol	0	0	0	217	13,099	0	0	0	0	13,316	0.67
methyl acetate	0	0	0	0	0	0	0	0	0	0	0.00
ethyl acetate	0	0	0	0	15,728	0	0	0	0	15,728	0.79
isopropyl acetate	0	0	0	0	0	0	0	0	0	0	0.00
n-butyl acetate	0	0	0	0	9,518	0	0	0	0	9,518	0.48
s-butyl acetate	0	0	0	0	9,518	0	0	0	0	9,518	0.48
dichloromethane	0	0	0	0	8,738	0	0	0	0	8,738	0.44
1,1,1 trichloroethane	0	0	0	0	24,552	0	0	0	0	24,522	1.24
tetrachloroethylene	0	0	0	0	17,275	0	0	0	0	17,275	0.87
trichloroethylene	0	0	0	0	23,730	0	0	0	0	23,730	1.20
methylcyclohexane	0	0	0	0	4,514	0	0	0	0	4,514	0.23
styrene	0	0	0	0	7,894	0	0	0	0	7,894	0.40
n-butanol	0	0	0	0	31,355	0	0	0	0	31,355	1.58
diacetone alcohol	0	0	0	0	19,078	0	0	0	0	19,078	0.96
formic acid	0	0	0	0	0	0	0	0	46	46	0.00
acetic acid	0	0	0	0	0	0	0	0	46	46	0.00
methylchloride	0	0	0	852	0	0	0	0	0	852	0.04
CFCs	0	0	0	0	5,215	0	0	0	0	5,215	0.26
proan-2-ol	0	0	0	0	8,969	0	0	0	0	8,969	0.45
2-methylpropanol	0	0	0	0	22,070	0	0	0	0	22,070	1.11
butan-2-ol	0	0	0	0	19,034	0	0	0	0	19,034	0.96
butanone	0	0	0	0	16,586	0	0	0	0	16,586	0.84
2-methylpentan-4-one	0	0	0	0	22,987	0	0	0	0	22,987	1.16
cyclohexone	0	0	0	0	11,919	0	0	0	0	11,919	0.60
cyclohexanol	0	0	0	0	4,514	0	0	0	0	4,514	0.23
butylglycol	0	0	0	0	18,729	0	0	0	0	18,729	0.94
methoxypropanol	0	0	0	0	4,926	0	0	0	0	4,962	0.25
3,5-dimethylethylbenz.	0	0	0	0	10,730	0	0	0	0	10,730	0.54
3,5-diethyltoluene	0	0	0	0	10,730	0	0	0	0	10,730	0.54
unknown	0	68,600	16,569	12,263	33,614	0	2,760	632	0	134,439	6.78
Total	652,000	175,000	143,000	56,000	787,000	0	133,995	34,000	3,000	1,983,995	100.00

Table 3.19: Tropospheric budgets of NO_y, CO and O_3 from a two dimensional model[182]. Results are for a simulation appropriate to mid-1980s emission levels and climate.

Category[a]	NO_y[b]	CO	O_3
Emissions	+40.0	+898	0
Net chemical effect	0.0	-809	+2074
Dry/wet deposition	-40.2	-70	-2777
Stratospheric flux[c]	+0.2	-19	+692
Total	0	0	(-11)

[a] Fluxes given as Tg N a^{-1}, Tg C a^{-1}, and Tg O_3 a^{-1}.

[b] NO_y = NO_x + HNO_3 + HNO_4 + PAN + organic nitrates

[c] A positive value indicates transport from the stratosphere.

APPENDIX 3.1

Dry Deposition of Photochemical oxidants

The destruction of ozone at the earth's surface plays an important role in the budget of tropospheric ozone and strongly influences concentrations and the lifetime of ozone in the boundary layer. Dry deposition is the direct absorption/destruction of a gas at the Earth's surface. This distinguishes it from removal by precipitation, although water surfaces as well as vegetation and soil may participate. Deposition requires the transfer of ozone through the lower part of the boundary layer to the surface followed by decomposition or chemical reaction on contact with the surface. The first part of the deposition process depends on the nature of the boundary layer whereas the second part depends of the chemical reactivity of the surface with respect to ozone. Dry deposition depletes ozone concentrations directly, but also depletes ozone indirectly through the removal of the important precursor species e.g. NO_2, HNO_3 and PAN.

In studies of dry deposition the rate of transfer, or deposition velocity, v_g is often expressed by

$$V_g(z) = \frac{F}{c(z)} = \frac{1}{r_a + r_b + r_s} \qquad (3.8)$$

where F is the downward flux of the pollutant, which is related to the eddy diffusivity, and c is its concentration at height z. Dry deposition processes are best understood by considering a resistance analogue. In direct analogy with electrical resistance theory, the major resistances to deposition are represented by three resistors in series. These are:

r_a, the aerodynamic resistance, which describes the resistance to transfer towards the surface through normally turbulent air;

r_b, the boundary layer resistance, which describes the transfer through a laminar boundary layer at the surface;

r_s, the surface canopy resistance, which describes the resistance to update by the surface itself. This can vary enormously, from virtually zero for gases such as HNO_3 vapour which attach irreversibly to surfaces, to very high values for gases of low solubility which are not utilised by plants.

Whether the flux is downward or upward, it is driven by a concentration gradient, dc/dz. By analogy with Fick's first law of molecular diffusion:

$$F = K\frac{dc}{dz} \qquad (3.9)$$

where K is the eddy diffusivity coefficient for gas transport in the vertical. The principle of *similarity* may be used for finding the eddy diffusivity coefficient by assuming that the eddy diffusivity coefficient for gases is the same as eddy diffusivity coefficient for another property, e.g. sensible heat or momentum. Under certain conditions the eddy diffusivity coefficient for momentum, heat, water vapour and gases may be assumed to be identical[184]. Prediction of fluxes is then possible by combining profiles of appropriate atmospheric properties.

The main advantage of the resistance analogy is that the total resistance may be sub-divided into the atmospheric and surface components. That is, those parts of the total flux which are constant for all gases under a given set of meteorological conditions, can be separated from those parts which are constant for a given gas under all meteorological conditions. The aerodynamic resisitance, r_a, may be obtained directly from measurements of wind velocity and temperature gradients over flat homogeneous terrain

where u is the wind velocity and u. is the friction velocity. The friction velocity, u., is

$$r_a = \frac{u(z)}{u_*^2} \qquad\qquad (3.10)$$

directly proportional to the rate of change of wind speed with height. Correction terms are often applied to the above equation to account for non-neutral stability[185].

The boundary layer resistance, r_b, appears in the literature[186] as the Stanton number, B, where

$$r_b = \frac{1}{Bu} \qquad\qquad (3.11)$$

The Stanton number is obtained from experiments, and these show that it is a function of the Schmidt number (which in turn is a function of the kinematic viscosity of air and the diffusivity of the species) and the roughness Reynolds number[187]. Various formulations of B^{-1} have been calculated for different surfaces and wind conditions[188, 189, 190, 191, 192] and are discussed in detail in the relevant references.

Once the atmospheric resistance terms have been subtracted from the total resistance the remaining term represents, r_s, the surface resistance. The surface resistance is the sum of several processes operating in parallel, each representing uptake by a different component of the absorbing surface[192]. For vegetation this may be simplified into a term representing absorption by the soil beneath the canopy.

There are several methods routinely used to determine resistances or dry deposition velocities in the field, each with its own advantages and disadvantages[193, 194, 195, 196]. The three basic methods of determining fluxes of trace gases are: Box methods, eddy correlation and flux gradient techniques. Box methods[197, 198, 199] are quick and simple to carry out, but yield data concerning a small area of ground, typically 1 m^2. These methods are inherently limited since they may alter the local environment, thus making it unrepresentative. It is often difficult to deploy a chamber for long-term continuous measurements, or deploy enough chamber replicates to obtain statistically reliable results. Micrometeorological techniques (i.e. eddy correlation and flux gradient) do not disturb the environment around a plant canopy. Such techniques use continuous gas concentration measurements and time-averaged micrometeorological measurements at a point to provide an area-integrated, ensemble average of the exchange rates between the surface and the atmosphere.

Over an extensive uniform stand of vegetation there is a section of the turbulent boundary layer where fluxes of heat, momentum and water vapour and any other entrained gas are constant with height[200]. Bulk rates of exchange between canopy and air flowing over it can be determined by measuring vertical fluxes in this part of the boundary layer. In suitable conditions, micrometeorological methods permit the vertical flux of species to be determined by considering the turbulent transfer of that species.

Table 3.20: Ozone dry deposition determined by the profile technique.

Surface	Deposition velocity, cm s^{-1}	Reference
Grassland	0.58 (day)	206
	0.29 (night)	206
	0.06-1.0	190
	0.23 (wet)	201
	0.51 (damp)	201
	0.69 (dry)	201
	0.1 (night)	203
	1.0 (midday)	203
	0.06-1.0	205
Short wet grass (3 cm)	0.39	208
Short dry grass (3 cm)	0.77	208
Long wet grass (8 cm)	0.62	208
Long dry grass (8 cm)	0.85	208
Heather/grass	0.29	204
Grass/grain	0.6 (day)	207
	0.17 (night)	207
Soyabean	0.6 (day)	207
	0.17 (night)	207
Barley	0.2-1.99	208
Snow	0.08	201
Forest	1.0	207

Two of the more popular flux gradient techniques are the aerodynamic or profile method and the Bowen ratio method. The profile method is based on vertical gradients of windspeed, temperature and concentration[201]. In order to calculate deposition velocities this method assumes that in turbulent flow, where mechanical mixing dominates, an eddy transfers atmospheric properties equally so that exchange of all gases, vapours, heat and momentum are expected to be similar. Details of deposition velocities determined by this technique are given in Table 3.20[202, 203, 204, 205, 206, 207, 208]. Eddy diffusivities are evaluated as a function of momentum flux, surface roughness, and atmospheric stability, which are derived from supporting meteorological data[209]. The profile method is straightforward in neutral stability, when profiles of only windspeed and concentration are required. Outside such conditions the wind profile must be corrected, empirically, for the effects of thermal buoyancy. The accepted corrections appear valid over short crops but are less accurate over tall rough crops.

The Bowen ratio determination of flux is derived from the energy balance of the underlying surface[184]. Eddy diffusivity coefficients for the transport of gases are assumed to be the same as those for for sensible heat or water vapour and are determined from measurements of vertical gradients of temperature, humidity and ozone concentration as well as the net radiation and soil heat flux. For a flux to be estimated with any accuracy using this technique substantial net radiation fluxes (> 50 W m^{-2}) are required. At night, during Winter or in cloudy conditions the available net radiation is often too small for accurate flux determination.

Both the profile and Bowen ratio method are indirect methods of determining the flux since they rely on the measurement of mean potentials and their gradients in the atmosphere. Eddy correlation is a direct method[200] and requires simultaneous measurement of rapid fluctuations of vertical windspeed in the constant flux region of the surface boundary layer and the associated fluctuations in gas concentration, temperature and humidity. Only a stress on the horizontal wind due to a horizontal surface, together with a mean vertical velocity of zero, is assumed. The technique requires instruments with fast response times, usually less than 1 s, and a fast logging system. Table 3.21 summarizes measurements made by the eddy correlation technique[210, 211, 212, 213, 214, 215, 216, 217, 218, 219, 220, 221, 222].

In all micrometeorological techniques there are strict requirements on the nature of the fetch and the position of the sampling equipment. The fetch must be uniform for a sufficient distance upwind to ensure that a layer, in which the flux of material is constant with height, is deep enough to surround the sampling equipment. For the profile method it is expedient to sample over a range of heights within this layer, in order to reduce the measurement error on the gradient. Stationary meteorological conditions are needed if the meterological measurements are to be meaningful: windspeed, wind direction, temperature thermal stability and atmospheric concentration of the species under investigation should not change significantly during data sampling. If the eddy correlation technique is adopted, then the flux of any interfering compounds, that is, compounds which can react rapidly with the gas under study, should also not vary. In practice some relaxation of these rigorous demands are allowed[223].

Table 3.21: Ozone dry deposition velocity determined by the eddy correlation technique.

Surface	Deposition Velocity,cm s^{-1}	Surface Resistance, s cm^{-1}	References
Grass	-(day)	2.0	221
	-(night)	2.9	221
	0.32-1.17	-	203
	(avge.=0.57)	-	203
	0.04-0.93	-	218
	(avge.=0.47)	-	218
	0.09-0.43	-	204
	0.36 (day)[a]	2.65	220
	0.22 (night)[a]	4.61	220
Maize	0.2-0.8	0.5-0.4	226
Maize/soyabeans	-	4-11	214
Wheat	0.19	2.1	212
Barley	1.0	-	208
Sunflowers	0.01-1.5	-	213
deciduous forest	0.65-1.70	-	191
	1.0 (day)	-	219
Spruce forest	1.8 (day)	-	211
	70 (night)	-	211
Pine forest	-	0.5	215
Wet soil	0.01	10	222
Soil in deciduous forest	0.05	22-28	217
Grass/bare soil	0.7 (day)	-	216
	0.3 (night)	-	216
Water	0.01	90	222
Oceans	-	18	215
Snow	0.03	34	222

[a] Potential oxidant (O_3 + NO_2) deposition.

Deposition studies, carried out in many different natural environments, indicated that ozone deposition changes with time of day. For vegetation this diurnal variation results from stomata opening in the morning and closing in the evening. Often such diurnal cycles in the field are obscured by diurnal changes in atmospheric resistances. These are generally larger at night in response to smaller wind velocities and stable stratification of the air close to the ground. Dew and rainwater may influence deposition. Since ozone is weakly soluble in water it has been suggested that ozone uptake is inhibited when foliage is wet because stomata are covered with water[201, 224, 225, 226]. However, recent measurements, on individual red maple leaves, indicate substantially enhanced deposition[227]. Although decreased stomatal resistance was observed, the increase in ozone deposition resulted from the presence of compounds in aqueous form that readily scavenge ozone. The effect of rainwater is likely, therefore, to be a function of the composition of the rainwater (see chapter 4, below).

Few studies of PAN and H_2O dry deposition have been reported. Due to its high solubility and reactivity in aqueous medium, H_2O_2 surface resistance is expected to be small. Hence it should deposit in a similar manner to HNO_3 (Ref. 228). For PAN a range of deposition velocities have been measured. Garland and Penkett[229] conducted laboratory measurements of PAN deposition velocities to grass and soil, and obtained an average value of 0.25 cm s[-1]. Hill[230] studied PAN deposition to an alfalfa canopy and reported a deposition velocity of 0.75 cm s[-1]. More recently the observed PAN deposition velocity averaged 0.54 cm s[-1] during nighttime for vegetative surfaces[231]. Like ozone, PAN has a very large surface resistance for deposition to water. The deposition velocity for PAN to water has been estimated to be 0.008 cm s[-1] (Ref. 232).

1. Chameides, W.L., Lindsay, R.W., Richardson, J. and Kiang C.S. (1988) The Role of Biogenic Hydrocarbons in Urban Photochemical Smog, Science, 241, 1473-1475.

2. Fishman, J. (1991) Probing Planetary Pollution from Space, Environ. Sci. Technol., 25, 612-620.

3. Lovelock, J.E. (1988) *Ages of Gaia*, Oxford University Press, Oxford, U.K., 252.

4. Lovelock, J.E. (1989) Geophysiology, the science of Gaia, Rev. Geophys., 27, 215-222.

5. Kirchner. J.W. (1989) The Gaia Hypothesis: can it be tested, Rev. Geophys., 27, 223-235.

6. Margulis, L. and Guerrero, R. (1991) Kingdoms in Turmoil, New Scientist, 23 March 1991, 46-50.

7. Woese, C.R. and Wolfe, R.S. (1985) *The Bacteria, Archaebacteria*, 8, Academic Press, New York,

8. Whitman, W.B. (1985) Methanogenic Bacteria, in Woese, C.R. and Wolfe, R.S., *The Bacteria*, Academic Press, New York.

9. Cicerone, R.J. and Oremland, R.S. (1988) Biogeochemical Aspects of Atmospheric Methane, Global Biogeochemical Cycles, 2, 299-327.

10. Wolin, M.J. and Miller, T.L. (1987) Bioconversion of Organic Carbon to CH_4 and CO_2, Geomicrobiol. J., 5, 239-259.

11. Stryer, L. (1988) *Biochemistry*, 3rd ed., W.H. Freeman and Co., New York, USA, pp 1089.

12. Oremland, R.S., (1988) The biogeochemistry of methanogenic bacteria, in *Biology of Anaerobic Microorganisms*, A.J.B. Zehnder (ed.), J. Wiley and Sons, New York, USA, pp 641-702.

13. Crutzen, P.J., Aselmann, I. and Seiler, W. (1986) Methane Production by Domestic Animals, Wild Ruminants, other Herbivorous Fauna, and Humans, Tellus, 38B, 271-284.

14. Anthony, C. (1982) *The Biochemistry of Methylotrophs*, Academic Press, NY, pp 431.

15. Keller, M., Mitre, M.E. and Stallard, R.F. (1990) Consumption of Atmospheric Methane in Soils of Central Panama: Effects of Agricultural Development, Global Biogeochem. Cycles, 4, 21-27.

16. Oremland, R.S., and Culbertson, C.W. (1992) Importance of methane-oxidizing bacteria in the methane budget as revealed by the use of a specific inhibitor, Nature, 356, 421-423.

17. Seiler, W., Conrad, R. and Scharffe, D. (1984) Field Studies of Methane Emission from Termite Nests into the Atmosphere and Measurement of Methane Uptake by Tropical Soils, J. Atmos. Chem., 1, 171-186.

18. Striegl, R.G., McConnaughey, T.A., Thorstenson, D.C., Weeks, E.P., and Woodward, J.C. (1992) Consumption of atmospheric methane by desert soils, Nature, 357, 145-147.

19. Svensson, B.H. and Rosswall, T. (1984) In situ Methane Production from Acid Peat in Plant Communities with Different Moisture Regimes in a Subarctic Mire, Oikos, 43, 341-350.

20. Sebacher, D.I., Harriss, R.C., Bartlett, K.B., Sebacher, S.M. and Grice, S.S. (1986) Atmospheric Methane Sources: Alaskan Tundra Bogs, an Alpine Fen, and a Subarctic Marsh, Tellus, 38B, 1-10.

21. Crill, P.M., Bartlett, K.B., Harriss, R.C., Gorham, E., Verry, E.S., Sebacher, D.I., Madzas, L. and Somner, W. (1988) Methane Flux from Minnesota Peatlands, Global Biogeochem. Cycles, 2, 371-384.

22. Wilson, J.O., Crill, P.M., Bartlett, K.B., Sebacher, D.I., Harriss, R.C. and Sass, R.L. (1989) Seasonal Variations of Methane Emissions from a Temperate Swamp, Biogeochem., 8, 55-71.

23. Harriss, R.C. and Sebacher, D.I. (1991) Methane Flux in Forested Swamps of the Southeastern United States, Geophys. Res. Lett., 8, 1002-1004.

24. Burke, R.A., Barber, R.J., Sackett, W.M. (1988) Methane Flux and Stable Hydrogen and Carbon Isotope Composition of Sedimentary Methane from the Florida Everglades, Global Biogeochem. Cycles, 2, 329-340.

25. Harriss, R.C., Sebacher, D.I., Bartlett, K.B., Bartlett, D.S. and Crill, P.M. (1988) Sources of Atmospheric Methane in the South Florida Environment, Global Biogeochem. Cycles, 2, 231-243.

26. Miller, L.G. and Oremland, R.S. (1988) Methane Efflux from the Pelagic Regions of Four Lakes, Global Biogeochem. Cycles, 2, 269-278.

27. Bartlett, K.B., Crill, P.M., Sebacher, D.I., Harriss, R.C., Wilson, J.O. and Melack, J.M. (1988) Methane Flux from the Central Amazonian Floodplain, J. Geophys. Res., 93, 1571-1582.

28. Chanton, J.P. and Martens, C.S. (1988) Seasonal Variations in Ebullitive Flux and Carbon Isotopic Composition of Methane in a Tidal Freshwater Estuary, Global Biogeochem. Cycles, 2, 289-298.

29. Goreau, T.J., and de Mello, W.Z. (1988) Tropical Deforestation - some effects on atmospheric chemistry, Ambio, 17, 275-281.

30. Whiting, G.J., Chanton, J.P., Bartlett, D.S., Happell, J.D. (1991) Relationships between CH_4 emission, biomass, and CO_2 exchange in a subtropical grassland, J. Geophys. Res., 96, 13,067-13,071.

31. Chanton, J.P., and Dacey, J.W.H. (1991) Effects of vegetation on methane flux, reservoirs and carbon isotopic composition, in Trace Gas Emissions by Plants, T.D. Sharkey, E.A. Holland, and H.A. Mooney, Academic Press, London, UK. pp 65-86.

32. Whiting, G.J., and Chanton, J.P. (1992) Plant dependent CH_4 emission in a sub-Arctic Canadian fen, Global Biogeochem. Cycles, 6, 225-231.

33. Fung, I., John, J., Lerner, J., Matthews, E., Pratter, M., Steele, L.P., Fraser, P.J. (1991) Three-dimensional Model Synthesis of the Global Methane Cycle, J. Geophys. Res., 96, 13,033 - 13,065

34. Nisbet, E.G. (1989) Some Northern Sources of Atmospheric Methane: Production, history and future implications, Can. J. Earth Sci., 26, 1603-1611.

35. Whalen, S.C. and Reeburgh, W.S. (1988) A Methane Flux Time Series for Tundra Environments, Global Biogeochem. Cycles, 2, 399-409.

36. Moore, T., Roulet, N., and Knowles, R. (1990) Spatial and Temporal Variations of Methane Flux from Subarctic/Northern Boreal Fens, Global Biogeochem. Cycles, 4, 29-46.

37. Devol, A.H., Richey, J.E., Forsberg, B.R., Martinelli, L.A. (1990) Seasonal Dynamics in Methane Emissions from the Amazon River Floodplain to the Troposphere, J. Geophys. Res., 95, 16417-16426.

38. Bartlett, K.B., Crill, P.M., Bonassi, J.A., Richey, J.E., Harriss, R.C. (1990) Methane Flux from the Amazon River Floodplain: Emissions during Rising Water, J. Geophys. Res., 95, 16773-16788.

39. Chanton, J.P., Crill, P., Bartlett, K., and Martens, C. (1989) Amazon capims (floating meadows): a source of ^{13}C enriched methane to the troposphere, Geophys. Res. Lett., 16, 799-802.

40. Wassmann, R., Thein, U.G., Whiticar, M.J., Rennenberg, H., Seiler, W., and Junk, W.J. (1992) Methane emissions from the Amazon floodplain: characterization of production and transport, Global Biogeochem. Cycles, 6, 3-13.

41. Cicerone, R.J., Shetter, J.D. and Delwiche, C.C. (1983) Seasonal variation of methane flux from a California rice paddy, J. Geophys. Res., 88, 1022-1024.

42. Schutz, H., Holzapfel-Pschorn, A., Conrad, R., Rennenberg, H., and Seiler, W. (1989) A 3-year continuous record on the influence of daytime, season, and fertiliser treatment on methane emission rates from an Italian rice paddy, J. Geophys. Res., 94, 16,405-16,416.

43. Dai, A. (1988) M.S. Thesis, Institute of Atmospheric Physics, Academica Sinica, Beijing, China.

44. Khalil, M.A.K., Rasmussen, R.A., Wang, M-X., and Ren, L. (1991) Methane emissions from rice fields in China, Environ. Sci. Technol., 25, 979-981.

45. Holzapfel-Pschorn, A. and Seiler, W. (1986) Methane Emission during a Cultivation Period from a Rice Paddy, J. Geophys. Res., 91, 11,803-11,814.

46. Sass, R.L., Fisher, F.M., Harcombe, P.A., Turner, F.T. (1990) Methane Production and Emission in a Texas Rice Field, Global Biogeochem. Cycles, 4, 47-68.

47. Cicerone, R.J., Delwiche, C.C., Tyler, S.C., and Zimmerman, P.R. (1992) Methane emissions from California rice paddies with varied treatments, Global Biogeochem. Cycles, 6, 233-248.

48. Cicerone, R.J. and Shetter, J.D. (1981) Sources of Atmospheric Methane: Measurement in Rice Paddies and a Discussion, J. Geophys. Res., 86, 7203-7209.

49. Sass, R.L., Fisher, F.M., Wang, Y.B., Turner, F.T., and Jund, M.F. (1992) Methane emission from rice fields: the effect of floodwater management, Global Biogeochem. Cycles, 6, 249-262.

50. Sheppard, J.C., Westberg, H., Hoppes, I.F., Ganesea, K., Zimmerman, P. (1982) Inventory of Global Methane Sources and their Production Rates, J. Geophys. Res., 87, 1305-1312.

51. Bingemer, H.G., and Crutzen, P.J. (1987) The Production of Methane from Solid Wastes, J. Geophys. Res., 92, 2181-2187.

52. Warneck, P. (1988) *Chemistry of the Natural Atmosphere*, Academic Press, San Diego, USA.

53. Haynes, R.J. (1986) *Mineral Nitrogen in the Plant-soil System*, Academic Press, London, pp 483.

54. Firestone, M.K. and Davidson, E.A. (1989) Microbiological Basis of NO and N_2O Production and Consumption in Soil, in Andreae, M.O. and Schimel, D.S. (Eds.) *Exchange of Trace Gases between Terrestrial Ecosystems and the Atmosphere*, John Wiley and Sons, U.K., 7-21.

55. Galbally, I.E. (1989) Factors Controlling NO_x Emissions from Soils, in Andreae, M.O. and Schimel, D.S. (Eds) *Exchange of Trace Gases between Terrestrial Ecosystems and the Atmosphere*, John Wiley and Sons, U.K., 23-37.

56. Robertson, G.P. (1989) in Procter, J. (Ed.) *Mineral Nutrients in Tropical Forest and Savanna Ecosystems*, Oxford: Blackwell Scientific, UK.

57. Malhi, S.S., and McGill, W.B. (1982) Nitrification in three Alberta soils: effect of temperature, moisture and substrate concentration, Soil Biol. Biochem., 14, 393-399.

58. Williams, E.J. and Fehsenfeld, F.C. (1991) Measurement of Soil Nitrogen Oxide Emissions at Three North American Ecosystems, J. Geophys. Res., 96, 1033-1042.

59. Magalhaes, A.M.T., Chalk, P.M., Rudra, A.B. and Nelson, D.W. (1985) Formation of Methyl Nitrite in Soil Treated with Nitrous Acid, Soil, Sci. Soc. Am. J., 49, 623-625.

60. Tortoso, A.C. and Hutchinson, G.L. (1990) Contributions of Autotrophic and Heterotrophic Nitrifiers to Soil NO and N_2O Emissions, Appl. Environ. Microbiol., 56, 1799-1805.

61. Johansson, C. and Gabally, I.E. (1984) The Production of Nitric Oxide in a Loam under Aerobic and Anaerobic Conditions, Appl Environ. Microbiol., 47, 1284-1289.

62. Anderson, I.C., Levine, J.S., Poth, M.A. and Riggan, P.J. (1988) Enhanced Biogenic Emissions of Nitric Oxide and Nitrous Oxide Following Biomass Burning, J. Geophys. Res., 93, 3893-3898.

63. Galbally, I.E. and Roy, C.R. (1978) Loss of Fixed Nitrogen from Soils by Nitric Oxide Exhalation, Nature, 275, 734-735.

64. Williams, E.J., Parrish, D.D. and Fehsenfeld, F.C. (1987) Determination of Nitrogen Oxide Emissions from Soils: Results from a Grassland Site in Colorado, United States, J. Geophys. Res., 92, 2173-2179.

65. Delany, A.C., Fitzjarrald, D.R., Lenshow, D.H., Pearson, R., Wendel, G.J. and Woodruff, B. (1986) Direct Measurements of Nitrogen Oxides and Ozone Fluxes over Grassland, J. Atmos. Chem., 4, 429-444.

66. Johansson, C., Rodhe, H. and Sanheuza, E. (1988) Emission of NO in a Tropical Savanna and Cloud Forest during the Dry Season, J. Geophys. Res., 93, 7180-7192.

67. Johansson, C. (1984) Field Measurements of Emissions of Nitric Oxide from Fertilized and Unfertilized Forest Soils, J. Atmos. Chem., 1, 429-442.

68. Williams, E.J., Parrish, D.D., Buhr, M.P. and Fehsenfeld, F.C. (1988) Measurement of Soil NO_x Emissions in Central Pennsylvania, J. Geophys. Res., 93, 9539-9546.

69. Johansson, C., and Sanhueza, E. (1988) Emission of NO from savanna soils during rainy season, J. Geophys. Res., 93, 14,193-14,198.

70. Slemr, F., and Seiler, W. (1984) Field measurements of NO and NO_2 emissions from fertilised and unfertilised soils, J. Atmos. Chem., 2, 1-24.

71. Went, F.W. (1955) Air Pollution, Sci. Am., 192, 63-72.

72. Ipatiev, W. and Whittorf, N. (1897) J. Prakt. Chem., 55, 1.

73. Richards, J.H. and Hendrickson, J.B. (1964) *The Biosynthesis of Steroids, Terpenes and Acetogenins*, W.A. Benjamin inc., New York, pp 416.

74. Went, F.W. (1960) Blue Hazes in the Atmosphere, Nature, 187, 641-643.

75. Wayne, R.P (1991) *Chemistry of Atmospheres*, Oxford Science Pub., OUP, UK. pp 447.

76. Zimmerman, P.R. (1979) Testing of Hydrocarbon Emissions from Vegetation. EPA/450/4-79/004, U.S. Environmental Protection Agency, Atlanta, GA.

77. Atkinson, R. (1990) Gas-phase tropospheric chemistry of organic compounds: a review, Atmos. Environ., 24A, 1-41.

78. Mgaloblishvili, M.P., Khetsuriana, N.D., Kalandadze, A.N. and Sanadze, G.A. (1979) Localization of Isoprene Biosynthesis in Poplar Leaf Chloroplasts, Sov. Plant Physiol., 25, 837-842.

79. Evans, R.C., Tingey, D.T., Gumpertz, M.L. and Burns, W.F. (1982) Estimates of Isoprene and Monoterpene Emission Rates in Plants, Bot. Gaz., 143, 304-310.

80. Evans, R.C., Tingey, D.T. and Gumpertz, M.L. (1985) Interspecies Variations in Terpenoid Emissions from Englemann and Sikta Spruce Seedlings, Forest Sci., 31, 132-142.

81. Tingey, D.T., Evans, R.C. and Gumpertz, M.L. (1981) Effects of Environmental Conditions on Isoprene Emissions from Live Oak, Planta, 152, 565-570.

82. Monson, R.K. and Fall, R. (1989) Isoprene Emission from Aspen Leaves, Plant Physiol., 90, 267-274.

83. Sanadze, G.A. and Kalandadze, A.N. (1966) Light and Temperature Curves of the Evolution of C_5H_8, Sov. Plant. Physiol., 13, 411-413.

84. Rasmussen, R.A. and Jones, C.A. (1973) Emission of Isoprene from Leaf Discs of Hamamelis, Phytochem., 12, 15-19.

85. Tingey, D.T. (1991) The Effect of Environmental Factors on the Emission of Biogenic Hydrocarbons from Live Oak and Slash Pine, in *Atmospheric Biogenic Hydrocarbons*, 1, Ann Arbour, MI.

86. Sanadze, G.A. (1969) Light-dependent Excretion of Molecular Isoprene, Prog. Photosynth. Res., 2, 701-706.

87. Jones, C.A. and Rasmussen, R.A. (1975) Production of Isoprene by Leaf Tissue, Plant Physiol., 55, 982-987.

88. Intergovernmental Panel on Climate Change (1990) *Climate Change*, Houghton, J.T., Jenkins, G.J. and Ephraums J.J. (Eds.), Cambridge University Press, Cambridge, U.K. pp 364.

89. Zimmerman, P.R. (1979) Determination of Emission Rates of Hydrocarbons from Indigenous Species of Vegetation, EPA/904/9-77/028, US Environmental Protection Agency, Atlanta, GA, USA.

90. Arey, J., Corchnoy, S.B. and Atkinson, R. (1991) Emission of Linalool from Valencia Orange Blossoms and its Observation in Ambient Air, Atmos. Environ., 25A, 1377-1381.

91. Bufler, U. and Wegmann, K. (1991) Diurnal Variations of Mono-terpene Concentrations in Open-top Chambers and in the Welzheim Forest Air, F.R.G., Atmos. Environ., 25A, 251-256.

92. Yokouchi, Y. and Ambe, Y. (1984) Factors affecting the Emission of Monoterpenes from Red Pine (pinus densiflora), Plant Physiol., 75, 1009-1012.

93. Yokouchi, Y., Okaniwa, M., Ambe, Y., and Fuwa, K. (1983) Seasonal variation of monoterpenes in the atmosphere of a pine forest, Atmos. Environ., 17, 743-750.

94. Yokouchi, Y., and Ambe, Y. (1988) The diurnal variation of atmospheric isoprene and monoterpene hydrocarbons in an agricultural area in summertime, J. Geophys. Res., 93, 3751-3759.

95. Hewitt, C.N., Kok, G.L. and Fall, R. (1990) Hydroperoxides in Plants Exposed to Ozone Mediate Air Pollution Damage to Alkene Emitters, Nature, 344, 56-58.

96. Lamb, B., Westberg, H. and Allwine, G. (1986) Isoprene Emission Fluxes Determined by an Atmospheric Tracer Technique, Atmos. Environ., 20, 1-8.

97. Lamb, B., Westberg, H. and Allwine, G. (1985) Biogenic Hydrocarbon Emissions from Deciduous and Coniferous Trees in the USA, J. Geophys. Res., 90, 2380-2390.

98. Arnts, R.R., Seila, R.L., Kuntz, R.L., Mowry, F.L., Kuoerr, K.R. and Dudgeon, A.C. (1978) Measurements of α-pinene Fluxes from a Loblolly Pine Forest, in *Proc. 4th Int. Conf. on Sensing of Environmental Pollutants*, 829-833, Amer. Chem. Soc., Washington, D.C., USA.

99. Kanakidou, M., Bonsang, B. and Lambert, G. (1989) Light Hydrocarbon Vertical Profiles and Fluxes in a French Rural Area, Atmos. Environ., 23, 921-927.

100. Zimmerman, P.R., Greenberg, J.P. and Westberg, C.E. (1988) Measurements of Atmospheric Hydrocarbons and Emission Fluxes in The Amazon Boundary Layer, J. Geophys. Res., 93, 1407-1416.

101. Ayers, G.P. and Gillett, R.W. (1988) Isoprene Emissions from Vegetation and Hydrocarbon Emissions from Bushfires in Tropical Australia, J. Atmos. Chem., 7, 177-190.

102. Rambler, M.B., Margulis, L. and Fisher, R. (1989) *Global Ecology*, Academic Press, London, UK, 204 pp.

103. Crutzen, P.J. and Andreae, M.O. (1990) Biomass Burning in the Tropics: Impact on Atmospheric Chemistry and Biogeochemical Cycles, Science, 250, 1669-1678.

104. Myers, N. (1989) *Deforestation Rates in Tropical Forests and Their Climatic Implications*, Friends of the Earth, London, U.K.

105. Kaufman, Y.J., Tucker, C.J. and Fung, I. (1990) Remote-Sensing of Biomass Burning in the Tropics, J. Geophys. Res., 95, 9927-9939.

106. Watson, C.E., Fishman, J. and Reichle, H.E. (1990) The Significance of Biomass Burning as a Source of Carbon Monoxide and Ozone in the Southern Hemisphere Tropics: A Satellite Analysis, J. Geophys. Res., 95, 16,443-16,450.

107. Kirchhoff, V.W.J.H., Marinho, E.V.A., Dias, P.L.S., Pereira, E.B., Calheiros, R., André, R. and Volpe, C. (1991) Enhancements of CO and O_3 from Burnings in Sugar Cane Fields. J. Atmos. Chem., 12, 87-102.

108. Dignon, J., Atherton, C.S., Penner, J.E. and Walton, J.J. (1991) NO_x Pollution from Biomass Burning: A Global Study. Presented at *11th Conference on Fire and Forest Meteorology*, Missoula, MT, April 1991.

109. Hegg, D.A., Radke, L.F., Hobbs, P.V., Rasmussen, R.A. and Riggon, R.J. (1990) Emissions of Some Trace Gases from Biomass Fires, J. Geophys. Res., 95, 5669-5675.

110. Crutzen, P.J. (1989) The Role of the Tropics in Atmospheric Chemistry, in Dickson, R.E. (Ed.), *Geophysiology of Amazonia*, Wiley, NY, 107-130.

111. Kuhlbusch, T.A., Lobert, J.M., Crutzen, P.J. and Warneck, P. (1991) Molecular Nitrogen Emissions from Denitrification during Biomass Burning, Nature, 351, 135-137.

112. Lobert, J.M., Scharffe, D.H., Hao, W.M. and Crutzen, P.J. (1990) Importance of Biomass Burning in the Atmospheric Budgets of Nitrogen-Containing Gases, Nature, 346, 552-554.

113. Robinson, J.M. (1989) On Uncertainty in the Computation of Global Emissions from Biomass Burning, Climatic Change, 14, 243-262.

114. Clements, H.B. and McMahon, C.K. (1980) Nitrogen Oxides from Burning Fossil Fuels Examined by Thermogravimetry and Evolved Gas Analysis, Thermochimica Acta, 35, 133-139.

115. Finlayson-Pitts, B.J. and Pitts, J.N. (1986) *Atmospheric Chemistry*, Wiley and Sons, New York, 1098 pp.

116. Seinfeld, J.H. (1989) Urban Air Pollution: State of the Science, 243, 745-752.

117. Faeth, G.M. (1979) in Chigier, N. (Ed.) *Energy and Combustion Science*, Student Edition 1, Pergamon, Oxford, U.K., 149-182.

118. Dryer, F.L. (1991) in Bartok, W. and Sarofim, A.F. (Eds.) *Fossil Fuel Combustion*, John Wiley and Sos, NY, 121-214.

119. Haynes, B.S. (1991) in Bartok, W. and Sarofim, A.F. (Eds.) *Fossil Fuel Combustion*, J. Wiley and Sons, NY, 261-326.

120. Warnatz, J. (1983) Hydrocarbon oxidation at high temperatures, Ber. Bunsennges. Phys. Chem., 87, 1008-1022.

121. Smoot, G.D. and Horton, M.D. (1979) in Chigier, N. (Ed.) *Energy and Combustion Science*, Student Edition 1, Pergamon, Oxford, U.K., 203-228.

122. Heywood, J.B. (1979) in Chigier, N. (Ed.) *Energy and Combustion Science*, Student Edition 1, Pergamon, Oxford, U.K., 229-258.

123. Glassman, I. (1987) *Combustion*, 2nd Edition, Academic Press, FLA.

124. Barnard, J.A. and Bradley, J.N. (1985) *Flame and Combustion*, 2nd Edition, Chapman and Hall, NY.

125. Bowman, C.T. (1991) in Bartok, W. and Sarofim, A.F. (Eds.) *Fossil Fuel Combustion*, J. Wiley and Sons, NY, 215-260.

126. Bachmaier, F., Eberius, K.H., and Just, T. (1973) Formation of nitric oxide and the detection of hydrogen cyanide in premixed hydrocarbon-air flames at 1 atmosphere, Combust. Sci. Technol., 7, 77-84.

127. Baulch, D.L., Cobos, C.J., Cox, R.A., Esser, C., Frank, P., Just, T., Kerr, J.A., Pilling, M.J., Troe, J., Walker, R.W., and Warnatz, R.W. (1992) Evaluated kinetic data for combustion modelling, J. Phys. Chem. Ref. Data, 21, 411-737.

128. U.S. Code of Federal Regulations. (1983) Title 40, Part 86, Control of Air Pollution From New Motor Vehicles etc.

129. Stump, F., Tejada, S., Ray, W., Dropkin, D. and Black, F. (1989) The Influence of Ambient Temperature on Tailpipe Emissions from 1984-1987 Model Year Light-duty Gasoline Motor Vehicles, Atmos. Environ., 23, 307-320.

130. Stump, F.S., Tejada, S., Ray, W., Dropkin, D., Black, F., Snow, R., Crews, W., Siudak, P., Davis, C.O. and Carter, P. (1990) The Influence of Ambient Temperature on Tailpipe Emissions - II, Atmos. Environ., 24A, 2105-2112.

131. Hoekman, S.K. (1992) Speciated Measurements and Calculated Reactivities of Vehicle Exhaust Emissions from Conventional and Reformulated Gasolines, Environ. Sci. Technol., 26, 1206-1216.

132. Williams, R.L., Lipari, F. and Potter, R.A. (1990) Formaldehyde, Methanol and Hydrocarbon Emissions from Methanol-Fuelled Cars, J. Air Waste Manage. Assoc., 40, 747-756.

133. Rice, R.W., Sanyal, A.K., Elrod, A.C. and Bata, R.M. (1991) Exhaust Gas Emissions of Butanol, Ethanol and Methanol- gasoline Blends, J. Engin. Gas Turbines and Power, 113, 377-381.

134. Stump, F., Knapp, K.T. and Ray, W.D. (1990) Seasonal Impact of Blending Oxygenated Organics with Gasoline on Motor Vehicle Tailpipe and Evaporative Emissions, J. Air Waste Manage. Assoc., 40, 872-880.

135. Horn, J.C. and Hoekman, S.K. (1989) Methanol-fueled Light-duty Vehicle Exhaust Emissions, Proc. Air Waste Manage. Assoc., Meeting in Anaheim, Calif., June 25-30, 1989, paper 89-9.3.

136. Gabele, P.A. and Knapp, K.T. (1993) A Characterization of Emissions from an Early Model Flexible-fuel Vehicle, J. Air Waste Manage., 43, 851-858.

137. Chang, T.Y., Hammerle, R.H., Japar, S.M., Salmeen, I.T. (1989) Alternative Transportation Fuels and Air Quality, Environ. Sci. Technol., 25, 1190-1197.

138. Calvert, J.G., Heywood, J.B., Sawyer, R.F., and Seinfeld, J.H. (1993) Achieving acceptable air quality: some reflections on controlling vehicle emissions, Science, 261, 27-45.

139. Smith, L.R. and Paskind, J. (1989) Characterization of Exhaust Emissions from Trap-equipped Light-duty Diesels, SAE Tech. Paper Ser., No. 891972.

140. Hickman, A.J. and Graham, M.A. (1993) Performance Related Exhaust Emissions from Heavy-duty Diesel Engines, Sci. Tot. Environ., 134, 211-223.

141. Latham, S. and Hickman, A.J. (1990) Exhaust Emissions from Heavy-duty Diesel Engined Vehicles, Sci. Tot. Environ., 93, 139-145.

142. Eberhard, G.A., Ansari, M. and Hoekman, S.K. (1989) Emissions and Fuel Economy Test Results for Methanol and Diesel-fueled Buses, Proc. Air Waste Manage., Meeting in Anaheim, Calif., 25-30 June 1989, paper 89-9.4.

143. Westerholm, R., Almen, J., Li, H., Raumug, U., Rosen, A. (1992) Exhaust Emissions from Gasoline Fuelled Light-duty Vehicles in Different Driving Conditions, Atmos. Environ., 26B, 79-90.

144. Potter, C.J., Bailey, J.C., Savage, C.A., Schmidl, B., Simmonds, A.C. and Williams, M.L. (1988) The Measurement of Gaseous and Particulate Emissions from Light-duty and Heavy-duty Motor Vehicles under Road Driving Conditions, SAE Tech. Paper Ser., No. 880313.

145. Bailey, J.C., Schmidl, and Williams, M.L. (1990) Speciated Hydrocarbon Emissions from Vehicles Operated over the Normal Speed Range on the Road, Atmos. Environ., 24A, 43-52.

146. Schurmann, D. and Staab, J. (1990) On-the-road Measurements of Automotive Emissions, Sci. Tot. Environ., 93, 147-157.

147. Nichols, R.A. (1990) J. Air Waste Manage. Assoc., 40, 238.

148. Halberstadt, M.L. (1990) J. Air Waste Manage. Assoc., 40, 238-239.

149. Seitz, L.E. (1989) California Methods for Estimating Air Pollutant Emissions from Motor Vehicles, Proc. Air Waste Manage., Meeting in Anaheim, Calif., 25-30 June 1989, Paper 89-7.4.

150. Bailey, J.C. and Eggleston, S. (1993) The Contribution of Gasoline Fuelled Vechicle Exhaust to the U.K. Speciated Hydrocarbon Inventory, Sci. Tot. Environ., 134, 263-271.

151. Black, F.M. (1988) Motor Vehicles as Sources of Compounds Important to Tropospheric and Stratospheric Ozone, EPA/600/D-88/067. EPA, Research Triangle Park, NC, USA.

152. Cadle, S.H., Knapp, K.T., Carlock, M., Lloyd, A.C., Gibbs, R.E. and Pierson, W.R. (1991) CRC-APRAC Vehicle Emissions Modelling Workshop, J. Air Waste Manage. Assoc., 41, 817-820.

153. Graedel, T.E., Bates, T.S., Bouwman, A.F., Cunnold, D., Dignon, J., Fung, I., Jacob, D.J., Lamb, B.K., Logan, J.A., Marland, G., Middleton, P., Pacyna, J.M., Placet, M., and Veldt, C. (1993) A Compilation of Inventories of Emissions to the Atmosphere, Global Biogeochem. Cycles, 7, 1-26

154. Oort, A.H. (1983) *Global Atmospheric Circulation Statistics, 1958-1973*, NOAA Prof. Paper 14, U.S. Dept. of Commerce, Rockville, MD, 180 pp.

155. Shea, D. (1986) *Climatological Atlas: 1950-1979, Surface Air Temperature, Precipitation, Sea-level Pressure, and Sea-surface Temperature (45S - 90 N)*, Tech. note NCAR/TN-269+STR, Nat. Centre for Atmos. Res., Boulder, Co., USA.

156. Hansen, J., and Lebedef, S. (1987) Global trends of measured surface air temperature, J. Geophys. Res., 92, 13,345-13,372.

157. Jones, P.D., Wigley, T.M.L., and Wright, P.B. (1986) Global temperature variations between 1861 and 1984, Nature, 322, 430-434.

158. Eischeid, J.K., Diaz, H.F., Bradley, R.S., and Jones, P.D. (1991) *A Comprehensive Precipitation Datat set for Global Land Areas*, DoE/ER-69017T-H1, US Dept. of Energy, Washington, DC 20585, USA.

159. Hahn, C.J., Warren, S.G., London, J., Jenne, R.L., and Chervin, R.M. (1988) *Climatological Data for Clouds over the Globe from Surface Observations*, NDP026, Carbon Dioxide Information Analysis Centre, Oak Ridge National Laboratory, Oak Ridge, Tenn., USA.

160. Rossow, W.B., and Schiffer, R.A. (1991) ISCCP cloud data products, Bull. Am. Met. Soc., 72, 2-20.

161. Couch, G.R. (1988) *Lignite Resources and Characteristics*, IEA Coal Research Report No. IEACR/13, London, UK.

162. Matthews, E. (1983) Global vegetation and land use: new high resolution databases for climate studies, J. Clim. Appl. Meteorol., 22, 474-487.

163. Olson, J., and Watts, J. (1982) *Map of major world ecosystem complexes*, Environmental Sciences Div., Oak Ridge National Laboratory, Oak Ridge, TN, USA.

164. Wilson, M.F., and Henderson-Sellers, A. (1985) A global archive of land cover and soils data for use in general circulation climate models, J. Clim., 5, 119-143.

165. Zobler, L. (1986) A world datafile for global climate modeling, NASA Tech. Mem. 87802, 32 pp.

166. Matthews, E., and Fung, I. (1987) Methane emission from natural wetlands: global distribution, area, and environmental characteristics of sources, Global Biogeochem. Cycles, 1, 61-87.

167. Aselmann, I., and Crutzen, P.J. (1989) Global distribution of natural freshwater wetlands and rice paddies, their net primary productivity, seasonality and possible methane emissions, J. Atmos. Chem., 8, 307-358.

168. Tucker, C.J., Fung, I.Y., Keeling, C.D., and Gammon, R.H. (1986) Expansion and contraction of the sahara desert from 1980-1990, Science, 253, 299-301.

169. Lerner, J., Matthews, E., and Fung, I. (1988) Methane emissions from animals: a global high resolution data base, Global Biogeochem. Cycles, 2, 139-156.

170. Matthews, E., Fung, I., and Lerner, J. (1991) Methane emission from rice cultivation: Geographic and seasonal distribution of cultivated areas and emissions, Global Biogeochem. Cycles, 5, 3-24.

171. *Seydlitz Weltatlas* (1984) Cornelson and Schroedel, Geographische Verlagsgesellschaft, Berlin, pp 232.

172. Mannini, A., Daniel, M., Kirchner, A., and Soud, H. (1990) *World Coal-fired Power Stations*, IEA Coal Research Report No. IEACR/28, London.

173. NAEI, National Atmospheric Emissions Inventory (1993) Warren Spring Laboratories, Gunnels Wood Road, Stevenage, Hertfordshire, SG1 2BX, UK.

174. QUARG, The Quality of Urban Air Review Group. (1993) First Report. Air Quality Division, Department of the Environment, Romney House, 43 Marsham Street, London SW1P 3PY, U.K.

175. Amann, M. (1990) Recent and Future Developments of Emissions of Nitrogen Oxides in Europe, Atmos. Environ., 24A, 2759-2765.

176. Simpson, D. and Styve, H. (1992) The Effects of VOC Protocol on Ozone Concentrations in Europe, EMEP MSC-W Note 4-92, MSC-W, The Norwegian Meteor Inst., PO Box 43 - Blindern, N-0313, Oslo 3, Norway.

177. Badr, O., and Probert, S.D. (1993) Oxides of nitrogen in the Earth's atmosphere, Appl. Energy, 46, 1-67.

178. Colbeck, I. and MacKenzie, A.R. (1993) Ozone Formation in Urban Plumes, in *Oxidants in the Environment*, J.O. Nriagu (ed.), J. Wiley and Sons, NY, USA.

179. Scheff, P.A. and Wadden, R.A. (1993) Receptor Modelling of Volatile Organic Compounds, 1, Emission Inventory and Validation, Environ. Sci. Technol., 27, 617-625.

180. Prinn, R., Cunnold, D., Simmonds, P., Alyea, F., Boldi, R., Crawford, A., Fraser, P., Gutzler, D., Hartley, D., Rosen, R. and Rasmussen, R. (1992) Global Average Concentration and Trend for Hydroxyl Radicals deduced from ALE/GAGE Trichloromethane (methyl chloroform) data for 1978-1990, J. Geophys. Res., 97, 2445-2462.

181. Crutzen, P.J. (1991) Methane's Sinks and Sources, Nature, 350, 381-382.

182. Law, K.S. and Pyle, J.A. (1993) Modelling Trage Gas Budgets in the Troposphere, 1, Ozone and Odd-nitrogen, I, J. Geophys. Res., 98, 18,377-18,400

183. Law, K.S. and Pyle, J.A. (1993) Modelling Trace Gas Budgets in the Troposphere, 2, CH_4 and CO, J. Geophys. Res., 98, 18,400-18,412.

184. Hargreaves, K.J., Fowler, D., Storeton-West, R.L. and Duyzer, J.H. (1992) The exchange of nitric oxide, nitrogen dioxide and ozone between pasture and the atmosphere, Environ. Pollut., 75, 53-59.

185. Garland, J.A. (1977) The dry deposition of sulphur dioxide to land and water surfaces, Proc. Roy. Soc. Lond., A354, 245-268.

186. Owen, P.R. and Thompson, W.R. (1963) Heat transfer across rough surfaces, J. Fluid Mech., 15, 321-334.

187. Brutsaert, W. (1975) The roughness length for water vapor, sensible heat and other scalars, J. Atmos. Sci., 32, 2028-2031.

188. Thom, A.S. (1972) Momentum, mass and heat exchange of vegetation, Quart. J. Roy. Met. Soc., 98, 124-134.

189. Chamberlain, A.C. (1966). Transport of gases to and from grass and grass-like surfaces, Proc. Roy. Soc., 290, 236-265.

190. Delany, A.C. and Davies, T.D. (1983) Dry deposition of NO_x to grass in rural East Anglia, Atmos. Environ., 17, 1391-1371.

191. Padro, J., den Hartog, G. and Neumann, H.H. (1991) An investigation of the ADOM dry deposition module using summertime O_3 measurements above a deciduous forest, Atmos. Environ., 25A, 1689-1704.

192. Wesely, M.L. and Hicks, B.B. (1977) Some factors that affect the deposition rates of sulfur dioxide and similar gases on vegetation, J. Air Pollut. Control Assoc., 27, 1110-1116.

193. Baldocchi, D.D., Hicks, B.B. and Meyers, T.P. (1988) Measuring biosphere-atmosphere exchanges of biologically related gases with micrometeorological methods, Ecology, 69, 1331-1340.

194. Fowler, D. and Duyzer, J.H. (1989) Micrometeorological techniques of the measurement of trace gas exchange, in Andreae, M.O. and Schimel, D.S. (eds.) *Exchange of Trace Gases Between Terrestrial Ecosystems and the Atmosphere*, John Wiley and Sons, NY, 189-207.

195. McMahon, T.A. and Denison, P.J. (1979) Empirical atmospheric deposition parameters -a survey, Atmos. Environ., 13, 571-585.

196. Sehmel, G.A. (1980) Particle and gas deposition : a review, Atmos. Environ., 14, 983-1011.

197. Galbally, I.E. and Roy, C.R. (1980) Destruction of ozone at the earth's surface, Quart. J. Roy. Met. Soc., 106, 599-620.

198. Rondon, A., Johansson, C. and Granat, L. (1993) Dry deposition of nitrogen dioxide and ozone to coniferous forests, J. Geophys. Res., 98, 5159-5172.

199. Simmons, A.J. and Colbeck, I. (1990) Resistance of various building materials to ozone deposition, Environ. Technol., 11, 973-978.

200. Monteith, J.L. and Unsworth, M.H. (1990) *Principles of Environmental Physics*, Second Edition, Arnolds, London.

201. Colbeck, I. and Harrison, R.M. (1985) Dry deposition of ozone: Some measurements of deposition velocity and of vertical profiles to 100 metres, Atmos. Environ., 19, 1807-1818.

202. Dolske, D.A. and Gatz, D.F. (1985) A field intercomparison of methods for the measurement of particle and gas dry deposition, J. Geophys. Res., 90, 2075-2084.

203. Droppo, J.G. Jnr. (1985) Concurrent measurements of ozone dry deposition using eddy correlation and profile flux measurements, J. Geophys. Res., 90, 2111-2118.

204. Duyzer, J.H. and Bosveld, F.C. (1988) Measurements of dry deposition fluxes of O_3, NO_3, SO_2 and particles over grass/heathland vegetation and the influence of surface inhomogeneity, KNMI TNO Report No. R88/111.

205. Duyzer, J.H., Meyer, G.M. and van Aalst, R.M. (1983) Measurement of dry deposition velocities of NO, NO_2 and O_3 and the influence of chemical reactions, Atmos. Environ., 17, 2117-2120.

206. Garland, J.A. and Derwent, R.G. (1979) Destruction at the ground and the diurnal cycle of concentration of ozone and other gases, Quart. J. Roy. Met. Soc., 105, 169-183.

207. Robinson, E., Lamb, B. and Chockalingam, M.P. (1982) Determination of dry deposition rates for ozone, US Environ. Protect. Agency, report no. EPA/600/3-82-042.

208. Simmons, A.J. (1992) *Studies of Ozone Deposition*, Ph.D. Thesis, University of Essex, U.K.

209. Hicks, B.B., Wesely, M.L. and Durham, J.L. (1980) Critique of methods to measure dry deposition, United States EPA 600/9-80-050.

210. Eastman, J.A., Wesely, M.L. and Stedman, D.H. (1980) On the mechanisms that control vertical ozone flux to vegetative surfaces, International Ozone Symposium, Boulder, Colorado, 462-470.

211. Enders, G. (1992) Deposition of ozone to a mature spruce forest : measurements and comparison to models, Environ. Pollut., 75, 61-67.

212. Evilsizor, B.A., Stedman, D.H. and Stocker, D.W. (1987) Measurement of NO_x dry deposition, Committee no. AP - 12, Final report to the co-ordinating council.

213. Gusten, H., Ernst, N. and Heinrich, G. (1992) Continuous eddy flux measurements of ozone over agricultural land, *EUROTRAC Annual Report 1991*, Part 4, *BIATEX*, Fraunhofer Institute (IFU), Garmisch-Partenkirchen.

214. Hicks, B.B., Matt, D. R. and McMillen, R.T. (1989) A micrometeorological investigation of surface exchange of O_3, SO_2 and NO_2 : a case study, Boundary-Layer Meteorol., 47, 321-336.

215. Lenschow, D.H., Pearson, R. and Stankov, B.B. (1982) Measurements of ozone vertical flux to ocean and forest, J. Geophys. Res., 87, 8833-8837.

216. Massmann, W.J. (1993) Partitioning ozone fluxes to sparse grass and soil and the inferred resistances to dry deposition, Atmos. Environ., 27A, 167-174.

217. Meyers, T.P. and Baldocchi, D.D. (1993) Trace gas exchange above the floor of a deciduous forest, 2: SO_2 and O_3 deposition, J. Geophys. Res., 98, 12631-12638.

218. Neumann, H.H. and den Hartog, G. (1985) Eddy correlation measurements of atmospheric fluxes of ozone, sulphur and particulates during the Champaign intercomparison study, J. Geophys. Res., 90, 2097-2110.

219. Padro, J. (1993) Seasonal contrasts in modelled and observed dry deposition velocities of O_3, SO_2 and NO_2 over three surfaces, Atmos. Environ., 27A, 807-814.

220. Stocker, D.W., Stedman, D.H, Zeller, K.F., Massman, W.J. and Fox, D.G. (1993) Fluxes of nitrogen oxides and ozone measured by eddy correlation over a shortgrass prairie, J. Geophys. Res., 98, 12619-12630.

221. Wesely, M.L. (1983) Turbulent transport of ozone to surfaces common in the Eastern half of the United States, in Schwartz, S.E. (ed.), *Trace Atmospheric Constituents: Properties, Transformations and Fates*, J. Wiley and Sons, NY, 346-369.

222. Wesely, M.L., Cook, D.R. and Williams, R.M. (1981) Field measurements of small ozone fluxes to snow, wet bare soil and lake water, Boundary-Layer Met., 20, 459-471.

223. Nicholson, K.W. (1988) The dry deposition of small particles: a review of experimental measurements, Atmos. Environ., 22, 2653-2666.

224. Hicks, B.B., Baldocchi, D.D., Meyers, T.P., Hosker, R.P. and Matt, R. (1987) A preliminary multiple resistance routine for delivering dry deposition velocities from measured quantities, Water Air Soil Pollut., 36, 311-330.

225. Wesely, M.L. (1989) Parameterization of surface resistances to gaseous dry deposition in regional-scale numerical models, Atmos. Environ., 23, 1293-1304.

226. Wesely, M.L., Eastman, D.R., Cook, D.R. and Hicks, B.B. (1978) Daytime variations of ozone eddy fluxes to maize, Boundary-Layer Met., 15, 361-373.

227. Fuentes, J.D. and Gillespie, T.J. (1992) A gas exchange system to study the effects of leaf surface wetness on the deposition of ozone, Atmos. Environ., 26A, 1165-1173.

228. Walcek, C.J. (1987) A theoretical estimate of O_3 and H_2O_2 dry deposition over the northeast United States, Atmos. Environ., 21, 2649-2659.

229. Garland, J.A. and Penkett, S.A. (1976) Absorption of Peroxy Acetyl Nitrate and Ozone by Natural Surfaces, Atmos. Environ., 10, 1127-1131.

230. Hill, A.C. (1977) Vegetation: A sink for atmospheric pollutants, J. Air Pollut. Control Assoc., 21, 341-346.

231. Shepson, P.B., Bottenheim, J.W., Hastie, D.R. and Venkatram, A. (1992) Determination of the relative ozone and PAN deposition velocities at night, Geophys. Res. Lett., 19, 1121-1124.

232. Kames, J., Schweighoefer, S. and Schurath, U. (1991) Henry's Law Constant and Hydrolysis of Peroxylacetyl Nitrate (PAN), J. Atmos. Chem., 12, 169-180.

CHAPTER 4

ATMOSPHERIC CHEMISTRY OF PHOTOCHEMICAL OXIDANTS

Given that 21 % of the earth's atmosphere is composed of molecular oxygen it comes as no surprise that oxidation is the dominant chemical process occurring in the troposphere. Trace atmospheric species that are released in any but the most oxidised of forms are subject to low temperature combustion which yields (in the absence of significant deposition of intermediates) compounds such as CO_2, H_2O, HNO_3 and H_2SO_4. However, direct reaction with O_2 is slow under tropospheric conditions for most trace species: it is

Figure 4.1

Schematic diagram of the oxidation of organic compounds in the troposphere. Organic compounds are the fuel for NO_x-catalysed ozone production. From Colbeck[2].

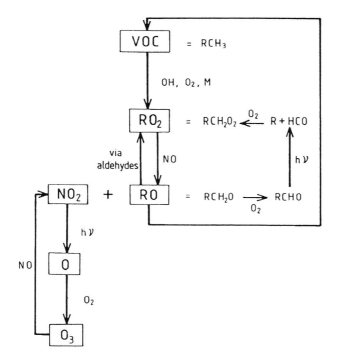

the more reactive radical moieties and ozone, produced by photochemical reactions, that are the atmosphere's principal oxidising agents[1, 2] (Figure 4.1). The rates of reaction of trace components with radicals and ozone will determine the ultimate chemical fate of the compounds from a large number of thermodynamically feasible routes (including, in theory at least, reaction with molecular oxygen). Investigating these possible fates is the role of the chemical kineticist, whose work is generally carried out in the controlled conditions of the kinetics laboratory. The results of such idealised studies, in which physical and fluid dynamical effects have been carefully avoided or minimised, form the basis of this chapter. Photolysis is dealt with first since it is the source of the energy required to initiate tropospheric oxidation: later sections discuss the interaction of these reactive species via bimolecular, termolecular, or heterogeneous processes; first in the remote troposphere, and then near sources of pollution.

The real troposphere is, however, far from the idealised state of the laboratory. As we have seen in chapter 3, emissions of trace species are highly inhomogeneous in both time and space. The troposphere is a chemical reaction vessel of vast proportions. Emitted compounds will not necessarily come into instant contact with oxidising agents. Some allowance must be made for mixing on a macroscopic as well as microscopic scale: i.e., the chemistry of the troposphere is highly coupled to the fluid dynamics of the troposphere. A discussion of the relevant fluid dynamics is outside the scope of this book, but should not be forgotten when considering the implications of the chemistry discussed here to the troposphere as a whole.

The preferred values for the absorption cross-sections, quantum yields and rate coefficients for the reactions discussed can be found in refs. 3, 4, 5, 6, 7, 8, and 9. These are the references which should be consulted for the purpose of building quantitative chemical mechanisms. For convenience, photolysis rates and thermal reaction rates are included alongside most reactions. The photolysis rates have been calculated using a two-stream approach, and give the noon-time rates of photolysis at the surface (1000 mbar), at 45 °N, on June 21st, under clear skies and 350 DU of ozone. The thermal rate coefficients have been calculated for a temperature of 298 K and a pressure of 1000 mbar. The first order rate coefficients have units of s^{-1}. Second order rate coefficients have units of cm^3 molec s^{-1}. Third order rate coefficients are given as effective second order rate coefficients, by including the concentration of the third body in the calculation of the rate coefficient. Third order rate coefficients therefore also have units of cm^3 molec s^{-1}.

4.1 PHOTOLYTIC PROCESSES

The absorption of a photon of light by a molecule may raise it to an electronically excited state that is unstable with respect to dissociation:

$$AB + h\upsilon \ (\lambda > \lambda_{crit}) \ \rightarrow \ AB^* \ \rightarrow \ A + B \qquad\qquad [4.1]$$

where λ_{crit} is the critical wavelength corresponding to a photon with sufficient energy to effect dissociation. The critical wavelength is not the only constraint on the process, however: it will only be efficient where a continuous potential energy surface connects products and reactants (i.e. it is *adiabatic*[10]). In particular the principle of *spin conservation* can be applied. This states that absorption of a photon does not change the *spin multiplicity* of a state, and that the spin multiplicities of the products are "paired" (i.e. singlet + singlet or triplet + triplet. For an explanation of states, spins, etc. see ref. 11, for example). In addition, the energy required for dissociation may come in part from the vibrational and rotational states of the reactant, so that the yield of product will not be an abrupt step-function of λ around λ_{crit}.

The intermediate in reactions such as reaction [4.1] will be an electronically excited reactant molecule. As an alternative to dissociation, this intermediate may undergo a number of other photochemical and photophysical processes such as re-emission, isomerisation, rearrangement, reaction with another molecule, dissociation to different products, radiationless transition, fluorescence and phosphorescence[12, 13]. Thus, for a given dose of radiation only a percentage of photon absorptions will result in reaction [4.1]. The overall *quantum yield* for the formation of product A in reaction [4.1] is thus:

$$\phi_A(\lambda_1) = \left(\frac{Number\ of\ molecules\ of\ A\ formed}{Number\ of\ photons\ absorbed} \right) at\ \lambda = \lambda_1 \qquad\qquad (4.1)$$

There is also a finite chance that incident radiation will not be absorbed at all. This effect gives rise to the *absorption cross-section*, σ (cm^2 $molec^{-1}$; we use absorption cross-sections per molecule, base *e*, throughout), in the well-known Beer-Lambert law for a given wavelength and temperature,

$$\log \frac{I}{I_o} = -\sigma_{\lambda,T}[A]l \qquad\qquad (4.2)$$

where I_o and I are incident and transmitted light intensities, respectively, [A] is the number concentration of the absorbing molecule (molecules cm^{-3}), and l is the path length in cm. The rate of loss of reactant AB in reaction [4.1], at wavelength λ, can thus be written as

$$-\left(\frac{d[AB]}{dt}\right)_{\lambda,T} = \sigma_{\lambda,T}\ \phi_{\lambda,T}\ F_{\lambda}\ [AB] \qquad (4.3)$$

where F is the photon intensity (photons $cm^{-3}\ s^{-1}$) within a given volume of atmosphere. F is a function of many variables, solar zenith angle, overhead ozone column, cloud cover and atmospheric aerosol loading[13]. Calculation of F is a problem in radiation transfer, the discussion of which is outside the scope of this work. Equation (4.3) has obvious resemblances to the classical rate equation for a unimolecular reaction, and an integrated photolytic rate constant is defined as[13]

$$J_{AB} = \int_{\lambda_1}^{\lambda_2} \sigma_{\lambda,T}\ \phi_{\lambda,T}\ F_{\lambda}\ d\lambda \qquad (4.4)$$

for the spectral region of interest. Figure 4.2 summarises schematically the absorption cross-sections for several important tropospheric species[14].

Ozone Photolysis

The O_2-O bond in ozone has an energy of only 101 kJ mol^{-1}. This implies from the Planck relation, $E=hc/\lambda$, that any radiation with wavelength less than 1185 nm (most of the spectrum of sunlight at ground level, see Figure 4.2) is sufficiently energetic to cleave the O_2-O bond[10]:

$$O_3(^1A_1) + h\upsilon\ (\lambda < 1180\ nm) \rightarrow O(^3P) + O_2(^3\Sigma g^-)$$

$$J_2 = 4.2\ x\ 10^{-4} \qquad [4.2]$$

At much shorter wavelengths the excess energy of the photon can be converted into electronic excitation in the products. The dominant process follows the principle of spin conservation, forming two singlet excited states:

$$O_3(^1A_1) + h\upsilon\ (\lambda < 310\ nm) \rightarrow O(^1D) + O_2(^1\Delta g)\quad J_3 = 2.9\ x\ 10^{-5} \qquad [4.3]$$

There is evidence indicating that, even at wavelengths shorter than the threshold, the primary quantum yield of $O(^1D)$ and $O_2(^1\Delta g)$ is not unity[15, 16, 17]. Furthermore the primary quantum yield of $O_2(^1\Delta g)$ remains high at $310 \leq \lambda \leq 334$ nm, although $\phi(O(^1D))$ becomes negligibly small. In this region, therefore, photodissociation appears to occur by

110

Figure 4.2

Schematic of the absorption coefficients of trace species that are important in tropospheric chemistry, and of the solar flux reaching the ground. From Cox[14].

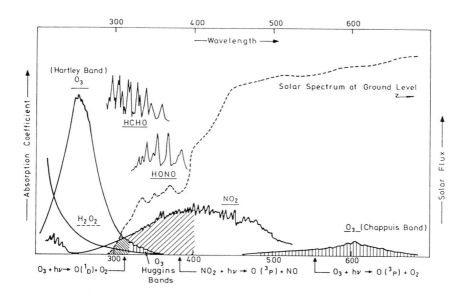

a "spin-forbidden" path[10]. Hence, care must be taken when inferring concentrations of either O_3 or $O(^1D)$ from measurements that can be made of the $O_2(^1\Delta g \rightarrow {}^3\Sigma g^-)$ airglow in the upper atmosphere, as has been proposed[18]. Once formed, $O(^1D)$ may return to its ground state via collisional energy transfer (thermal degradation) or abstract a proton from H_2O or CH_4.

The electronically excited molecular oxygen ($^1\Delta g$) might also be a candidate for a tropospheric oxidising agent, particularly of unsaturated hydrocarbons. Rate constants are of the order of 10^{-18} cm^3 molec^{-1} s^{-1} for these reactions, however. If we compare this value with the rate coefficients for the reaction of an alkene with OH and O_3 (around 10^{-12} and 10^{-18} cm^3 molec^{-1} s^{-1}, respectively); it is obvious that $O_2(^1\Delta g)$ would have to be present in concentrations similar to, or greater than, that of O_3 itself to have any significant effect on tropospheric chemistry[13]. In fact, concentrations of around 10^8 molec cm^{-3} (10 pptv at the surface) seem more likely[19]

The Photolysis of inorganic nitrogen compounds

Ozone is generated in the troposphere as a result of the photolysis of nitrogen dioxide:

$$NO_2 + h\nu \ (\lambda < 420 \ nm\ ) \rightarrow NO + O(^3P) \hspace{2cm} J_4 = 8.6 \ x \ 10^{-3} \hspace{1cm} [4.4]$$

At $\lambda \leq 300$ nm the quantum yield of NO is approximately unity, but in the region of tropospheric interest ($300 \leq \lambda \leq 400$ nm) it is about 0.8. This drop in efficiency is attributed to collisional deactivation. The absorption cross-section of NO_2 rises steadily from 1.3 x 10^{-19} cm^2 molec^{-1} at 300 nm, to about 6 x 10^{-19} cm^2 molec^{-1} for $\lambda \leq 400$ nm.

Photolysis can occur outside the wavelength region ($\lambda < 397.9$ nm) within which the photons have sufficient energy to cleave the N-O bond. Photolysis in this energy deficient region ($397.9 < \lambda < 420$ mm) is attributed to the utilisation of rotational and possibly vibrational energy stored in the molecule. The exact mechanism has yet to be confirmed. Davidson et al.[20] find that the absorption cross-section of NO_2 is only a weak function of temperature, in contrast to earlier work (e.g. ref. 21). Direct measurements of photolysis rates by sunlight under carefully controlled conditions[22] support the conclusion that photolysis rates at +30°C and -70°C are very similar, with $J_4(30°C)/J_4(-70°C) \approx 1.07$.

Nitric oxide is not photolabile in the troposphere; other oxides of nitrogen are:

$$NO_3 \ + h\nu \ (\lambda < 4600nm) \rightarrow NO + O_2 \hspace{1.5cm} J_5 = 2.4 \ x \ 10^{-2} \hspace{1cm} [4.5]$$

$$NO_3 \ + h\nu \ (\nu < 580 \ nm) \rightarrow NO_2 + O(^3P) \hspace{1cm} J_6 = 2.0 \ x \ 10^{-2} \hspace{1cm} [4.6]$$

$$N_2O_5 + h\nu \ (\lambda < 406nm) \ \rightarrow NO_3 + NO_2 \hspace{1cm} J_7 = 4.4 \ x \ 10^{-5} \hspace{1cm} [4.7]$$

$$HONO + h\nu \ (\lambda < 400nm) \rightarrow HO \ + NO \hspace{1cm} J_8 = 1.8 \ x \ 10^{-3} \hspace{1cm} [4.8]$$

$$HNO_3 + h\nu \ (\lambda < 320nm) \ \rightarrow HO + NO_2 \hspace{1cm} J_9 = 5.9 \ x \ 10^{-7} \hspace{1cm} [4.9]$$

Note that the wavelength limit for nitric acid photolysis is close to the short wavelength limit for penetration into the troposphere, and that it is very slow under these conditions[23]. In contrast, reactions [4.5], [4.6], [4.7] and [4.8] are relatively efficient, so that these oxides of nitrogen generally have pronounced diurnal variations in concentrations with a minimum around midday[24]. Note also that, given an efficient source of nitrous acid in the troposphere, reaction [4.8] may be an important source of OH.

The Photolysis of Organic Species

Unsubstituted hydrocarbons are not photolabile in the troposphere. Simple monoalkenes, for example, have π-π^* electronic transitions corresponding to the absorption of light with wavelengths between 160 and 210 nm. Aromatic compounds have similar transitions corresponding to light with wavelengths of about 230 nm.

Organic compounds containing nitrogen are an important reservoir of NO_x. Their formation is described in § 4.2. In the atmosphere. the most abundant organic compound containing nitrogen is PAN[25], peroxy acetyl nitric anhydride, more commonly called peroxy acetylnitrate[a, 26]. At tropospheric wavelengths, the absorption cross-section for PAN itself is less than 1×10^{-21} cm^2 molec^{-1}, and that of peroxypropionic nitric anhydride (PPN) even smaller[13] (cf. the absorption cross-section for NO_2 around 300nm, above). At longer wavelengths no absorption occurs, so that photolysis of PAN is not an important destructive mechanism in the troposphere. Alkyl nitrates and peroxynitrates form as minor products in the reaction between alkyl peroxy radicals and NO or NO_2 (refs. 14 and 27; see § 4.2 and 4.3), particularly for alkyl chains composed of more than 4 carbon atoms. Again, absorption cross-sections tail-off for wavelengths greater than 300 nm, giving integrated tropospheric rate constants for photolysis of the order of 10^{-6} - 10^{-7}s^{-1} (refs. 28 and 29). This implies that photolysis, i.e.

$$RONO_2 + h\nu \ (\lambda \approx 340nm) \rightarrow RO + NO_2 \qquad\qquad J_{10} \approx 10^{-6} \qquad [4.10]$$

and attack by OH (k $\approx$ 1 x 10^{-12} cm^3 molec^{-1} s^{-1}; § 4.2) are of roughly equal importance as tropospheric loss mechanisms. Table 4.1 lists estimated atmospheric photolysis rates for some alkyl nitrates and alkyl peroxynitrates [30, 31, 32, 33, 34].

In addition to PAN compounds another class of peroxy compounds, the hydroperoxides (ROOH), have begun to be closely studied with respect to "clean-air" chemistry[35]. The simplest hydroperoxide H_2O_2, has a weak absorption for $\lambda \leq 360$ nm and dissociates to give OH with a quantum yield of unity[36]

$$H_2O_2 + h\nu \ (\lambda \leq 360 \text{ nm}) \rightarrow 2 \text{ OH} \qquad\qquad J_{11} = 6.9 \text{ x } 10^{-6} \qquad [4.11]$$

In middle latitudes the small absorption cross-section for this process means that a more important loss mechanism is scavenging by cloud water. Similarly, organic peroxides absorb weakly in the near u.v.[9] but are probably more reactive towards attack by OH to form aldehydes (§ 4.2), and to scavenging by cloud water.

The aldehydes, formed by atmospheric oxidation of volatile organic compounds (§

---

[a] Roberts has reviewed and criticised current nomenclature used for organic nitrates. In particular, the common term peroxyacyl nitrate should be replaced by peroxy acetyl nitric acid anhydride. However, use of the old term is so widespread that it will be some time before the new, chemically consistent, term becomes established. As an interim measure, we will usually use the term "PAN compounds" as a generic term for acyl compounds containing nitroperoxy substituents. The term "organic nitrate" then refers only to compounds containing the nitrooxy group, and the term "organic peroxynitrate" to non-acyl compounds containing nitroperoxy substituents.

Table 4.1: Photodissociation rates for $RONO_2$ and $ROONO_2$ compounds. From Roberts[26]. ZA = solar zenith angle.

Compound	$10^6 J(s^{-1})$	Conditions	Refs.
CH_3ONO_2	2.2	ZA=45°, surface	30
	1.5	ZA=0°, 980 m	31
$CH_3CH_2ONO_2$	1.9	ZA=0°, 980 m	31
	1.4	ZA=0°, surface	32
$CH_3CH_2CH_2ONO_2$	1.3	ZA=0°, surface	32
$(CH_3)_2CHONO_2$	3.9	ZA=0°, 980 m	31
$CH_3CH_2CH(ONO_2)CH_3$	3.5	ZA=0°, 980 m	31
	2.5	ZA=0°, surface	32
$(CH_3)_3CONO_2$	10.0	ZA=0°, 980 m	31
$CH_3C(O)CH_2ONO_2$	42.0	ZA=0°, 980 m	31
$HOONO_2$	9.6	troposphere	33
$CH_3C(O)OONO_2$	0.06-0.21	ZA=0-60°, surface	34
$CH_3CH_2C(O)OONO_2$	0.07-0.23	ZA=0-60°, surface	34

4.2), and emitted directly into the troposphere (chapter 3) are themselves photolabile, having absorption cross-sections of around 10^{-20} cm^2 molec^{-1} at wavelengths between 300 and 340 nm. There are two channels for the photodissociation of formaldehyde

$$HCHO + h\nu \ (\lambda < 360 \text{ nm}) \rightarrow H + HCO \qquad J_{12} = 1.7 \times 10^{-5} \qquad [4.12]$$

$$HCHO + h\nu \ (\lambda < 360 \text{ nm}) \rightarrow H_2 + CO \qquad J_{13} = 4.3 \times 10^{-5} \qquad [4.13]$$

The difference in reactivity between the products of the reaction means that it is important to know accurately the primary quantum yield of each channel. Reaction [4.12] produces reactive radicals which contribute to the oxidative capacity of the atmosphere by forming HO_2 (§ 4.2); this is the dominant channel for $300 \leq \lambda \leq 320$ nm (ref. 13). At wavelengths larger than 340 nm dissociation to the relatively stable molecular products (reaction [4.13]) occurs preferentially.

The photochemistry of higher aldehydes is complicated by the fact that there are generally several possible dissociative channels. Acetaldehyde, for example, has $\sigma = 1 - 4$ x 10^{-20} cm^2 molec^{-1} ($295 \leq \lambda \leq 325$ nm) and four possible reaction pathways[13]:

$$CH_3CHO + h\nu \ (\lambda < 330 \text{ nm}) \rightarrow CH_4 + CO \qquad J_{14} = 2.8 \times 10^{-15} \qquad [4.14]$$

$$CH_3CHO + h\nu \ (\lambda < 330 \text{ nm}) \rightarrow CH_3 + HCO \qquad J_{15} = 5.3 \times 10^{-5} \qquad [4.15]$$

$$CH_3CHO + h\nu \ (\lambda < 330 \text{ nm}) \rightarrow CH_3CO + H \qquad\qquad\qquad\qquad [4.16]$$

$$CH_3CHO + h\nu \ (\lambda < 330 \text{ nm}) \rightarrow CH_2CO + H_2 \qquad\qquad\qquad\qquad [4.17]$$

In laboratory studies H_2 is detected in trace quantities only, thus ruling out channels 16 and 17. Route 14 only becomes significant for wavelengths shorter than 280 nm, and only 15 occurs for longer wavelengths[6]. C_2H_5CHO photolyses to produce C_2H_5 and CHO only[6]. Aldehydes with longer carbon chains are generally assumed to follow the same pattern (e.g. ref. 23).

The photodegradation of biogenic compounds results in the production of aldehydes. Most important of those formed are methacrolein ($CH_2=C(CH_3)CHO$) and hydroxyacetaldehyde ($HOCH_2CHO$), both of which are photolabile[37, 38]:

$$CH_2=C(CH_3)CHO + hv \rightarrow CH_2=C(CH_3) + HCO \qquad [4.18]$$

$$HOCH_2CHO + hv \rightarrow HOCH_2 + HCO \qquad [4.19]$$

The products of reactions [4.18] and [4.19] are very reactive towards molecular oxygen and will form peroxy radicals that can convert NO to NO_2 (§ 4.2). Photolysis rate coefficients have not been evaluated for the individual species under atmospheric conditions: some fraction of the acetaldehyde rate coefficient is commonly used (e.g. ref. 38).

Ketones are emitted directly into the planetary boundary layer (chapter 3) and are formed in the oxidation of hydrocarbons with more than 3 carbon atoms (§ 4.2). In the troposphere, ketones can undergo photolysis. There are two channels for the photodissociation of acetone, for example:

$$(CH_3)_2CO + hv \ (\lambda < 340 \text{ nm}) \rightarrow CH_3CO + CH_3 \qquad [4.20]$$

$$(CH_3)_2CO + hv \ (\lambda < 340 \text{ nm}) \rightarrow 2 \ CH_3 + CO \qquad [4.21]$$

Typical mid-latitude photolysis rate coefficients are 3.1×10^{-6} s^{-1} and 2.4×10^{-7} s^{-1} for reactions [4.20] and [4.21], respectively[23]. Note that a subsequent product of reaction [4.20] will be PAN:

$$CH_3CO + O_2 + M \rightarrow CH_3C(O)O_2 + M \qquad k_{22} = 5.0 \times 10^{-12} \qquad [4.22]$$

$$CH_3C(O)O_2 + NO_2 + M \rightarrow CH_3C(O)O_2NO_2 + M \qquad k_{23} = 7.2 \times 10^{-12} \qquad [4.23]$$

an important reservoir compound for NO_x in the free troposphere. Nonsymmetric ketones dissociate such that formation of the more stable alkyl radical predominates. Hence butanone, for example, forms CH_3 and C_2H_5 radicals in the ratio 1:40 approximately[13]:

$$CH_3C(O)C_2H_5 + hv \ (\lambda < 340 \text{ nm}) \rightarrow CH_3CO + C_2H_5 \qquad J_{24} = 8.4 \times 10^{-7} \qquad [4.24]$$

The rate of photolysis under these conditions is similar in magnitude to the pseudo first-order rate constant for butanone's reaction with OH (assuming $[OH] \approx 1 \times 10^6$ cm^{-3}, implies $k' \approx 3 \times 10^{-6}$ s^{-1}).

An important intermediate in the photo-oxidation of biogenically emitted isoprene is methyl vinyl ketone (MVK). Most modelling studies of the tropospheric chemistry of BNMHC (biogenic non-methane hydrocarbons) omit the photolytic reactions of MVK and other ketonic intermediates that are outlined in Brewer *et al.*[39], e.g.:

$$CH_3C(O)CH=CH_2 + h\nu \ (\lambda < 320nm) + 2O_2 \rightarrow CH_3CO_3 + HCHO + HO_2 + CO \quad [4.25]$$

which is expected to be a loss mechanism for MVK comparable in size to reaction with OH (ref. 37). Note that the products of reaction [4.25] have been given as the more long-lived peroxy compounds which from rapidly in the oxygen-rich atmosphere, rather than the primary alkyl photolysis products.

In addition to the aldehydic and ketonic species formed in the degradation of complex volatile organic compounds (both biogenic and anthropogenic in origin), dicarbonyl species are produced. Methylglyoxal is, for example, a major intermediate in the photo-oxidation of isoprene. It absorbs in the regions $\lambda < 340$ nm and $355 < \lambda < 460$ nm (refs. 37 and 38)

$$CH_3C(O)CHO + h\nu \ (\lambda<460 \ nm) + 2O_2 \rightarrow CH_3C(O)O_2 + HO_2 + CO$$
$$J_{26} = 1.4 \times 10^{-4} \quad [4.26]$$

Biacetyl may then be assumed to follow a similar path:

$$(CH_3CO)_2 + h\nu \ (\lambda<460 \ nm) \rightarrow 2 \ CH_3CO \quad\quad\quad [4.27]$$

Glyoxal may cleave to form more stable molecular products[13]

$$OHC.CHO + h\nu \ (\lambda<460 \ nm) \rightarrow HCHO + CO \quad\quad J_{28} = 1.4 \times 10^{-5} \quad [4.28]$$
$$OHC.CHO + h\nu \ (\lambda<460 \ nm) \rightarrow 2 \ CO + H_2 \quad\quad J_{29} = 7.8 \times 10^{-5} \quad [4.29]$$

in the appropriate ratio, (reaction [4.28]):(reaction [4.29]), of 13:87. The importance of photolysis for methylglyoxal and biacetyl in the troposphere is not known accurately. Uncertainties in the photolysis of aldehydic, ketonic and dicarbonyl species could limit our ability to accurately evaluate the role of BNMHC on both regional and global scales.

Halogenated organic compounds enter the troposphere from biogenic and anthropogenic activity. Organic compounds containing bromine have been implicated in the observed spring-time decrease of surface ozone levels in the arctic[40]. Bromoform, emitted from marine algae, is photolysed to produce active bromine and bromine oxide radicals

$$CHBr_3 + h\nu \ (\lambda < 310 \ nm) \rightarrow Br + CHBr_2 \quad\quad\quad [4.30]$$
$$CHBr_2 + O_2 \rightarrow BrO_x + products \quad\quad\quad [4.31]$$

As indicated above, the exact mechanism has not been specified. Further reaction produces BrO which is also photolabile[19] (see §4.2 for further details)

$$BrO + h\nu \ (\lambda < 515 \ nm) \rightarrow Br + O \quad\quad\quad [4.32]$$

Condensed phase photolysis

Atmospheric species dissolved in the aqueous phase present in clouds and fogs will also be subject to a flux of actinic radiation. As the radiation impinges on the surface of an aerosol or cloud droplet, a fraction of it will be reflected back into the gaseous medium, so that actinic fluxes inside a single, isolated, droplet are generally assumed to be approximately 20 - 80 % of the clear-sky value[42]. Similarly, primary quantum yields are reduced because a greater percentage of photolysed fragments are held close together by the solvent, and recombine (the *solvent cage effect*). However, path lengths in clouds are much longer than in cloudless air because multiple scattering can occur[41, 42], so that photolysis rates may or may not increase for constituents which become dissolved. Two photolytic reactions in the condensed phase of importance to photo-oxidant chemistry are:

$$O_3(aq) + H_2O(l) + hv \rightarrow H_2O_2(aq) + O_2(aq) \qquad\qquad J_{33} = 3.2 \times 10^{-5} \qquad [4.33]$$

$$H_2O_2(aq) + hv \rightarrow 2\ OH(aq) \qquad\qquad J_{34} = 9.2 \times 10^{-6} \qquad [4.34]$$

where the photolysis rates given are for the equator in July, at 700 mb (ref. 43). Coupled with other aqueous phase reactions (§ 4.2), the photolytic reactions above may contribute to significant ozone destruction on a global scale. If such a mechanism is indeed as important as it now appears, it emphasises the hitherto neglected influence of aerosols and clouds on the composition of the troposphere that has already been realised in the case of ozone depletion in the polar stratosphere[44].

When a gas-phase molecule becomes adsorbed on a solid surface, the initial (chemisorbed) monolayer can be profoundly affected by the underlying solid. The interaction of sorbate and sorbent bonds will change the absorption spectrum of the sorbate[45], generally lowering the energy required to photolyse the species by 0.5-1.0 eV (50-100 kJ mol^{-1}, *cf.* a typical bond energy, 300 kJ mol^{-1}). An example of this dramatic change in absorption spectrum is given in Figure 4.3. Subsequent layers can become physisorbed on top of the initial monolayer[46]. The molecules in these upper layers are less strongly affected by interaction with the sorbent and so generally require more energy to photolyse. Zhou and White[45] conclude from their study of the photochemistry of small organic compounds on glass that:

(i)   the chemisorbed first layer has unique photochemistry which begins at a significantly lower energy than photolysis in the gas phase;

(ii)   surface photochemistry is compound specific (witness the different spectra of CH_3Br and CH_3Cl in Figure 4.3)

Figure 4.3

Comparison of gas phase and adsorbed optical absorption coefficients (photodissociation cross sections) of ethyl bromide and ethyl chloride[45]. Absorption coefficients of adsorbed species are for monolayer coverage on Ag{111}. Zero of the y-axis corresponds to an absorption coefficient of 10^{-20} cm^2.

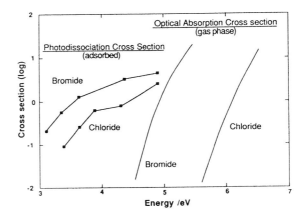

(iii)   multiple layers of physisorbed molecules react 3-5 times more slowly than the unique chemisorbed monolayer;

(iv)   the photochemistry does not follow the optical absorption cross-section of the gas phase molecule; and

(v)   photochemistry is not generally related to the *work function*[b, 47] of the adsorbent, and occurs on irradiation by photons with energy lower than the work function.

Clearly heterogeneous photolysis could contribute to atmospheric chemistry because, although the adsorptive capacity of the atmospheric aerosol will be small relative to the volume of the gas phase, the adsorbed molecules have properties very different from their gaseous counterparts. However, the actual heterogeneous photochemistry observed in the atmosphere will be a strong function of aerosol particle phase, composition, surface structure, and the amount of photolabile adsorbate on the surface (which will, in turn, be

---

[b] The work function of a solid is the minimum energy required for electrons to be ejected from its surface[47]. Since this minimum energy is sensitive to the condition of the adsorbent surface, the work function varies with adsorbate and coverage.

a function of the rate of reactant adsorption relative to the sorption rates of the many possible sorbates present in the air).

Table 4.2: Rates of some heterogeneous reactions involving photolysis, expressed as the rate of methane production, except [¶], which is given as the rate of CH_3Br loss. Note the various units in column 4.

Sorbate	Sorbent	Light Source	Specific Reaction rate	Refs.
Ethanoic acid	ZnO	1000 W Hg-Xe	484×10^{-10} mol m^{-2} h^{-1}	48
	TO$_2$	"	26.4×10^{-10} mol m^{-2} h^{-1}	"
	WO$_3$	"	33.8×10^{-10} mol m^{-2} h^{-1}	"
	MnO$_2$	"	traces	"
	MoO$_3$	"	0	"
	Fe$_2$O$_3$	"	29.4×10^{-10} mol m^{-2} h^{-1}	"
	MgO	"	$\approx 150 \times 10^{-10}$ mol m^{-2} h^{-1}	"
	SiO$_2$	"	14.2×10^{-10} mol m^{-2} h^{-1}	"
	γ-Al$_2$O$_3$	"	5.6×10^{-10} mol m^{-2} h^{-1}	"
CH$_3$Br	Ag{111}	100 W Hg	$\approx 3 \times 10^{-15}$ mol m^{-2} h^{-1} [¶]	49
C$_2$H$_2$Cl$_2$	Pd{111}	Hg broad band	Not reported	50, 51
	Pt{111}	"	"	"
2-pentanone	vycor glass	"	50×10^{-5} ml h^{-1}	51
2-butanone	"	"	30 ml h^{-1}	"
acetone	"	"	12 ml h^{-1}	"

Table 4.2 summarises some typical laboratory studies of heterogeneous photolysis[48, 49, 50, 51]: sorbents used range from pure transition metals to porous glass, whilst sorbates used have tended to be small organic molecules. Such compounds (*e.g.* CH$_3$COOH, CH$_3$Br, etc.) are found in the atmosphere, but small inorganic molecules such as O$_3$ and NO$_2$ are generally more abundant and so may dominate the surface layers of suitable particles (of course, the major components of air: O$_2$, N$_2$, Ar, CO$_2$, and H$_2$O must also be considered). The soluble mineral fraction of atmospheric aerosol is outlined in Table 4.3: sea-salt and carbonaceous matter (not shown) make up a large part of the overall range of compounds. The surface photochemistry of real aerosol particles is not known, though the sorption processes occurring prior to photolysis have been studied for some relevant species[52]. Of course, the hygroscopicity of much of the condensed material in the atmospheric aerosol means that the particles will deliquesce to form saturated liquid droplets at even moderate relative humidities. There will then be a layer of solution between any solid surface and the gas phase.

Table 4.3: Mass concentrations ($\mu g \ m^{-3}$) of the soluble mineral fraction of inorganic aerosol for various types of location.

Element or group	Urban	Continental	Marine
SO_4^{2-}	15.1	3.2	2.6
NO_3^-	6.4	0.9	0.05
Cl^-	2.0	0.1	4.6
Br^-	0.3	-	0.02
NH_4^+	5.9	1.3	0.2
Na^+	2.1	0.05	2.9
K^+	0.7	0.06	0.1
Ca^{2+}	1.7	0.2	0.2
Mg^{2+}	0.9	-	0.4
Al_2O_3	5.0	0.2	-
SiO_2	13.5	0.7	-
Fe_2O_3	4.6	0.1	0.06
CaO	-	0.1	-
sum	59.6	6.9	11.2

4.2 TROPOSPHERIC CHEMISTRY FAR FROM ANTHROPOGENIC SOURCES

As outlined in chapter 3, anthropogenic activity releases a huge variety of compounds into the planetary boundary layer and free troposphere. The magnitudes of the emissions will vary greatly around the globe, being largest in the densely populated and heavily industrialised regions of Europe, North America and the Far East. Far from these relatively localised areas, however, air masses will contain those trace species which are sufficiently long-lived to be transported from their point of release, intermediate compounds in the oxidation of these long-lived species, and any locally emitted species of biogenic origin (should the air parcel be at or near ground level). In practice this means that the chemistry of photo-oxidants far from anthropogenic sources is the chemistry of O_3, NO, CH_4, H_2O, PAN, organic peroxides and hydroperoxides, and locally emitted terpenes (in special cases sea-salt aerosol and halogenated compounds may also play a part; see *e.g.* ref. 40. In this section we discuss the photochemistry of the remote troposphere for two reasons: because it is easier to introduce the basic characteristics of the photochemistry in the simpler reactant mixture found in the remote troposphere; and because man's activities may already be changing the composition of the remote troposphere on a global scale.

Inorganic Chemistry

In the troposphere, atomic oxygen is generated by the photolysis of NO_2 and O_3 (see reactions [4.2], [4.3] and [4.4] above).

$$NO_2 + h\nu \rightarrow NO + O(^3P) \qquad\qquad J_4 = 8.6 \text{ x } 10^{-3} \qquad [4.4]$$

Ground-state atomic oxygen reacts with molecular oxygen to (re)form ozone in a trimolecular reaction:

$$O(^3P) + O_2 + M \rightarrow O_3 + M \qquad\qquad k_{35} = 1.2 \text{ x } 10^{-14} \qquad [4.35]$$

but ozone reacts with NO to re-form NO_2

$$O_3 + NO \rightarrow NO_2 + O_2 \qquad\qquad k_{36} = 1.8 \text{ x } 10^{-14} \qquad [4.36]$$

so that there is a photochemical cycling of NO and NO_2 (or $O(^3P)$ and O_3). A *photostationary state* is reached rapidly, such that

$$[O_3] = \frac{J_4[NO_2]}{k_{36}[NO]} \qquad\qquad (4.5)$$

i.e. production and loss rates will be brought into balance. The rapid interconversion of O_3 and $O(^3P)$, and of NO and NO_2, make it convenient to consider these compounds as forming families: $O_x = O_3 + O(^3P)$; and $NO_x = NO + NO_2$. It is also convenient to bracket the sum of O_3 and NO_2 together since they make up the total reservoir of "potential ozone". Potential ozone will be conserved in an air parcel if reactions [4.4], [4.35], and [4.36] are the only reactions controlling NO_x partitioning (the concentrations of $O(^3P)$ and OH are very small by comparison).

Other possible reaction pathways must now be considered. The photolysis of O_3 has been discussed in § 4.1. Any $O(^3P)$ formed is rapidly reconverted into O_3 by reaction [4.35]; excited $O(^1D)$ may follow the same path after thermal degradation (collision without reaction), or extract a proton from H_2O or CH_4:

$$O(^1D) + N_2 \rightarrow O(^3P) + N_2 \qquad\qquad k_{37} = 2.6 \text{ x } 10^{-11} \qquad [4.37]$$
$$O(^1D) + O_2(^3\Sigma g^-) \rightarrow O(^3P) + O_2(^1\Sigma g^+ \text{ or } ^1\Delta g) \qquad\qquad k_{38} = 4.0 \text{ x } 10^{-11} \qquad [4.38]$$
$$O(^1D) + H_2O \rightarrow 2\ OH \qquad\qquad k_{39} = 2.2 \text{ x } 10^{-10} \qquad [4.39]$$
$$O(^1D) + CH_4 \rightarrow OH + CH_3 \qquad\qquad k_{40} = 1.4 \text{ x } 10^{-10} \qquad [4.40]$$

Given reactant abundances of $2.1 \text{ x } 10^{19}$, $5.6 \text{ x } 10^{18}$, $3 \text{ x } 10^{17}$ and $4.3 \text{ x } 10^{13}$ molec cm^{-3} for N_2, O_2, H_2O and CH_4, respectively, these reactions have pseudo first order rate coefficients of $5.5 \text{ x } 10^8$, $2.2 \text{ x } 10^8$, $6.6 \text{ x } 10^7$, and $6.0 \text{ x } 10^4$ s^{-1} for reactions [4.37] to [4.40], respectively. This implies that for every $O(^1D)$ atom converted to OH about 12 atoms are thermally deactivated and reconverted to ozone. This minor route is, however, of central importance in photochemical oxidant chemistry because of the reactive nature of the hydroxyl radical. It will become obvious in what follows that the hydroxyl radical, OH, interconverts rapidly with the hydroperoxy radical, HO_2. It is again convenient to consider the two compounds as a family. The odd-hydrogen family, HO_x, is then given by OH + HO_2 (the concentration of free hydrogen radicals in the troposphere is very small, see reaction [4.61]).

The rapid cycling of potential ozone, between NO_2 and O_3 itself, can be interrupted if N-containing compounds other than NO and NO_2 are formed[53]

$OH + NO_2 + M \rightarrow HNO_3 + M$	$k_{41} = 1.2 \times 10^{-11}$	[4.41]
$O(^3P) + NO_2 + M \rightarrow NO_3 + M$	$k_{42} = 1.4 \times 10^{-12}$	[4.42]
$O_3 + NO_2 \rightarrow NO_3 + O_2$	$k_{43} = 3.2 \times 10^{-17}$	[4.43]
$NO_2 + NO_3 + M \rightarrow N_2O_5 + M$	$k_{44} = 6.4 \times 10^{-13}$	[4.44]
$NO + NO_3 \rightarrow 2 NO_2$	$k_{45} = 2.3 \times 10^{-11}$	[4.45]
$OH + NO + M \rightarrow HONO + M$	$k_{46} = 8.4 \times 10^{-12}$	[4.46]
$HO_2 + NO_2 + M \rightarrow HO_2NO_2 + M$	$k_{47} = 1.3 \times 10^{-12}$	[4.47]

Most of these sinks for NO_x are temporary, the products being thermally or photochemically labile (§ 4.1). Nitric acid, however, formed via reaction [4.41], is the most oxidised form of atmospheric nitrogen, and as such it is relatively stable, reacting only slowly with OH

$$OH + HNO_3 \rightarrow NO_3 + H_2O \qquad k_{48} = 1.5 \times 10^{-13} \qquad [4.48]$$

HNO_3 can be an important contributor to the overall acidity of an air mass[54] and it can be removed quickly by both dry and wet deposition[55, 56]. This is an example of the general point that the termination of radical chains processes in the troposphere commonly occurs by the formation of stable, highly oxidised, species which are subsequently deposited out of the system. Reactions of the nitrate radical have also been studied[57]: in general, reactions with closed-shell (molecular) species are not rapid whilst those with other radicals such as OH

$$OH + NO_3 \rightarrow HO_2 + NO_2 \qquad k_{49} = 2.0 \times 10^{-11} \qquad [4.49]$$

are much faster. However, for daytime tropospheric OH concentrations, loss of NO_3 through reaction [4.49] is some four orders of magnitude smaller than loss via photolysis[57] and only reaction [4.45] can compete. Other NO_3 chemistry is discussed below in the section on nocturnal oxidant chemistry.

Other inorganic processes are important in the production of acid deposition by the atmospheric oxidation of sulphur compounds, but this is outside the scope of this book. The chemistry of hydrogen peroxide is dealt with at the same time as that of the organic peroxides, below.

Aqueous phase chemistry

Aqueous phase mechanisms can be important in the oxidant photochemistry of the remote troposphere. Of course, the occurrence of aqueous-phase photochemistry presupposes that some fraction of the gas-phase reactants have been partitioned into the available liquid phase: for a droplet in equilibrium with its surroundings, Henry's law may be applied, i.e.

$$\gamma_A \, [A]_{(aq)} = H_A \, e_A \qquad (4.6)$$

where H_A is the effective Henry's law constant for species A in M atm^{-1}, e_A is its equilibrium partial pressure above the surface in atm, γ_A is the dimensionless activity coefficient of A in the droplet, and $[A]_{(aq)}$ is the aqueous phase concentration, in M (*i.e.* mol l^{-1}) at the surface (which is equal to the bulk concentration if the droplet is well-mixed). An "effective Henry's law coefficient", H_A^*, can be defined which takes account of the fact that A is removed from solution by dissociation:

$$H_A^* = \frac{H_A}{1 + \dfrac{K_a}{[H^+]_{eff}}} \qquad (4.7)$$

Here K_a is the dissociation constant of A in solution (mol l^{-1}), and $[H^+]_{eff}$ is the effective hydrogen ion concentration (mol l^{-1}). A summary of Henry's coefficients for gases of importance to photo oxidant chemistry is given in Table 4.4[58, 59, 60, 61, 62, 63, 64, 65]. Table 4.5 summarises some of the aqueous phase equilibria which are important in determining the dissolution of the gases in Table 4.4[66, 67, 68]. It should be borne in mind that the concentration of minor species in the aqueous phase is a strong function of the hydrogen ion concentration (equation (4.7)), and that the hydrogen ion concentration itself will depend on the acidity of the major components of the aqueous phase, such as H_2SO_4, and NH_4^+.

External conditions such as changes in temperature, changes in emission, or chemical reaction, may change the gas phase concentration on timescales shorter than the time required for the droplet to reach equilibrium. The rate at which A partitions into solution

Table 4.4: Some Henry's Law Coefficients for gases of importance to photochemical oxidant chemistry. From Lelieveld[42].

Equilibrium		H_A(298 K) M atm^{-1}	$-\Delta H/R$ K	References
$HO_2(g)$	$\rightleftharpoons$ $HO_2(aq)$	2.0×10^3	6600	58
$H_2O_2(g)$	$\rightleftharpoons$ $H_2O_2(aq)$	7.4×10^4	6615	59
$O_3(g)$	$\rightleftharpoons$ $O_3(aq)$	1.1×10^{-2}	2300	60
$HCHO(g)$	$\rightleftharpoons$ $HCHO(aq)$	6.3×10^3	6425	61
$HCOOH(g)$	$\rightleftharpoons$ $HCOOH(aq)$	3.7×10^3	5700	62
$CH_3OOH(g)$	$\rightleftharpoons$ $CH_3OOH(aq)$	2.2×10^2	5653	59
$CH_3OO(g)$	$\rightleftharpoons$ $CH_3OO(aq)$	2.0×10^3	6600	estimate
$HNO_3(g)$	$\rightleftharpoons$ $HNO_3(aq)$	2.1×10^5	8700	64
$NO(g)$	$\rightleftharpoons$ $NO(aq)$	1.9×10^{-3}	1480	64
$NO_2(g)$	$\rightleftharpoons$ $NO_2(aq)$	6.4×10^{-3}	2500	65
$NO_3(g)$	$\rightleftharpoons$ $NO_3(aq)$	15.0		63

is then governed by the slowest rate of mass transfer in the system. Mass transfer can be limited by diffusion of the gas-phase species to the droplet surface, by diffusion across the gas-liquid interface, or by diffusion within the droplet itself[69]. The characteristic timescales for bulk diffusion to the droplet surface, τ_{dg}, and for the establishment of interfacial equilibrium, τ_i, are:

$$\tau_{dg} = \frac{r^2}{4D_g} \tag{4.8}$$

$$\tau_i = D_l \left(\frac{4RTH_A^*}{\alpha \bar{c}} \right)^2 \tag{4.9}$$

where r is the radius of the droplet, D_g is the gas-phase diffusion coefficient, D_l is the liquid-phase diffusion coefficient, R is the gas constant, T is the ambient temperature, α is the mass accommodation coefficient of the surface, and $\bar{c}$ is the mean thermal velocity of a gaseous molecule at temperature, T. Diffusion within the droplet is characterised by the timescale, τ_{da}:

$$\tau_{da} = \frac{r^2}{D_l} \tag{4.10}$$

For droplets with radius greater than about 0.1 mm, internal circulations develop and the

Table 4.5: Some aqueous phase equilibria of importance to photo oxidant chemistry. From Lelieveld[42].

Equilibrium		$K_a(298\ K)$ mol l^{-1}	$-\Delta H/R$ K	References
$HO_2(aq)$	$\leftrightharpoons O_2^- + H^+$	3.5×10^{-5}		66
$H_2O_2(aq)$	$\leftrightharpoons HO_2^- + H^+$	2.2×10^{-12}	-3730	67
$HCOOH(aq)$	$\leftrightharpoons HCOO^- + H^+$	1.8×10^{-4}	-1510	68
$HNO_3(aq)$	$\leftrightharpoons NO_3^- + H^+$	15.4		58

timescale for internal mixing is rapid enough to be neglected in the estimation of mass transfer rates. The timescale for chemical reaction, τ_{cr}, is:

$$\tau_{cr} = \left(\frac{1}{[A]_{(aq)}} \frac{d[A]_{(aq)}}{dt} \right)^{-1} \qquad (4.11)$$

Under typical tropospheric conditions τ_{dg} and τ_{cr} are large with respect to τ_i and τ_{da}. If τ_{dg} is short compared to τ_{cr}, then the Henry's Law equilibrium condition can be used.

The chemistry occurring in a cloud droplet will depend on its composition, and its composition can vary widely depending on the airmass from which it is formed. Polluted continental air masses produce cloud droplets containing significant amounts of transition metal ions such as Fe^{n+} and Mn^{n+}. These transition metal ions can catalyse the production of hydrogen peroxide in the droplets, and the oxidation of S^{IV} compounds[70]. In cloud droplets formed in clean maritime airmasses, dissolution of ozone and HO_2 results in the permanent removal of ozone:

$HO_2(g) \leftrightharpoons H^+(aq) + O_2^- (aq)$ [4.50]

$O_2^- (aq) + O_3(aq) \leftrightharpoons O_2(aq,g) + O_3^- (aq)$ [4.51]

$O_3^- (aq) + H^+(aq) \rightarrow OH + O_2(aq,g)$ [4.52]

This process may be important in the global ozone budget of the troposphere[43].

The hydrolysis of N_2O_5 in, or on, liquid aerosol

$N_2O_5(g) + H_2O(l) \rightarrow 2\ HNO_3(aq)$ [4.53]

appears be important in the troposphere, as well as in the stratosphere[71, 72]. The large Henry's Law constant for HNO_3 (Table 4.4) means that most of the HNO_3 formed by reaction [4.53] remains in solution. Reaction [4.53] removes odd nitrogen (NO_2 and NO_3), odd oxygen (via NO_3) and potential ozone (NO_2) irreversibly, and so may alter the photochemical character of the airmass in which it occurs. More details of the effect of

N_2O_5 hydrolysis are given in the section on nocturnal oxidant chemistry, below.

If the tropospheric aerosol is dry - i.e., temperatures and/or humidities are below those required for deliquesence - dinitrogen pentoxide or nitrogen dioxide may react with surface inorganic sea-salts[73] :

$$N_2O_5(g) + NaBr(s) \rightarrow BrNO_2 + NaNO_3 \qquad [4.54]$$
$$N_2O_5(g) + NaCl(s) \rightarrow ClNO_2 + NaNO_3 \qquad [4.55]$$
$$2\ NO_2(g) + NaBr(s) \rightarrow BrNO + NaNO_3 \qquad [4.56]$$

liberating bromine and chlorine radicals which can act as photochemical initiators, or as ozone destruction catalysts (see below).

Oxidation of Methane

Methane is the most abundant hydrocarbon present in the earth's atmosphere. Figure 4.4 is a schematic of the overall oxidation of CH_4 to CO_2. Note the recycling of HO_x radicals in a NO_x -rich environment:

$$CH_4 + OH + O_2 \rightarrow CH_3O_2 + H_2O \qquad k_{57} = 6.5 \times 10^{-15} \quad [4.57]$$
$$CH_3O_2 + NO \rightarrow CH_3O + NO_2 \qquad k_{58} = 7.6 \times 10^{-12} \quad [4.58]$$
$$CH_3O + O_2 \rightarrow HCHO + HO_2 \qquad k_{59} = 1.9 \times 10^{-15} \quad [4.59]$$
$$HO_2 + NO \rightarrow NO_2 + OH \qquad k_{60} = 8.3 \times 10^{-12} \quad [4.60]$$

Net: $\quad CH_4 + 2\ NO + 2\ O_2 \rightarrow HCHO + 2\ NO_2 + H_2O$

so that oxidation has been effected without loss of the reactive species responsible for its initiation. In fact, the oxidation of NO to NO_2 will lead to an ozone production via 4 and 35 which will, in turn, provide more $O(^1D)$ for conversion to OH. In this sense the oxidation of alkanes in a NO_x-rich environment is *autocatalytic*.

Ravishankara[74] points out that very few measurements of the rate-limiting step in [4.57] (formation of CH_3) have been made under conditions similar to those found in the upper troposphere, so that there is some uncertainty in the overall conversion rate. This uncertainty must be resolved if a definitive explanation for the observed rate of methane concentration increase is to be given. At present it is not clear whether increased methane concentrations are due to any increasing flux into the atmosphere, or to a decreased rate of chemical destruction via the reactions given above and in Figure 4.4[75]. There are important feedback mechanisms between the magnitude of methane emissions and the chemical composition of the atmosphere into which it is emitted, particularly relating to methane's

Figure 4.4

Schematic outline of the methane oxidation cycle. From Ravishankara[74].

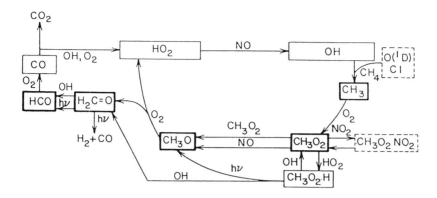

role as a "greenhouse" gas and the sensitivity of bacterial emissions to temperature (chapter 3). Methane and CO, the product of methane oxidation, are the major sinks for tropospheric OH. Future emissions of methane can be expected, therefore, to have an increasing lifetime to chemical destruction because OH is being increasingly removed by methane emissions.

In an atmosphere containing sufficient NO_x there are three routes of chemical transformation of the aldehyde formed above: photodissociation to carbon monoxide

HCHO + hv (λ < 360nm) $\rightarrow$ CO + H_2	J_{13} = 4.3 x 10^{-5}	[4.13]

photodissociation to HCO and subsequent oxidation

	HCHO + hv (λ < 360nm) $\rightarrow$ HCO + H	J_{12} = 1.7 x 10^{-5}	[4.12]
	H + O_2 + M $\rightarrow$ HO_2 + M	k_{61} = 7.0 x 10^{-13}	[4.61]
	HCO + O_2 $\rightarrow$ CO + HO_2	k_{62} = 6 x 10^{-12}	[4.62]
2x	(HO_2 + NO $\rightarrow$ NO_2 + OH)	k_{60} = 8 x 10^{-12}	[4.60]
2x	(NO_2 + hv(λ < 420nm) $\rightarrow$ NO + O(^{3}P))	J_4 = 8.6 x 10^{-3}	[4.4]
2x	(O(^{3}P) + O_2 $\rightarrow$ O_3)	k_{35} = 1.2 x 10^{-14}	[4.35]

Net: HCHO + 4 O_2 + 3 hv $\rightarrow$ CO + 2 OH + 2 O_3

where again there is an autocatalytic production of OH and O_3; and direct attack by OH

HCHO + OH $\rightarrow$ HCO + H_2O	k_{63} = 1 x 10^{-11}	[4.63]

$$HCO + O_2 \rightarrow CO + HO_2 \qquad\qquad k_{62} = 6 \times 10^{-12} \qquad [4.62]$$

followed by $NO \rightarrow NO_2$ conversion by HO_2 as above and further production of ozone. In the atmosphere, these reaction pathways will occur in parallel (and in parallel with physical processes such as dry deposition and scavenging by cloud droplets) and will compete for the available formaldehyde. Crutzen[76] has estimated that on average 50-60% of HCHO molecules follow [4.13]; 20-25% follow [4.12] and subsequent reactions; 20-30% are attacked by OH directly, as would be expected from the rate coefficients quoted above. The ultimate oxidation of CO is also accomplished via attack by OH

$$CO + OH \rightarrow CO_2 + H \qquad\qquad k_{64} = 2 \times 10^{-13} \qquad [4.64]$$

$$H + O_2 + M \rightarrow HO_2 + M \qquad\qquad k_{61} = 7.0 \times 10^{-13} \qquad [4.61]$$

where, once more, a sufficient source of NO will regenerate OH directly from reaction [4.60] and indirectly by the photolysis of ozone.

This autocatalytic behaviour is not found in NO_x-poor regions of the troposphere. In that case, peroxides undergo self-reaction to eliminate molecular oxygen:

$$HO_2 + O_3 \rightarrow OH + 2\,O_2 \qquad\qquad k_{65} = 2.0 \times 10^{-15} \qquad [4.65]$$

$$HO_2 + HO_2 \rightarrow H_2O_2 + O_2 \qquad\qquad k_{66} = 1.5 \times 10^{-12} \qquad [4.66]$$

$$HO_2 + CH_3O_2 \rightarrow CH_3OOH + O_2 \qquad\qquad k_{67} = 3.1 \times 10^{-12} \qquad [4.67]$$

$$CH_3O_2 + CH_3O_2 \rightarrow CH_3OH + HCHO + O_2 \qquad\qquad k_{68} \geq 2.1 \times 10^{-13} \qquad [4.68]$$

$$CH_3O_2 + CH_3O_2 \rightarrow 2\,CH_3O + O_2 \qquad\qquad k_{69} = 1.3. \times 10^{-13} \qquad [4.69]$$

$$CH_3O_2 + CH_3O_2 \rightarrow CH_3OOCH_3 + O_2 \qquad\qquad k_{70} \leq 3 \times 10^{-14} \qquad [4.70]$$

(the reaction between CH_3O_2 and O_3 has a rate coefficient of less than 2×10^{-17} cm^{-3} molec^{-1} s^{-1} (ref. 6) and so can be neglected). Note that a dependence on water vapour concentration has been discovered in studies of the kinetics of reaction [4.66] but not in the case of the other reactions[8, 35, 77]. The contribution from each of the three channels in the methyl peroxy self reaction is sensitive to temperature, with reaction [4.69] becoming more important as temperature is decreased[78].

The ratio of rate coefficients for reaction [4.65] to reaction [4.60] is 1 : 4150 at 298 K. Ozone concentrations must, therefore, exceed NO concentrations by a factor of ≈ 4000 for the peroxide-ozone reaction to become significant. For ozone concentrations typical of the non-urban troposphere (30-60 ppbv), this is equivalent to a threshold NO concentration of about 10 pptv. Large regions of the lower troposphere (Figure 4.5) do appear to have sufficiently low NO concentrations[24, 79] for radical-radical combination reactions to be important.

Figure 4.5

Two-dimensional model calculation showing the distribution of NO$_x$ (pptv) in the troposphere and lower stratosphere for June[79]. Values increase markedly above the tropopause ($\approx$ 13 km at the equator) due to the production of NO$_y$ from N$_2$O in the stratosphere.

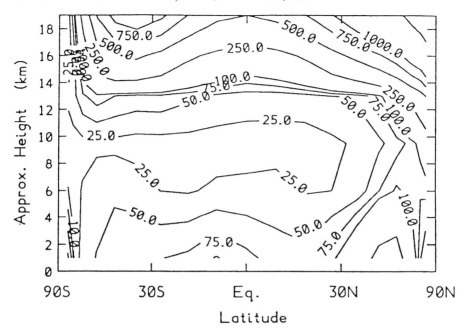

In NO-poor environments then, methane is still attacked by OH but subsequent loss of odd hydrogen (HO$_x$) can occur[76].

$$CH_4 + OH + O_2 \rightarrow CH_3O_2 + H_2O \qquad k_{57} = 6.5 \times 10^{-15} \quad [4.57]$$

$$CH_3O_2 + HO_2 \rightarrow CH_3OOH + O_2 \qquad k_{71} = 3.1 \times 10^{-12} \quad [4.71]$$

$$CH_3OOH + OH \rightarrow HCHO + OH + H_2O \qquad k_{72} = 1.8 \times 10^{-12} \quad [4.72]$$

Net: $CH_4 + OH + HO_2 \rightarrow HCHO + 2\ H_2O$

Methyl peroxide can also be photolysed slowly[80]

$$CH_3OOH + h\nu\ (\lambda < 632\ nm) \rightarrow CH_3O + OH \qquad J_{73} = 5.4 \times 10^{-6} \quad [4.73]$$

$$CH_3O + O_2 \rightarrow HCHO + HO_2 \qquad k_{59} = 1.3 \times 10^{-13} \quad [4.59]$$

which, taken together with reactions [4.57] and [4.71] gives the net reaction

Net: $CH_4 + O_2 \rightarrow HCHO + H_2O$

and no loss of odd hydrogen.

The oxidation of formaldehyde is also affected by the absence of NO since there is no longer an opportunity for HO_x recycling:

$$HCHO + h\nu \ (\lambda < 360nm) \rightarrow HCO + H \qquad\qquad J_{12} = 1.7 \times 10^{-5} \qquad [4.12]$$

$$HCO + O_2 \rightarrow CO + HO_2 \qquad\qquad k_{62} = 6 \times 10^{-12} \qquad [4.62]$$

$$H + O_2 + M \rightarrow HO_2 + M \qquad\qquad k_{61} = 7.0 \times 10^{-13} \qquad [4.61]$$

$$2x \quad (HO_2 + O_3 \rightarrow OH + 2O_2) \qquad\qquad k_{65} = 2 \times 10^{-15} \qquad [4.65]$$

Net: $\quad HCHO + 2\,O_3 + h\nu \rightarrow CO + 2\,OH + 2\,O_2$

or alternatively

$$HCHO + OH \rightarrow HCO + H_2O \qquad\qquad k_{63} = 1 \times 10^{-11} \qquad [4.63]$$

$$HCO + O_2 \rightarrow CO + HO_2 \qquad\qquad k_{62} = 6 \times 10^{-12} \qquad [4.62]$$

$$HO_2 + O_3 \rightarrow OH + 2\,O_2 \qquad\qquad k_{65} = 2 \times 10^{-15} \qquad [4.65]$$

Net: $\quad HCHO + O_3 \rightarrow CO + H_2O + O_2$

which destroys ozone without production of OH as above. The primary photolysis path of formaldehyde, to yield CO and H_2 directly, is not affected by NO_x levels. NO concentrations, then, control the effect that methane oxidation (and, by extension, that of other alkanes) has on ambient ozone levels. Crutzen[76] has estimated that, in the absence of heterogeneous processes, the oxidation of one molecule of CH_4 to CO in a *NO-poor* atmosphere leads to the net *destruction* of 3.5 HO_x and 1.7 O_3 molecules whereas, in *NO-rich* conditions, the same process results in the net *production* of 0.5 HO_x and 3.7 O_3 molecules. This should not be confused with the effect of intense *urban* NO emissions on ozone levels. In that case, NO emissions can reduce ozone concentrations in an air mass via reaction [4.36] and the photo-stationary state relationship.

Given a present estimate of the methane and CO source strengths ($\approx 1.9 \times 10^{10}$ g a^{-1}) and assuming sufficient NO throughout the atmosphere, up to 6×10^{15} g a^{-1} of ozone could be produced[76]. Although there is evidence of increasing surface ozone concentrations (chapter 6) it is not of this scale, a fact attributable to the NO-poor condition of much of the lower troposhere (Figure 4.5). It should be borne in mind, however, that increased NO emissions, particularly close to large biogenic sources of methane at present emitting into NO-poor environments, will have the double-edged effect of shutting off an ozone destroying cycle (the hydroperoxide reactions) and switching on an ozone producing one (the NO→NO_2 interconversion reactions).

Whilst the above discussion includes all the major steps in the oxidation of methane, one or two other points are also worth mentioning. Firstly, in NO_x-rich environments, peroxy radicals can undergo reaction with NO_2 as well as NO:

$$CH_3O_2 + NO_2 + M \rightarrow CH_3O_2NO_2 + M \qquad\qquad k_{74} = 2.6 \times 10^{-12} \qquad [4.74]$$

At lower-tropospheric temperatures methyl peroxynitrate is unstable with respect to thermal decomposition[26]:

$$CH_3O_2NO_2 \rightarrow CH_3O_2 + NO_2 \qquad\qquad k_{75} = 1.4 \qquad [4.75]$$

so that the compound (and its homologues) are unlikely to be important in the redistribution of NO_x and for tropospheric chemistry in general.

Similarly, an additional pathway exists in the reaction of alkyl peroxy radicals with nitric oxide[26, 8], resulting in the formation of alkyl nitrates ($RONO_2$). In general the fraction of these nitrates formed increases in proportion to the number of carbon atoms in the reactant alkyl peroxy species, and so are not a significant product for the reaction of CH_3O_2 with NO.

There are also complicating factors to be taken into account when discussing the reactivity of alkyl peroxy radicals in NO-poor air. In general few data exist for the reactions of these species with HO_2 (e.g. reaction [4.71]). This should be borne in mind when the chemistry of higher alkyl species is discussed below and in § 4.3. Atkinson[8] assigns an uncertainty of a factor of 2 to the rate coefficient for reaction [4.71] at 298 K. Furthermore, there is some uncertainty as to the products formed in the reaction. General accounts[14, 76, 81] usually follow the scheme given above but there is some recent evidence that a second channel may be important[35]:

$$CH_3O_2 + HO_2 \rightarrow CH_3OOH + O_2 \qquad\qquad k_{71} = 3.1 \times 10^{-12} \qquad [4.71]$$
$$CH_3O_2 + HO_2 \rightarrow HCHO + H_2O + O_2 \qquad\qquad k_{76} = 2.1 \times 10^{-12} \qquad [4.76]$$

where $k_{76}/(k_{71} + k_{76}) \approx 0.4$. If this is the case the effect is to transfer odd-hydrogen from a semi-permanent reservoir, CH_3OOH (since it is photolysed only slowly and has a lifetime for reaction with OH roughly equal to that for rainout) to a temporary reservoir, HCHO. As we have seen, aldehydes are photolabile in the troposphere yielding, in part, HCO and H which are rapidly converted to $CO + 2 HO_2$.

The sequence of reactions [4.57], [4.71] and [4.72] has also been simplified by assuming that the only products of the reaction of CH_3OOH with OH are HCHO, OH (from the decomposition of CH_2OOH) and H_2O. In fact, abstraction of the hydrogen from the hydroxyl group can compete with abstraction of a hydrogen from the methyl group[74]:

$$CH_3OOH + OH \rightarrow CH_3O_2 + H_2O \qquad\qquad k_{77} = 3.7 \times 10^{-12} \quad [4.77]$$

$$CH_3OOH + OH \rightarrow CH_2OOH + H_2O \qquad\qquad k_{72} = 1.8 \times 10^{-12} \quad [4.72]$$

$$CH_2OOH \rightarrow HCHO + OH \qquad\qquad\qquad\qquad\qquad\qquad [4.78]$$

with $k_{77}/(k_{77} + k_{72}) \approx 0.67$ at 298 K. The production of CH_3O_2, which is a source of CH_3OOH in reaction [4.71], means that methyl hydroperoxide can act as a catalyst for odd hydrogen destruction until it is removed slowly by reaction with OH at the methyl group, or photolysis:

$$CH_3OOH + OH \rightarrow CH_3O_2 + H_2O \qquad\qquad k_{77} = 3.7 \times 10^{-12} \quad [4.77]$$

$$CH_3O_2 + HO_2 \rightarrow CH_3OOH + O_2 \qquad\qquad k_{71} = 3.1 \times 10^{-12} \quad [4.71]$$

Net: $\quad OH + HO_2 \rightarrow H_2O + O_2$

The expanded sequence for methane oxidation, including the "semi-catalytic" effect of methyl peroxide, is then (using the ratio 1:2 for $k_{71}:k_{76}$ and $k_{72}:k_{77}$)

$$CH_4 + OH + O_2 \rightarrow CH_3O_2 + H_2O \qquad\qquad\qquad k_{57} = 6.5 \times 10^{-15} \quad [4.57]$$

$$3CH_3O_2 + 3HO_2 \rightarrow 2CH_3OOH + HCHO + H_2O + 3O_2 \qquad k_{79} = 5.2 \times 10^{-12} [4.79]$$

$$3\ CH_3OOH + 2\ OH \rightarrow 2\ CH_3O_2 + HCHO + 3\ H_2O \qquad k_{80} = 5.5 \times 10^{-12} \quad [4.80]$$

Net: $\quad CH_4 + CH_3OOH + 3\ OH + 3\ HO_2 \rightarrow 2\ HCHO + 5\ H_2O + 2\ O_2$

The expression is not as satisfactory as those derived earlier because a partially oxidised methyl species appears on the left hand side. Nevertheless, it does give some indication of the overall oxidative process from long-lived species to reactive intermediates such as HCHO, and stable products such as water. The loss of odd-hydrogen from the system is clear, 6 radicals being destroyed. The 2 molecules of formaldehyde produced could quickly yield 3 HO_x radicals on photolysis, however, assuming 50% photolysis to stable products CO and H_2. The effect, then, of this reaction cycle and formaldelyde photolysis will be the loss of 3 HO_x radicals per molecule of CH_4 and CH_3OOH. One might pursue this analysis further by eliminating CH_3OOH from the left hand side of the net oxidation reaction. CH_3OOH is equivalent to $CH_3O_2 + HO_2 - O_2$, since reaction [4.71] is the only source of CH_3OOH in this system. Similarly, CH_3O_2 can be equated with $CH_4 + OH + O_2 - H_2O$ if one neglects the catalytic cycling of CH_3O_2 and CH_3OOH. This gives a final net oxidation reaction for CH_4 of:

Net: $\quad CH_4 + 2\ OH + 2\ HO_2 \rightarrow HCHO + 3\ H_2O + O_2$

which, when compared to the original net oxidation reaction calculated from reactions

[4.57], [4.71] and [4.72] indicates a loss of 4 HO_x radicals rather than the 2 shown initially. Again, the HCHO produced could form 1.5 radicals of HO_x upon reaction.

Biogenic Non-Methane Hydrocarbons

Methane's long lifetime with respect to attack by OH means that it can be transported far from its source before being destroyed. In marked contrast to this, the terpenes are highly reactive and consequently have short lifetimes. Biogenic activity releases a great variety of non-methane hydrocarbons (BNMHC) into the atmosphere, as discussed in chapter 3. Below, we describe the degradation of isoprene, the most abundant compound, and α-pinene, a compound often chosen as a characteristic terpene in modelling studies. It should be noted, however, that the inclusion of BNMHC emissions and chemistry into models of all scales, from the urban[82] to the global[83], is still at an early stage. Chapter 3 outlined some of the uncertainties in the calculation of emission source strengths and compositions; the atmospheric chemistry is discussed below.

When released into the atmosphere isoprene is very reactive to OH and O_3 (as are the anthropogenic unsaturated hydrocarbons; § 4.3). An example mechanism is given in Table 4.6[84]. Under tropospheric conditions the dominant degradative route is by addition of OH. By analogy with substituted monoalkenes[13, 8], reaction at the double bonds would be expected to occur in the ratio 6:4 (C(1)=C(2)):(C(3)=C(4)). In general, however, a 50:50 split is used since this gives a better fit to smog chamber data[38]. Addition at either end of the C(1)=C(2) double bond results in the production of methyl vinyl ketone (MVK): *i.e.* either

$$OH + CH_2=C(CH_3)-CH=CH_2 + O_2 \rightarrow OOCH_2-C(CH_3)(OH)-CH=CH_2$$
$$k_{81} = 1.0 \times 10^{-10} \quad [4.81]$$

$$OOCH_2-C(CH_3)(OH)-CH=CH_2 + NO \rightarrow OCH_2-C(CH_3)(OH)-CH=CH_2$$
$$+ NO_2 (+1\% \text{ nitrate}) \quad k_{82} \approx 4 \times 10^{-12} \quad [4.82]$$

$$OCH_2-C(CH_3)(OH)-CH=CH_2 \rightarrow HOCH_2-C(CH_3)(O)-CH=CH_2 \quad [4.83]$$

$$HOCH_2-C(CH_3)(O)-CH=CH_2 \rightarrow CH_3C(O)-CH=CH_2 + CH_2OH$$
$$k_{84} = 7.9 \times 10^4 \quad [4.84]$$

or

$$OH + CH_2=C(CH_3)-CH=CH_2 + O_2 \rightarrow HOCH_2-C(CH_3)(OO)-CH=CH_2$$
$$k_{85} = 1.0 \times 10^{-10} \quad [4.85]$$

Table 4.6: An isoprene oxidation scheme[84]. Rate coefficients given are in units appropriate to the order of reaction shown (reactants in brackets are assumed to be in large excess) and have not been revised.

Reaction		k_{298} cm^3-molec-s
ISOP + OH (+ O$_2$)	$\rightarrow$ ISOO$_2$	2.36×10^{-11}
ISOO$_2$ + NO	$\rightarrow$ NO$_2$ + HO$_2$ + HCHO + 0.5 (MVK + MACR)	3.8×10^{-12}
ISOO$_2$ + NO	$\rightarrow$ nitrate	4.2×10^{-13}
ISOP + O$_3$	$\rightarrow$ 0.4 HCHO + 0.25 (MACR + MAOO) + 0.17 (MVK + MVKOO + CO + CH$_2$O$_2$) + 0.04 HO$_2$	7.3×10^{-15}
MVKOO + NO	$\rightarrow$ MVK + NO$_2$	4.2×10^{-12}
MVKOO + NO$_2$	$\rightarrow$ MVK + NO$_3$	7×10^{-13}
MVKOO + H$_2$O	$\rightarrow$ products	3.4×10^{-18}
MAOO + NO	$\rightarrow$ MACR + NO$_2$	4.2×10^{-12}
MAOO + NO$_2$	$\rightarrow$ MACR + NO$_3$	7×10^{-13}
MAOO + H$_2$O	$\rightarrow$ products	3.4×10^{-18}
MVK + OH (+ O$_2$)	$\rightarrow$ MVINT	3.4×10^{-12}
MVINT + NO	$\rightarrow$ NO$_2$ + 0.7 (RCO$_3$ + RCHO) + 0.3 (HCHO + MGLY + HO$_2$)	3.8×10^{-12}
MVINT + NO	$\rightarrow$ nitrate	4.2×10^{-13}
MGLY + OH (+ O$_2$)	$\rightarrow$ RCO$_3$ + CO + H$_2$O	1.7×10^{-11}
MACR + OH (+ O$_2$)	$\rightarrow$ MAOO$_2$	1.0×10^{-11}
MAOO$_2$ + NO (+ O$_2$)	$\rightarrow$ PPOO + NO$_2$ + CO$_2$	4.2×10^{-12}
PPOO + NO (+ O$_2$)	$\rightarrow$ ACETOO + NO$_2$	4.2×10^{-12}
ACETOO + NO (+ O$_2$)	$\rightarrow$ MGLY + NO$_2$ + HO$_2$	4.2×10^{-12}
MAOO$_2$ + NO$_2$ (+ M)	$\rightarrow$ MPAN	4.8×10^{-12}
MPAN (+ M)	$\rightarrow$ MAOO$_2$ + NO$_2$	4.0×10^{16}
MACR + OH (+ O$_2$)	$\rightarrow$ MRO$_2$	3.9×10^{-12}
MRO$_2$ + NO	$\rightarrow$ NO$_2$ + HO$_2$ + HCHO + MGLY	3.8×10^{-12}
MRO$_2$ + NO	$\rightarrow$ nitrate	4.2×10^{-13}
MVK + O$_3$	$\rightarrow$ 0.5 (HCHO + MGLY) + 0.2 (CH$_2$O$_2$ + HO$_2$ + MCRIG + CO) + 0.15 (RCHO + RCO$_3$)	2×10^{-15}
MACR + O$_3$	$\rightarrow$ 0.5 HCHO + 0.2 (CH$_2$O$_2$ + HO$_2$ + MCRIG) + 0.35 CO + 0.15 (CH$_3$O$_2$ + RCOOH)	5×10^{-15}
MCRIG + NO	$\rightarrow$ MGLY + NO$_2$	4.2×10^{-12}
MCRIG + NO$_2$	$\rightarrow$ MGLY + NO$_3$	7×10^{-13}
MCRIG + H$_2$O	$\rightarrow$ products	3.4×10^{-18}
CH$_2$O$_2$ + NO	$\rightarrow$ HCHO + NO$_2$	4.2×10^{-12}
CH$_2$O$_2$ + NO$_2$	$\rightarrow$ HCHO + NO$_3$	7×10^{-13}
CH$_2$O$_2$ + H$_2$O	$\rightarrow$ products	3.4×10^{-18}

$$\text{HOCH}_2\text{-C(CH}_3)(\text{OO)-CH=CH}_2 + \text{NO} \rightarrow \text{HOCH}_2\text{-C(CH}_3)(\text{O)-CH=CH}_2$$
$$+ \text{NO}_2 \text{ (+1\% nitrate)} \qquad k_{86} \approx 4 \times 10^{-12} \qquad [4.86]$$

$$\text{HOCH}_2\text{-C(CH}_3)(\text{O)-CH=CH}_2 \rightarrow \text{CH}_3\text{-C(O)-CH=CH}_2 + \text{CH}_2\text{OH}$$
$$k_{87} = 7.9 \times 10^4 \qquad [4.87]$$

so that it is not necessary to specify the end of the double bond at which the initial OH addition takes place (although reaction to form the more stable primary alkyl radical is more likely). This is also the case for reaction at the C(3)=C(4) double bond: methacrolein

(MACR) being formed from addition of OH at either end of the double bond. Note the formation of organic nitrates which, in the mechanism in Table 4.6, act as permanent sinks for NO_x and BNMHC; and that the CH_2OH radicals formed react rapidly with oxygen to form formaldehyde.

$$CH_2OH + O_2 \rightarrow HCHO + HO_2 \qquad\qquad k_{88} = 9 \times 10^{-12} \qquad [4.88]$$

However, this may not be the complete picture. Atkinson et al.[85] present evidence

Figure 4.6

Chemical reactions involved in the formation of 3-methylfuran from isoprene. From Atkinson et al.[85].

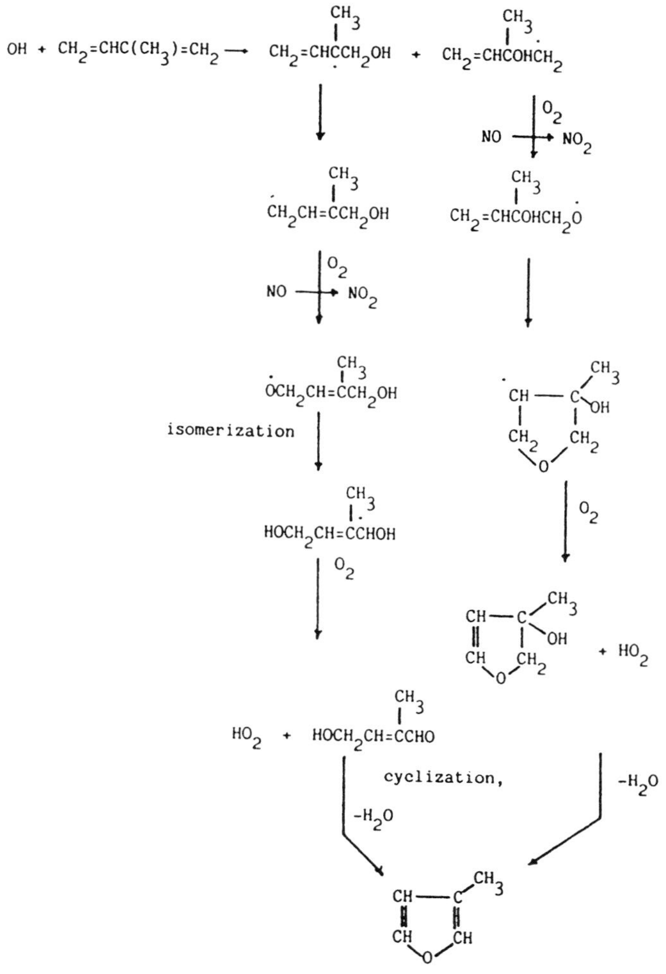

$$OH + CH_2=CHC(CH_3)=CH_2 \longrightarrow CH_2=CHCCH_2OH \;\; (CH_3) \;\; + \;\; CH_2=CHCOHCH_2 \;\; (CH_3)$$

$CH_2=CHCCH_2OH$ (with CH_3) $\downarrow$

$CH_2=CHCOHCH_2$ (with CH_3) : $NO \;|\; O_2 \rightarrow NO_2$

$CH_2CH=CCH_2OH$ (with CH_3) : $NO \;|\; O_2 \rightarrow NO_2$

$CH_2=CHCOHCH_2O$ (with CH_3)

$OCH_2CH=CCH_2OH$ (with CH_3) — isomerization

$CH—C$ ring (with CH_3, OH), CH_2, CH_2, O

$HOCH_2CH=CCHOH$ (with CH_3) $\downarrow O_2$

$\downarrow O_2$

$CH—C$ (with CH_3, OH), CH, CH_2, O $+ HO_2$

$HO_2 + HOCH_2CH=CCHO$ (with CH_3)

cyclization, $-H_2O$ / $-H_2O$

$CH—C$ (with CH_3), CH, CH, O (3-methylfuran ring)

135

of the formation of 3-methylfuran in the reaction, with a yield of about 4% of the total amount of isoprene reacted. Suggested mechanisms for the production of 3-methylfuran are given in Figure 4.6. In common with other aromatic species, 3-methylfuran has been found to be very reactive towards OH, having a rate coefficient of about 9×10^{-11} cm^3 $molec^{-1}$ s^{-1} at 296 K. This is more than an order of magnitude larger than those for methyl vinyl ketone and methacrolein (Table 4.6) so that this minor product may have a significant effect on numerical simulations when included in chemical schemes. Lastly, it should be noted that the reduction of peroxy radicals by NO is subject to the same limitations on NO availability as in the case of the CH_3O_2 radical and that, in NO-poor conditions, radical-radical recombinations and reactions with ozone will compete with the reactions in Table 4.6. To accommodate such NO-poor conditions, Killus and Whitten[37] include reactions of the type

$$HOCH_2\text{-}C(CH_3)(O_2)CH=CH_2 + HO_2 \rightarrow \text{product} \qquad k_{89} = 5 \times 10^{-12} \qquad [4.89]$$

where the unreactive product is presumably a hydroperoxide in the main. Pierotti et al.[86] have reported MACR/MVK ratios measured at rural sites in the USA which suggest that reactions such as those above are important in the clean continental boundary layer. Formaldehyde and the inorganic products of the reaction of isoprene with OH will undergo chemical transformation as outlined in our discussion of methane oxidation: for MVK and MACR the dominant reaction pathway is further reaction with OH (Table 4.6). Reactions with NO_3 and O_3 have been measured and appear to be of minor importance under atmospheric conditions. For both MVK and MACR, OH addition to the terminal carbon of the carbon-carbon double bond is the predominant addition reaction[87]. The Methacrolein product may then decompose in two ways:

$$HOCH_2\text{-}C(OO)(CH_3)CHO + NO \rightarrow HOCH_2\text{-}C(O)(CH_3)CHO$$

$$k_{90} \approx 4 \times 10^{-12} \qquad [4.90]$$

$$HOCH_2\text{-}C(O)(CH_3)CHO \rightarrow CH_3C(O)CHO + CH_2OH \quad k_{91} = 7.9 \times 10^4 \qquad [4.91]$$

$$HOCH_2\text{-}C(O)(CH_3)CHO \rightarrow HOCH_2C(O)CH_3 + HCO \quad k_{92} = 3.9 \times 10^5 \qquad [4.92]$$

producing either methylglyoxal ($CH_3C(O)CHO$) or hydroxyacetone ($HOCH_2C(O)CH_3$). Most model chemistry schemes have, to date, assumed that methylglyoxal production (91) predominates (e.g. refs. 38, 37, 39 and 88) but recent product studies have found that hydroxyacetone production predominates[89]. Alternatively, and occurring at a similar rate, OH can abstract an H-atom from the aldehydic functional group (Table 4.6) to form a peroxy acyl (RCO_3) radical. Further reaction of this species results in the production of methylglyoxal, or of a PAN analogue

$$CH_2=C(CH_3)-C(O)O_2 + NO_2 \rightleftharpoons CH_2=C(CH_3)-C(O)O_2NO_2 \qquad [4.93]$$

The rate coefficients for both forward reaction pathways are similar (Table 4.6) so that the product distribution will depend on the NO/NO$_2$ ratio (which will be controlled mainly by the ozone photo-stationary state) and the temperature dependent degradation of the PAN compound. Ultimately the acyl peroxy radical is destroyed by reaction with NO:

$$CH_2=C(CH_3)-C(O)O_2 + 3NO + 2\,O_2 \rightarrow CH_3C(O)CHO + 3NO_2 + HO_2 + CO_2$$
$$k_{94} = 1 \times 10^{-11} \qquad [4.94]$$

and the methyl glyoxal can go on to react further.

The abstraction of an aldehydic proton is not possible in the reaction of MVK with OH, and abstraction of an H-atom from the terminal methyl group is not efficient. Addition to the terminal carbon of the double bond is favoured and the reactions

$$OH + CH_3-C(O)CH=CH_2 + NO + 2\,O_2 \rightarrow CH_3C(O)CHO + HCHO + HO_2 + NO_2$$
$$k_{95} = 5.7 \times 10^{-12} \qquad [4.95]$$
$$OH + CH_3-C(O)CH=CH_2 + NO + O_2 \rightarrow HOCH_2CHO + CH_3CO + NO_2$$
$$k_{96} = 1.3 \times 10^{-11} \qquad [4.96]$$

occur in the ratio 3:7 (k_{95}:k_{96}), where reaction [4.95] is the result of addition of OH to the inner carbon of the double bond, and reaction [4.96] the result of addition to the outer carbon[90]. The production of glycolaldehyde (HOCH$_2$CHO) is favoured thermodynamically over production of methylglyoxal (CH$_3$C(O)CHO) in route [4.96], so that the production of methylglyoxal by this route can be assumed to be negligible. The product study of Tuazon and Atkinson[90] therefore supports the kinetic schemes used to date.

The other major reaction route for isoprene, and the unsaturated intermediates resulting from its degradation, is reaction with ozone (Table 4.6). The reactions of organic compounds with ozone have been reviewed by Atkinson and Carter[87], and updated rate constants are summarised by Atkinson[8]. The point is made in the updated review that the only products of the reactions of alkenes with ozone that are accurately known, are those of the reaction of ethene with ozone. This is as true for species of biogenic origin as for anthropogenic compounds. In terms of the chemistry of the atmosphere these uncertainties are unlikely to have a large effect: typically the percentages of isoprene, methyl vinyl ketone and methacrolein reacting with ozone rather than OH will be around 2%, 1% and 1%, respectively.

Further reaction of the hydrocarbon intermediates shown in Table 4.6 can be dealt with by discussing the reactivity of generic species. In fact, a generic treatment is made necessary in some instances because of the lack of data on specific compounds. The chemistry follows the general pattern outlined in Figure 4.1: peroxy radicals are produced which react with NO to produce NO_2 without loss of O_3 (other than via the minor O_3-unsaturated hydrocarbon reactions discussed above), or with NO_2 to produce PAN compounds. Reactions between peroxy radicals are not included. As discussed for the chemistry of methane, such reactions become important when NO levels fall below a few pptv, and they have now been included in most schemes designed for use with non-urban conditions (*e.g.* refs. 37, 88 and 91).

The chemistry of terpene compounds - a great variety of which are released by

Figure 4.7

Reactions of α-pinene with a) OH; and b) O_3. From Hatakeyama *et al.*[92].

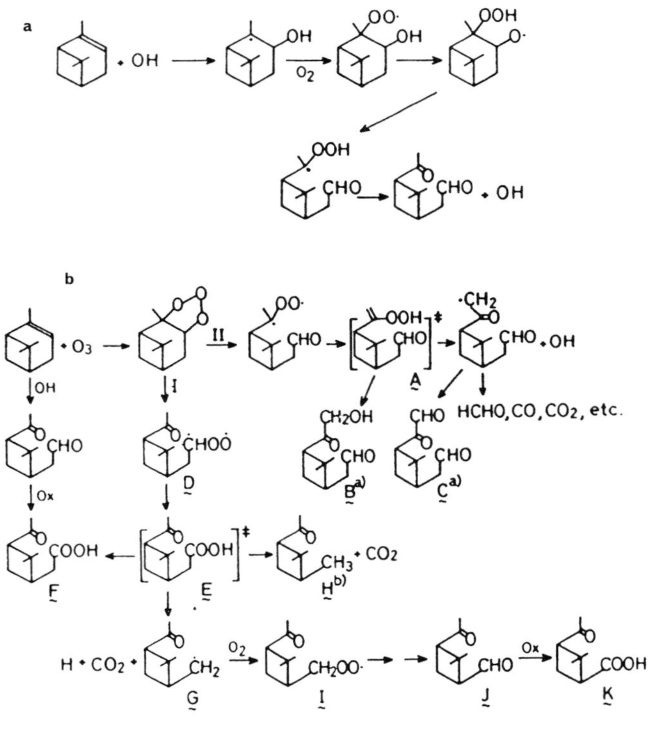

vegetation - is exemplified by that of α-pinene[92] (Figure 4.7). Exemplified is, perhaps, too strong a word: the reactivity and speciation of biogenic terpene emissions should not be regarded as well understood. Nonetheless, α-pinene is the monoterpene most commonly used in numerical simulations of tropospheric chemistry and some details are now emerging.

Table 4.7: An α-pinene oxidation scheme[38]. Rate coefficients given are in units appropriate to the order of the reaction (bracketed species are considered to be in large excess), and have not been revised.

Reaction		k_{298} cm^3-molec-s
PIN + OH (+ O$_2$)	$\rightarrow$ PINT	9.6×10^{-12}
PINT + NO (+ O$_2$)	$\rightarrow$ PCHO + NO$_2$ + HO$_2$	4.2×10^{-12}
PCHO + hv (+ 2 O$_2$)	$\rightarrow$ AL$_4$INT + HO$_2$ + CO	6.6×10^{-5}
AL$_4$INT + NO (+ O$_2$)	$\rightarrow$ ALD$_4$ + HO$_2$ + NO$_2$	4.2×10^{-12}
PCHO + hv	$\rightarrow$ MVK + RCHO	6.6×10^{-6}
PCHO + OH (+ O$_2$)	$\rightarrow$ PCO$_3$ + H$_2$O	1.4×10^{-11}
PCO$_3$ + NO$_2$	$\rightarrow$ PIAN	4.4×10^{-12}
PIAN	$\rightarrow$ PCO$_3$ + NO$_2$	4.0×10^{16}
PCO$_3$ + NO	$\rightarrow$ PO$_2$ + NO$_2$	4.2×10^{-12}
PO$_2$ + NO (+ O$_2$)	$\rightarrow$ ALD$_4$ + NO$_2$ + HO$_2$ + CO$_2$	4.2×10^{-12}
PIN + O$_3$	$\rightarrow$ BIR	4.0×10^{-15}
BIR + NO	$\rightarrow$ PCHO + NO$_2$	4.2×10^{-12}
BIR + NO$_2$	$\rightarrow$ PCHO + NO$_3$	6.4×10^{-13}
BIR + H$_2$O	$\rightarrow$ products	1.6×10^{-18}
ALD$_4$ + hv (+ 2 O$_2$)	$\rightarrow$ MPOX + HO$_2$ + CO	6.6×10^{-6}
MPOX + NO (+ O$_2$)	$\rightarrow$ product + HO$_2$ + NO$_2$	4.2×10^{-12}
ALD$_4$ + OH (+ O$_2$)	$\rightarrow$ ACO$_4$ + H$_2$O	6.2×10^{-12}
ACO$_4$ + NO$_2$	$\rightarrow$ PAN$_2$	4.4×10^{-12}
PAN$_2$	$\rightarrow$ ACO$_4$ + NO$_2$	4.0×10^{16}
ACO$_4$ + NO (+ O$_2$)	$\rightarrow$ MPOX + NO$_2$ + CO$_2$	4.2×10^{-12}

Rate constants at 298 K for the reaction of α-pinene with OH and O$_3$ are 5.37 x 10^{-11} and 8.5 x 10^{-17} cm^3 molec^{-1} s^{-1} (refs. 8 and 93) respectively, so that around 75% of the available α-pinene can be expected to react with OH. This process, shown in Figure 4.7a, results in cleavage of the 6-membered ring and formation of pinonaldehyde. Reaction with O$_3$ (Figure 4.7b) gives rise to several products which contain 4-membered rings and oxygenate moieties (aldehydic, ketonic, carboxylic functional groups) at opposite ends of the molecule. The biradical species formed by the reaction of ozone with α-pinene (and other terpenes) has been implicated in the oxidation of SO$_2$ to sulphuric acid aerosol near the surface of trees[94].

In all the products shown in Figure 4.7 the reactive moieties are sufficiently far apart for the species to react in a manner indicative of the most reactive functional group. This is the assumption adopted in the (necessarily speculative) mechanism of Lloyd et al.[38]

which is given in Table 4.7. Pinonaldehyde, formed by the reaction of α-pinene with OH and O_3 (in contrast to the mechanism of Hatakeyama et al.[92]) is photolysed in two ways (§ 4.1) and attacked by OH. Further degradation then occurs via reaction with NO, NO_2, and H_2O as for the isoprene scheme (Table 4.6) where, again, reactions between peroxy radicals have been omitted. Note that the scheme (Table 4.7) produces MVK: although different, the actual α-β unsaturated ketone formed in the reaction is sufficiently like MVK chemically to be treated as identical in the model, thus removing a variable from the set required for numerical solution of the mechanism.

An important feature of the degradation of terpenes is the production of organic aerosol. Several of the products shown in Figure 4.7 can become supersaturated with respect to aerosol formation at relatively low gaseous concentrations. Hatakeyama et al.[92] report that, for initial α-pinene concentrations between 20 and 120 ppbv and an excess of ozone, the carbon mass-based yields of aerosol remain constant at around 18 %. This will be an upper bound to the formation of organic aerosol particles under ambient conditions ([α-pinene] < 1 ppbv) since sufficient product must be formed to supersaturate the environment if condensation is to occur. Pandis et al.[95] report no significant aerosol formation from isoprene photochemical oxidation carried out in a smog chamber, and a maximum aerosol yield of 8 % (as available carbon) from β-pinene. Once formed, organic aerosols will prevent further reaction of the constituent species other than on the surface of the particles. Eventually coagulation and sedimentation of the particles will remove them from the system entirely, thus preventing the oxidation of some fraction of the total amount of carbon emitted. The process may be important, if accurate chemical schemes and CO budgets are to be produced.

Halocarbon Compounds

Continuing concern about the influence that anthropogenic chlorofluorocarbons (CFC) have on the stratospheric ozone layer (see, e.g. ref. 44) has led to the proposal that less environmentally damaging alternatives be put into production. In general, this implies the use of compounds which will degrade in the troposphere rather than in the stratosphere, since it is the inertness of CFCs to reaction in the troposphere which allows them to accumulate in the lower atmosphere and slowly percolate into the stratophere where they are photolabile. The most likely candidates to replace CFCs are the hydrochlorofluoro-carbons (HCFC) and hydrofluorocarbons (HFC) which are reactive in the troposphere by virtue of the C-H bonds they contain[96]: *e.g.*

$$CF_3CHCl_2 + OH \rightarrow CF_3CCl_2 + H_2O \qquad\qquad k_{97} = 3.7 \times 10^{-14} \qquad [4.97]$$

This is much faster than the equivalent reactions for CFCs which are the order of 10^{-18} cm^{-3} molec^{-1} s^{-1} at 298 K. Subsequent oxidation of the HCFC fragment is then expected to follow a pattern similar to that described above for methane[97] (Figure 4.8).

Figure 4.8

Photochemical oxidation of HCFC-124. From Atkinson[97]. Initial reaction (not shown) is with OH to form the haloalkyl radical.

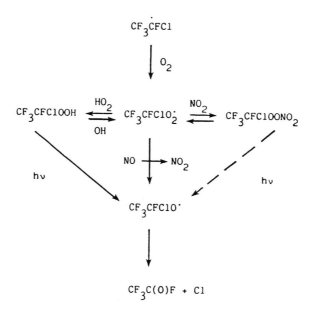

The effect of HCFCs on tropospheric ozone will depend on the rate of halocarbon peroxy radical formation, the rate of the reaction of halocarbon peroxy radicals with NO, and on the HCFC source strength. Table 4.8 summarises the rate data for the reaction of alternative fluorocarbons with OH to form halocarbon peroxy radicals[98]. On the whole, HCFCs and HFCs are about as reactive as methane. Furthermore, they occur in concentrations ranging from pptv to ppbv (10^7 - 10^{10} molec cm^{-3}) in remote and urban air, respectively[98]. Given the rate data, HCFCs would only effect urban ozone levels if concentrations greatly exceeded that of methane, which is present in ppmv concentrations, since methane does not make a significant contribution to surface ozone production on the

141

Table 4.8: Rate constants for the reaction of CFC replacements with the OH radical, and tropspheric lifetimes. After Niki[98].

Compound	Code	$k_{298}\times10^{15}$ $cm^3\ molec^{-1}\ s^{-1}$	Lifetime yr
HFCs			
CH_3CHF_2	152a	37.0	1.7
CH_2FCF_3	134a	4.8	13.2
CHF_2CF_3	125	2.5	25.4
HCFCs			
$CHClF_2$	22	4.7	3.5
CH_3CClF_2	142b	3.8	16.7
CH_3CHClF	124	10.0	6.3
CH_3CCl_2F	141b	8.0	7.9
$CHCl_2CF_3$	123	34.0	1.7

timescale of one day. HCFCs and HFCs are not, therefore, expected to significantly change urban air quality with respect to oxidant concentrations. On a global scale too, HCFC and HFC levels are estimated to be about six orders of magnitude below that of methane, so that their contribution to surface ozone concentrations will be negligible. A more important concern arising from HCFC and HFC emissions is the production of toxic and acidic substances such as CF_3COOH during their photochemical oxidation.

Other processes occurring in the remote troposphere

As mentioned in our earlier section, heterogeneous processes have been implicated in an observed decrease in surface ozone concentration seen to occur in the Arctic in springtime. A full discussion of the chemical and microphysical processes that lead to, and result from, atmospheric aerosol, is outside the scope of this book. The reader is directed to Seinfield[69] and Hinds[99] for the background information not given here.

Springtime surface ozone concentrations in the Arctic are strongly anti-correlated with the level of Br-containing compounds as measured by filter techniques[40] (filterable Br, f-Br). Although the detailed mechanism for this process has yet to be established, the following observations outline the behaviour which any proposed theory must obey:

i) Ozone depletion is rapid (with a timescale of less than 1 day) and occurs throughout the Arctic;

ii) f-Br is maximum during the spring months and minimum in the summer;

iii) the f-Br source appears to be marine.

At least two chemical schemes have been put forward to explain the springtime minimum in surface ozone concentrations. "Active bromine" is formed from either photolysis of bromoform[40], and/or heterogeneous reaction on sea-salt aerosol to form bromine nitrate[100]

Table 4.9: Bromine chemistry implicated in the decrease of ozone in the springtime Arctic boundary layer. Kinetic data for the heterogeneous reactions are given as estimated rates (molec cm^{-3} s^{-1}).

Reaction		k_{298} $cm^3/molec/s$	References
$CHBr_3 + hv$	$\rightarrow Br + CHBr_2$	6.0×10^{-7}	40
$CHBr_2$	$\rightarrow nBrO_x$	fast	40
$Br + O_3$	$\rightarrow BrO + O_2$	7.0×10^{-13}	40
$2\ BrO + hv$	$\rightarrow 2\ Br + O_2$	2.5×10^{-12}	40
$BrO + NO$	$\rightarrow Br + NO_2$	2.5×10^{-11}	40
$BrO + hv$	$\rightarrow Br + O$	5.0×10^{-3}	40
$BrO + NO_2\ (+ M)$	$\rightarrow BrNO_3\ (+ M)$	2.0×10^{-11}	40
$BrNO_3 + hv$	$\rightarrow BrO + NO_2$	4.0×10^{-4}	40
$Br + HCHO$	$\rightarrow HBr + CHO$	7.0×10^{-13}	40
$Br + HO_2$	$\rightarrow HBr + O_2$	8.0×10^{-13}	40
$HBr + OH$	$\rightarrow Br + H_2O$	1.0×10^{-11}	40
$BrO + HO_2$	$\rightarrow BrOH + O_2$	5.0×10^{-12}	40
$BrOH + hv$	$\rightarrow Br + OH$	fast	40
$N_2O_5 + NaBr(s)$	$\rightarrow BrNO_2 + NaNO_3(s)$	1×10^2	100
$2\ BrNO_2$	$\rightleftharpoons Br_2 + 2\ NO_2$	-	100
$2\ NO_2 + NaBr(s)$	$\rightarrow BrNO + NaNO_3(s)$	8×10^{-3}	105

(Table 4.9). The active bromine then depletes ozone via

$$Br + O_3 \rightarrow BrO + O_2 \qquad\qquad k_{98} = 1.2 \times 10^{-12} \qquad [4.98]$$

which is catalytic in BrO. Note that the reacton of BrO with ozone may also be important since the reaction

$$BrO + O_3 \rightarrow Br + 2O_2 \qquad\qquad k_{99} < 5 \times 10^{-15} \qquad [4.99]$$

(ref. 9) will have a pseudo-first order rate constant ($[O_3]{\approx}1 \times 10^{12}$ molec cm^{-3}) of the same magnitude as the spring-time arctic photolysis rate coefficient for BrO ($J{\approx}5 \times 10^{-3}s^{-1}$; ref. 40). The formation of nitryl bromide by the heterogeneous reaction of NO_2 with sea-salt aerosol can also result in the formation of active bromine[100]., i.e.

$$N_2O_5(g) + NaBr(s) \rightarrow BrNO_2 + NaNO_3 \qquad\qquad [4.54]$$

$$2\ NO_2(g) + NaBr(s) \rightarrow BrNO + NaNO_3 \qquad\qquad [4.56]$$

The relative contribution of inorganic bromine and organic bromine to the active bromine concentration is a subject of much debate[100, 101]. It has been established, however, that there is a significant biogenic source of bromoform in the Arctic[102, 103] and the Antarctic[104] boundary layers. The absolute rate of active bromine formation from reactions will be a function of aerosol characteristics such as size distribution and surface area, as well as N_2O_5 concentration, and so careful prediction of the atmospheric aerosol loading would be needed to properly quantify the process. To date this has not been attempted.

As an interesting extension to the discussion above, Sturges (ref. 80) has measured Bromine compounds at remote coastal sites in the Western Isles, off the northwest coast of Scotland, and found that a significant fraction of the total particulate bromine yield cannot be accounted for in terms of sea-salt and anthropogenic contributions[105]. It is suggested that the chemistry occurring during Arctic spring occurs throughout the marine boundary layer and is a global source of reactive bromine and sink of ozone. The exact magnitude of such a sink has not been determined.

Alkyl halides have also been implicated in the unexpectedly fast oxidation of dimethyl sulphide (DMS) in the marine troposphere. DMS is produced by algae in the oceans[106]; its oxidation may be the source of "background" sulphur-acid and non-sea salt particulate sulphate levels which cannot be attributed to anthropogenic emissions[107]. Production of cloud condensation nuclei by this pathway, leading to cloud formation and an increase in the earth's albedo, has been put forward as an example of biotic control of climate and is cited as support for the controversial Gaia hypothesis[108].

Atmospheric oxidation of DMS is believed to be predominantly by reaction with OH in the gas phase:

$$DMS + OH \rightarrow products \qquad\qquad k_{100} = 6.3 \times 10^{-12} \qquad\qquad [4.100]$$

to yield a variety of products including SO_2, methyl sulphonic acid, formaldehyde and dimethyl sulphoxide (DMSO) from a combination of parallel and sequential reactions that have yet to be deduced unambiguously[109]. If we define the lifetime for DMS in the troposphere as $\tau = 1/k[OH]$, and use a maximum OH concentration for the remote troposphere of 10^6 molec cm^{-3} (refs. 110, 111 and 112), we find that the esti- mated life time is about 44 h. However, measurements suggest that the atmospheric lifetime of DMS is often much shorter than this, sometimes less than 1 h (ref. 113). It appears that an unknown sink process is having a significant effect.

Chameides and Davies[114] suggested that IO_x radicals may be the missing oxidants.

Figure 4.9

Overview of the possible gas phase reactions involving iodine in the lower atmosphere. From Chatfield and Crutzen[115]. In the figure, CH₃I is considered to be the most important source of radical iodine, but I₂ or HOI may also be important.

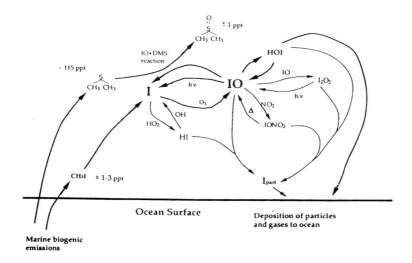

Formation is via the photolysis of methyl iodide[115] (Figure 4.9),

$$CH_3I + h\nu \rightarrow CH_3 + I \qquad\qquad J_{101} = 3 \times 10^{-6} \qquad\qquad [4.101]$$

where J is the 12 h average at the surface for a latitude of 30°N, followed by rapid reaction with ozone.

$$I + O_3 \rightarrow IO + O_2 \qquad\qquad k_{102} = 8 \times 10^{-13} \qquad\qquad [4.102]$$

It is unusual for an unoxidised radical, such as I, to undergo reaction with a species other than O_2 (which is, of course, in great excess). This is explained by the fact that the IO bond in IO_2 is extremely weak, in keeping with the observed behaviour of ClO_2. IO is cycled back to I by reaction with NO or by reaction with DMS. The reaction of IO with DMS results in net ozone destruction, whilst the reaction of IO with NO does not. We can make a first guess at the effect of IO on ozone concentrations in the remote boundary layer, therefore, by comparing its lifetimes for destruction by reaction with NO, and with DMS[116, 117]. The ratio of rate coefficients $k_{NO}/k_{DMS} \approx 1000$ at 298 K. A typical value for NO in the rural troposphere is 10 pptv; DMS appears to occur in concentrations of around 100 pptv above many ocean surfaces[106]. The effective first order loss rates for IO by reaction with NO and DMS, then, will be in the ratio $(k_{NO}[NO])/(k_{DMS}[DMS]) \approx 100$, implying that IO chemistry can play only a small part in the oxidation of DMS, and in the destruction of tropospheric ozone, in the marine boundary layer.

To summarise, then, one can conclude that the broad features of tropospheric chemistry, far from urban and industrial sources, have now been worked out. Explicit descriptions of the gas-phase chemistry of methane, carbon monoxides, the oxides of nitrogen, and ozone can be used in models of the atmosphere. Some other areas do remain unclear, however. The speciation and reactivity of biogenic non-methane hydrocarbons have yet to be established conclusively, particularly with respect to small- and regional-scale inventories. Our understanding of the chemistry of the springtime depletion of surface ozone in Arctic remains speculative, and no confident predictions can be made about the effect of such chemistry on the global tropospheric ozone budget. Lastly, we consider that the role of heterogeneous processes in tropospheric ozone chemistry remains largely unexplored, though the results of the work that has been done, that of Lelieveld and Crutzen[43], for example, suggest that there could be a significant effect.

4.3 OXIDANT CHEMISTRY IN URBAN AREAS

The principal differences between oxidant chemistry in urban areas compared and

Figure 4.10

Peak ozone concentrations as a function of initial NO_x and VOC concentrations [118] from the US EPA standard EKMA model.

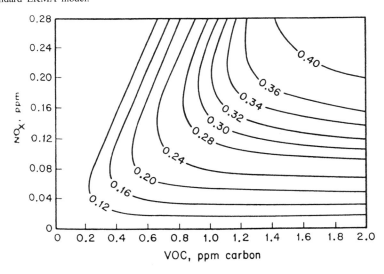

oxidant chemistry in rural areas are the strength of sources and diversity of precursors (chapter 3). The sheer number of precursors and intermediates makes it impossible to give an explicit description of all the chemistry occurring. For this reason the description below is general in nature, and draws on the discussion of different types of reactivity given in the previous sections. Figure 4.1 still stands. The oxidation of hydrocarbons is coupled closely to the NO_x - O_3 photostationary state: oxidation of VOC compounds by OH produces RO_2 radicals (where R is some organic moiety) which convert NO to NO_2 without destroying O_3. The concentration of O_3 will therefore increase, providing more OH via reactions [4.3] and [4.39], and so increase the rate at which VOC is oxidised. It is this feedback that gives urban airshed chemistry its characteristic non-linearity (Figure 4.10)[118]. For maximum ozone production, a balance of NO_x and VOC concentrations are required: too little NO_x and insufficient OH is produced to oxidised the VOC; too little VOC and RO_2 concentrations are not high enough to alter the position of the photostationary state significantly. Hence, towards the top-left of Figure 4.10 ozone formation is *hydrocarbon-limited*, and towards the bottom-right ozone formation is *NO_x-limited*. The non-linearity of ozone formation has been demonstrated experimentally by Kelly and Gunst[119]. They carried out a statistical analysis of the response of ambient Los Angeles air to injections of known quantities of NO_x, hydrocarbons, or clean air. Maximum ozone concentrations, $[O_3]_{max}$, were found to obey the following empirical formula:

$$[O_3]_{max} = 129 + 2.8[NO_x] - 6.8\frac{[NO_x]^2}{[HC]} + 19(T_{av} - 21.1) \qquad (4.12)$$

which clearly shows in mathematical form the non-linearity shown in Figure 4.10.

　　　We concentrate below on describing the relative reactivities of the various organic compounds, discussing degradation pathways only for those classes of compound that have not been previously discussed. The rate coefficients for the OH and NO_3 initiation reactions are given in Table 4.10 for many of the common hydrocarbons found in urban air[120]. The chemistry of higher alkanes in the troposphere follows the pattern described for methane. Initial attack by OH occurs preferentially to form the most stable alkyl radical. Hence tertiary and secondary hydrogens are the preferred hydrogens for abstraction by OH. Anthropogenic alkene chemistry follows a similar pattern to that outlined for the biogenic dialkene, isoprene. Initial reaction is radical addition of OH, followed by addition of O_2 to

Table 4.10: The reactivities of common hydrocarbons to OH and NO_3. Concentrations are the median values from Ref. 120 for the 25 most common hydrocarbons in urban air in the USA. Rate coefficients are for 298 K.

Compound	Concn. ppbC	k_{OH} (x10^{12})	k_{NO3} (x10^{16})
isopentane	45.3	3.9	1.6
n-butane	40.3	2.54	0.65
toluene	33.8	5.96	0.3
propane	23.5	1.15	-
ethane	23.3	0.27	0.01
n-pentane	22.0	3.94	0.8
ethylene	21.4	8.52	2.1
m, p-xylene	18.1	23.6,14.3	1.0,3.0
2-methylpentane	14.9	5.6	-
isobutane	14.8	2.34	1
acetylene	12.9	0.9	≤ 0.2
benzene	12.6	1.23	0.2
n-hexane	11.0	5.61	1.5
3-methylpentane	10.7	5.7	-
1,2,4-trimethylbenzene	10.6	32.5	15
propylene	7.7	26.3	94
2-methylhexane	7.3	-	-
o-xylene	7.2	13.7	3
2,2,4-trimethylpentane	6.8	3.68	-
methylcyclopentane	6.4	-	-
3-methylhexane	5.9	-	-
2-methylpropene, 1-butene	5.9	51.4,31.4	3100,120
ethylbenzene	5.9	7.1	6
m-ethyltoluene	5.3	19.2	-
n-heptane	4.7	7.15	1.4
isoprene	-	101	5900
α-pinene	-	53.7	58000

a hydroxy substituted alkyl peroxy radical. This peroxy radical then reacts with NO or HO_2 in NO-rich and NO-poor conditions, respectively. A minor reaction with NO_2 to form alkyl peroxy nitrates, will also occur. The chemistry of carbonyl compounds has also been discussed for the case of carbonyls produced in the degradation of isoprene and other BNMHC. The photolytic processes have already been discussed in § 4.1.

PAN compounds

The photochemical oxidation of carbonyl compounds leads to the production of peroxy-acetyl radicals, *e.g.*:

$$CH_3CHO + OH\ (+ O_2)\ \rightarrow\ CH_3C(O)O_2 + H_2O \qquad\qquad k_{103} = 1.6\ x\ 10^{-11}\quad [4.103]$$
$$CH_3C(O)O_2 + NO_2 + M\ \rightarrow\ CH_3C(O)O_2NO_2 + M \qquad k_{23} = 3.6\ x\ 10^{-12}\quad [4.23]$$
$$CH_3C(O)O_2NO_2 + M\ \rightarrow\ CH_3C(O)O_2 + NO_2 + M \qquad k_{104} = 1.8\ x\ 10^{-4}\quad [4.104]$$

The formation of PAN compounds is a chain terminating process which competes with the chain propagating reaction of peroxy radicals with NO:

$$CH_3C(O)O_2 + NO\ \rightarrow\ CH_3C(O)O + NO_2 \qquad\qquad k_{105} > 1\ x\ 10^{-11}\quad [4.105]$$
$$CH_3C(O)O + O_2\ \rightarrow\ CH_3O_2 + CO_2 \qquad\qquad k_{106} \approx 2\ x\ 10^{-12}\quad [4.106]$$
$$CH_3O_2 + NO\ \rightarrow\ CH_3O + NO_2 \qquad\qquad k_{58} = 7.6\ x\ 10^{-12}\quad [4.58]$$
$$CH_3O + O_2\ \rightarrow\ HCHO + HO_2 \qquad\qquad k_{59} = 1.9\ x\ 10^{-15}\quad [4.59]$$

Net: $CH_3C(O)O_2 + 2\ NO_2 + 2\ O_2\ \rightarrow\ HCHO + CO_2 + HO_2 + 2\ NO_2$

The ratio of rate coefficients for reactions [4.23] and [4.105] is insensitive to temperature between 283 K and 313 K (ref. 121). The equilibrium formed by reactions [4.23] and [4.104] is very sensitive to temperature and pressure[122, 123]; PAN formation becoming more favourable at low temperatures and pressures. Hence, although thermal decomposition is the only significant loss route for PAN in near the surface at ambient temperatures, reaction with OH becomes the more important route above about 7 km:

$$CH_3C(O)O_2NO_2 + OH\ \rightarrow\ products \qquad\qquad k_{107} = 1.1\ x\ 10^{-13}\quad [4.107]$$

If PAN compounds are lofted rapidly on formation, therefore, they can have greatly increased lifetimes and so be transported long distances from their source region, providing a supply of NO_x to the remote troposphere.

Aromatic compounds

The atmospheric chemistry of aromatic compounds is not well documented. It is known that the only degradation pathway of importance is reaction with OH (The discussion below is for monocyclic aromatic compounds only; details of the reactivity of polyaromatic hydrocarbons (PAH) can be found in ref. 13. PAH are not emitted in sufficient quantity to affect photo-oxidant production significantly). For alkyl substituted aromatics attack can be at an alkyl or aromatic centre: *e.g.*

$$C_6H_5\text{-}CH_3 + OH\ \rightarrow\ C_6H_5\text{-}CH_2 + H_2O \qquad\qquad k_{108} = 7.2\ x\ 10^{-13}\quad [4.108]$$

or

$$C_6H_5\text{-}CH_3 + OH\ \rightarrow\ (\underline{o},\ \underline{m},\ \underline{p})C_6H_5(OH)CH_3 \qquad\qquad k_{109} = 5.3\ x\ 10^{-12}\quad [4.109]$$

For toluene, $k_{108}/(k_{108} + k_{109}) \approx 0.92$. In general values for other substituted aromatics are lower than this[8]. The rest of the abstraction pathway is the same as for other alkyl radicals:

$$C_6H_5\text{-}CH_2 + O_2 \rightarrow C_6H_5\text{-}CH_2O_2 \qquad\qquad k_{110} \approx 1 \times 10^{-13} \qquad [4.110]$$

$$C_6H_5\text{-}CH_2O_2 + NO \rightarrow C_6H_5\text{-}CH_2O + NO_2 \qquad k_{111} \approx 4 \times 10^{-12} \qquad [4.111]$$

$$C_6H_5\text{-}CH_2O + O_2 \rightarrow C_6H_5\text{-}CHO + HO_2 \qquad k_{112} \approx 8 \times 10^{-15} \qquad [4.112]$$

$$C_6H_5\text{-}CH_2O_2 + NO \ (+ M) \rightarrow C_6H_5\text{-}CH_2ONO_2 \ (+ M) \qquad k_{113} \approx 4 \times 10^{-13} \qquad [4.113]$$

where $k_{111} : k_{113} \approx 10:1$.

The subsequent steps of the addition mechanism are not well defined, however. It had been thought that, in common with other organic radicals, initial reaction would be with O_2, eg.

$$C_6H_5(OH)CH_3 + O_2 \rightarrow C_6H_4(OH)CH_3 + HO_2 \qquad k_{114} \leq 2 \times 10^{-16} \qquad [4.114]$$

or

$$C_6H_5(OH)CH_3 + O_2 \rightarrow C_6H_5(OH)(O_2)CH_3 \qquad k_{115} \leq 2 \times 10^{-16} \qquad [4.115]$$

$$C_6H_5(OH)(O_2)CH_3 + NO \rightarrow NO_2 + \text{ring cleavage products}$$

$$k_{116} \approx 4 \times 10^{-12} \qquad [4.116]$$

and mechanisms based on this assumption have been common in the literature (*e.g.* refs. 23 and 124). Recent work[125, 125] has shown the production of ring retaining compounds following the reaction of aromatic species with OH, and points out that the rate constants for the reaction of the methyl cyclohexadienyl radical with O_2 and NO_2 are in the ratio

Figure 4.11

The oxidation of benzene.

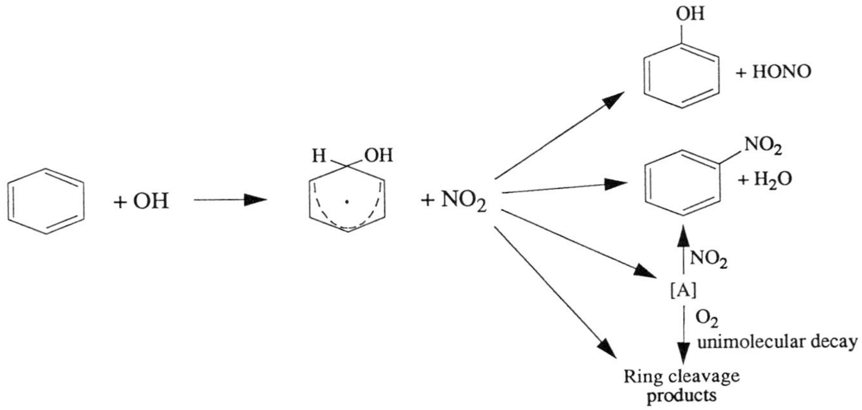

150

$k(O_2)/k(NO_2) < 10^{-5}$ approximately (a reaction with NO has also been observed[126] but is expected to be negligible under atmospheric conditions). Given atmospheric concentrations, which have a ratio of $[O_2]/[NO_2] > 10^4$, it is apparent that reaction with NO_2 will be important and may be dominant. A pathway has been put forward[127] for the benzene - OH reaction, to account for the observed products (Figure 4.11), where [A] is as yet unidentified. It is important to ascertain the relative importance of the two channels in Figure 4.11 because, although the same products may result, the number of NO - NO_2 conversions per molecule of aromatic hydrocarbon reacted may be different. This will, in turn, influence our estimate of the ozone-forming potential of such species.

To summarise, the atmospheric degradation of monocyclic aromatic hydrocarbons results in the formation of aromatic hydroxy compounds, aromatic aldehydes, aromatic nitrates, and the products formed on ring cleavage. The ring cleavage products include (in the case of toluene): glyoxal; methylglyoxal; PAN; $CH_3C(O)C(O)CH=CH_2$; $CH_3C(O)$-$CH=CH_2$; $CH_3CH=CHCH=CH_2$; $OHC-C(OH)=CHCHO$; and $CH_3C(O)CH=CHCH=CHCHO$ (ref. 8). The exact yield of each of the ring cleavage products is not known at present[8] and the products of about 40 % of the overall reaction of toluene with OH remain to be quantified. This is also the case for the benzene-OH reaction. Of course, in the atmosphere, all the ring cleavage products will undergo further reaction unless removed by some deposition process. Those details that have been established about the reactivity of the intermediate oxidation products are reviewed in detail by Atkinson[8]. The chemistry of glyoxal and methylglyoxal has been discussed above (§ 4.1 and 4.2). It is clear, however, that aromatic oxidation in the polluted atmosphere results in substantial conversion of NO to NO_2, so that the potential of aromatics emission to produce ozone, is large. The initial reactivity of the more common aromatic hydrocarbons towards OH is given in Table 4.10: m-xylene and p-xylene are about five times more reactive than toluene, which is itself about five times more reactive than benzene.

Alkynes

Alkynes can occur in urban atmospheres at ppbv levels. They react predominantly with OH (ref. 8). For acetylene the following mechanism is proposed

$OH + HC\equiv CH + M \rightarrow HOCH=CH + M$ $k_{117} = 5.5 \times 10^{-13}$ [4.117]

$HOCH=CH \rightarrow CH_2\text{-}CHO + O_2 \rightarrow (CHO)_2 + OH$ [4.118]

$HOCH=CH + O_2 \rightarrow HOCH=CHOO$ [4.119]

$$HOCH=CHOO \rightarrow HCO + HCOOH \qquad [4.120]$$

$$HOCH=CHOO + NO \rightarrow HOCH-CHO + NO_2 \qquad [4.121]$$

$$HOCH-CHO + O_2 \rightarrow (CHO)_2 + HO_2 \qquad [4.122]$$

leading to the production of biacetyl and formic acid. Note that HO_x is regenerated in reactions [4.118] and [4.122], and that step [4.121] converts NO to NO_2 (hence promoting ozone formation). The fraction of the overall reaction occurring via the isomerisation mechanism, without NO to NO_2 conversion, is not yet known. Higher alkynes react to give analogous acids and di-aldehydes though about 35 % of the propyne reaction appears to be via an unknown pathway[8]. The subsequent chemistry of dicarbonyls has been discussed earlier (§ 4.1 and 4.2). The major loss process for the carboxylic acids will be deposition, both wet and dry.

Alcohols

The possible introduction of methanol-fuelled vehicles into the United States' car fleet has stimulated interest in the reactivity of alcohols in the urban atmosphere. Their predominant degradation pathway is by reaction with OH:

$$CH_3OH + OH \rightarrow H_2O + CH_3O \qquad k_{123} = 1.4 \times 10^{-13} \quad [4.123]$$

$$CH_3OH + OH \rightarrow H_2O + CH_2OH \qquad k_{124} = 7.9 \times 10^{-13} \quad [4.124]$$

(ref. 8) where the ratio $k_{123}/(k_{123}+k_{124})$ is approximately 15 %. Subsequent reaction of the radical products is rapid

$$CH_3O + O_2 \rightarrow HCHO + HO_2 \qquad k_{59} = 1.9 \times 10^{-15} \quad [4.59]$$

$$CH_2OH + O_2 \rightarrow HCHO + HO_2 \qquad k_{88} = 9.0 \times 10^{-12} \quad [4.88]$$

and the temporary sink, formaldehyde, will further degrade by photolysis and reaction with OH.

Ethers

Ethers occur in the urban troposphere as a consequence of their use as anti-knock agents in lead-free gasoline. A significant part of the carbon content of ethers is rapidly converted to relatively inert esters[128, 129, 130]. Methyl *t*-butyl ether, for example, forms *t*-butyl formate:

$$CH_3OC(CH_3)_3 + OH + NO + 2\ O_2 \rightarrow 0.6\ HCOOC(CH_3)_3 + 0.4\ CH_3OC(CH_3)_2CHO$$

$$+ HO_2 + NO_2 + H_2O \qquad k_{125} = 3 \times 10^{-12} \quad [4.125]$$

which has an atmospheric lifetime of several days, and 2-methoxy 2-methyl propanal which will react rapidly with OH:

$$CH_3OC(CH_3)_2CHO + OH + 2\ NO + 3\ O_2 \rightarrow (CH_3)_2CO + HCHO + CO_2$$
$$+ HO_2 + 2\ NO_2 + H_2O \qquad k_{126} = 2.7 \times 10^{-11} \quad [4.126]$$

The relative inertness of the products of ether oxidation allows transport and dilution to reduce the possibility of local photochemical oxidant pollution episodes. Once transported to the remote troposphere, photochemical oxidant production is usually limited by the availability of NO_x, so the addition of unreactive ether oxidation products will make little difference to the potential for ozone formation.

Nocturnal oxidant chemistry

Photolytic production of NO and OH is not possible at night. All the NO_x present is initially converted to NO_2, which can then react with ozone to produce the nitrate radical:

$$O_3 + NO_2 \rightarrow NO_3 + O_2 \qquad\qquad k_{43} = 3.2 \times 10^{-17} \quad [4.43]$$

This reaction has a small rate coefficient, but the concentrations of both reactants are often high enough over urban areas to make the rate of NO_3 significant. Further reaction of the NO_3 with NO_2 produces N_2O_5, which can be hydrolysed in, or on, aqueous aerosol.

$$NO_2 + NO_3 + M \rightarrow N_2O_5 + M \qquad\qquad k_{44} = 6.4 \times 10^{-13} \quad [4.44]$$
$$N_2O_5(g) + H_2O(l) \rightarrow 2\ HNO_3(aq) \qquad\qquad\qquad [4.53]$$

Nitric acid is almost inert in the lower troposphere, as we have seen, so that the effect of the above reactions is to remove odd nitrogen without removal of reactive hydrocarbons. The photochemistry occurring subsequent to this kind of process might be expected therefore, to be different from that occurring before the process.

There is a more direct effect, however. The nitrate radical can act as a nocturnal oxidant, reacting with hydrocarbons, particularly with alkenes (Table 4.10), and producing significant amounts of peroxy radicals. A significant amount of the alkene loading of an airmass may be oxidised in this way[131, 132]. Alkenes react with NO_3 by addition to form the most stable nitrooxy substituted alkyl radical. This radical then reacts very rapidly with oxygen to form the equivalent alkyl peroxy radical. In daylight, peroxy radicals react with NO or other peroxy radicals, especially HO_2, depending on the NO_x loading of the airmass. At night, however, NO is not available for reaction even within polluted airmasses, and further reaction is with other peroxy radicals

$$R_1R_2CHO_2 + HO_2 \rightarrow R_1R_2CHOOH + O_2 \qquad k_{127} \approx 5 \times 10^{-12} \quad [4.127]$$
$$2\ R_1R_2CHO_2 \rightarrow 2\ R_1R_2CHO + O_2 \qquad 10^{-16} \leq k_{128} \leq 10^{-12} \quad [4.128]$$

$$2\ R_1R_2CHO_2 \rightarrow R_1R_2CHOH + R_1R_2CO + O_2 \qquad 10^{-16} \leq k_{129} \leq 10^{-12} \qquad [4.129]$$

with the nitrate radical itself

$$R_1R_2CHO_2 + NO_3 \rightarrow R_1R_2CHO + NO_2 + O_2 \qquad\qquad [4.130]$$

or with NO_2 to form peroxynitrates

$$R_1R_2CHO_2 + NO_2\ (+\ M) \rightarrow R_1R_2CHO_2NO_2\ (+\ M) \qquad k_{131} \approx 10^{-12} \qquad [4.131]$$

The hydroperoxides produced in the above reactions are most likely to be deposited into cloud droplets or wet surfaces, but the peroxynitrates will probably remain in the gas-phase until they are photolysed the following day. The overall effect of nocturnal chemistry is, therefore, to slowly oxidise unsaturated hydrocarbons to peroxides and nitrates, and to remove odd nitrogen. Both these effects will tend to make the subsequent day's photochemistry less liable to produce ozone. This may be one reason why air parcels that are loaded with pollutants as they pass over urban centres at night do not produce large ozone concentrations even after many hours of illumination[139].

Relative Reactivity

Rather than continuing the catalogue the reactions of individual organic species *ad infinitum* (perhaps, bearing in mind the health effects of photochemical air pollution, one should say *ad nauseam*), an overall impression can be formed by considering the relative degradation rates or the relative ozone forming rates of the compounds. The use of relative degradation rates foregoes any mechanistic details and assumes that reaction of a hydrocarbon molecule, HC, with OH inevitably leads to ozone formation, *i.e.*

$$HC + OH + O_2 \rightarrow RO_2\ (+\ H_2O) \qquad\qquad [4.132]$$
$$RO_2 + NO \rightarrow RO + NO_2 \qquad\qquad [4.133]$$
$$NO_2 + h\nu + O_2 \rightarrow O_3 + NO \qquad\qquad [4.4]$$

The amount of ozone produced per mole of hydrocarbon will depend, therefore, on its concentration and reactivity with OH. The relative reactivity of species, j, in a mix of i hydrocarbons is then

$$R_j = \frac{k_j[HC]_j}{\sum_i k_i[HC]_i} \qquad\qquad (4.13)$$

Using the typical abundances of hydrocarbons given and their reactivities, relative degradation rates are straightforward to calculate (Table 4.11).

Table 4.11: Relative removal rates for the 25 most abundant compounds based on median concentrations. From Niki[98]; concentrations are from Table 4.10, converted to molec cm^{-3} at 1000 mbar.

Compound	k (x10^{-12}), 298 K, cm^3 molec^{-1} s^{-1}	Concn. (x10^{10}) molec cm^{-3}	Loss Rate (x10^{-2}) s^{-1}	% of total Loss Rate
Methane	0.0077	5748.0	44.3	3.0
Isopentane	3.9	22.7	88.5	5.9
n-butane	2.7	25.2	68.0	4.5
Toluene	6.4	12.1	77.4	5.2
Propane	1.1	19.6	21.5	1.4
Ethane	0.3	29.1	8.7	0.6
n-pentane	4.1	11.0	45.1	3.0
Ethylene	8.8	26.8	235.8	15.7
m-xylene	20.6			
p-xylene	16.8	5.7	95.8	6.4
2-methylpentane	5.5	6.2	34.1	2.3
Isobutane	2.2	9.3	20.4	1.4
Acetylene	0.9	16.1	14.5	1.0
Benzene	1.0	5.3	5.3	0.35
n-hexane	5.3			
2-ethyl-1-butene	32.7	4.6	150.4	10.0
3-methylpentane	5.5	4.5	24.8	1.6
1,2,4-trimethylbenzene	40.0	2.9	116.0	7.9
propylene	24.6	6.4	157.4	10.5
2-methylhexane	5.5	2.6	14.3	1.0
o-xylene	14.2	2.3	32.7	2.1
2,2,4-trimethylpentane	3.6	2.1	7.6	0.5
methylcyclopentane	5.2	2.3	12.0	0.8
3-methylhexane	7.1	2.1	14.9	1.0
2-methylpropene	52.3			
1-butene	42.1	3.7	155.8	10.4
Ethylbenzene	8.0	1.8	14.4	1.0
m-ethyltoluene	17.1	1.5	25.6	1.7
n-heptane	7.4	1.7	12.6	0.8

Alternatively, the complete cycle of oxidation and ozone production can be investigated using smog chambers or computer models. Experimental measurements of ozone formation due to the addition of a single hydrocarbon to a standard VOC - NO$_x$ "air" mixture must be the benchmark to which all theoretical estimates of reactivity are compared since no assumptions are made about the chemistry occurring in the system. If the results of computer simulations do, in fact, tally with these experimental observations then the

Table 4.12: Experimental and calculated incremental reactivities (IR(0)), fraction reacted (F), and limiting incremental reactivities (IR(0)/F), after 6 h irradiation in a surrogate NO_x-air mixture[134].

Compound	Experimental IR(0)	Calculated IR(0)	Fraction reacted, F	IR(0)/F
benzaldehyde	-1.6	-1.22	0.37	-3.3
toluene	-0.23	-0.23	0.20	-1.16
propene	-0.03	0.03	0.92	0.03
t-but-2-ene	0.3	-0.11	1.00	-0.11
n-butane	0.05	0.048	0.087	0.55
ethanol	0.01	0.014	0.094	0.15

Notes:

$$IR(0) = (\lim, \Delta[organic] \rightarrow 0) \ dR/d[organic]$$

$$R = [O_3(organic \ incremented)] - [O_3(no \ increment)]$$
$$-[NO(organic \ incremented)] - [NO(no \ increment)]$$

The experiments were carried out at VOC/NO ratios of 40.
Negative values denote the inhibition of ozone formation.

---

mechanism used in the model may be considered to be verified, at least in terms of its legislatively significant results. There have been many smog chamber studies of hydrocarbon reactivity, but Carter and Atkinson[133] were the first to discuss the topic in terms of the addition of single hydrocarbon species to a surrogate air mixture. They used a surrogate air with VOC/NO $\approx$ 40; Table 4.12 gives some results from the study. Incremental reactivities, which are potentials for ozone formation per atom of carbon, are complex functions of both hydrocarbon structure and the reaction time (not shown here). For example, after 6 hours of reaction, benzaldehyde, toluene and propene all inhibit ozone formation in the surrogate air that was used in the study. This complex relation between ozone formation, hydrocarbon structure, and atmospheric composition invalidates the assumptions used to rank hydrocarbon species solely by their degradation rate (above). Happily, however, the over-riding trends are consistent.

In terms of degradation rate, and of incremental reactivity, alkenes have the greatest ozone forming potential, followed by aromatic compounds, and alkanes have the smallest ozone forming potential. It should be borne in mind, however, that, in the troposphere as a whole, alkanes (particularly CH_4) will be much more important due to the sheer quantity emitted.

Experimental studies such as that of Carter and Atkinson[133] are clearly central to our understanding of ozone formation potentials but they demand painstaking and time-consuming attention to reproducibility in the constitution of the surrogate air and hydrocarbon samples. A simpler way to obtain incremental reactivities is to use computer models. Prior knowledge of degradation pathways is assumed in this approach, so that it is necessary to compare the results of a simulation with that from experimental studies for some (arbitrary) number of representative hydrocarbon species. Once this has been done, and the results of the model shown to be in reasonable agreement with experiment, then the results of simulations using different, but structurally related, hydrocarbons can be put forward with some confidence.

Pursuing just such a method, Carter and Atkinson have extended their experimental results using an urban-air box model[134]. The behaviour of sixteen compounds was examined: CO, ethane, n-butane, n-octane, n-pentadecane, ethene, propene, *trans*-but-2-ene, benzene, toluene, m-xylene, formaldehyde, acetaldehyde, benzaldehyde, methanol and ethanol. The ozone forming potential of these species per carbon atom, was found to be a complex function of VOC/NO_x ratio and the scenario used. The variation in reactivity could not be explained simply in terms of the OH reaction rate, particularly for those species with alternative degradation pathways (*e.g.* formaldehyde) and when PAN formation became important at large VOC/NO_x ratios. Ozone forming potentials generally peaked when $VOC/NO_x \approx 6$. This maximum effect of hydrocarbon injection occurs at VOC/NO ratios somewhat below the optimum ratio for ozone production, and corresponds to the injection of hydrocarbons into air masses in which ozone formation is hydrocarbon-limited (Figure 4.10).

Hough and Derwent[135] allocate ozone production between hydrocarbons using an explicit chemical mechanism in a trajectory model. these results are specific to the scenario used (conditions typical of photochemical episodes in SE England) and do not take account of non-linear effects in the model response explicitly. Nevertheless, the calculations do show the large difference in ozone formation per molecule of hydrocarbon, with aromatic compounds being the most efficient in this respect. Similar reactivity tests have been carried out as part of the study of the impact that methanol fuelled vehicles could have on air quality[136, 137]. Results indicate that methanol can be substantially less efficient at forming ozone than the current mixture of hydrocarbons within the urban airshed, but that the effect is a function of the VOC/NO_x ratio[137].

Model sensitivity studies may also give useful information on the subject of relative reactivities. Because most mechanisms, or the sensitivity experiments themselves, are to some degree "lumped" or otherwise compressed (§ 4.4), results generally reflect the ozone formation potential of generic species. For example, Derwent and Hov[138] used the same model as that used by Hough and Derwent[135], together with latin hypercube sampling of input parameter values, to gauge the sensitivity of model output to (amongst others) oxygenate and aromatic hydrocarbon emissions. No correlation with NO_x levels was possible but it was shown that, for the scenario used, variation of oxygenate emissions did not significantly alter ozone formation. In contrast, varying aromatic hydrocarbon emissions did cause a significant increase in ozone formation.

MacKenzie *et al.*[139] studied a similar scenario using the model of Cocks and Fletcher[140]. Interactions between parameters were estimated by using a factorial design experiment. Increasing the initial concentrations of alkenes, aromatics and alkanes significantly increased the maximum ozone levels. The effect of each category of hydrocarbon on ozone concentrations varied, alkenes being most effective, and alkanes least. All three categories interacted significantly with initial concentrations of NO_x, as would be expected from the results of Carter and Atkinson[133]. Clearly the topic of relative reactivity is one that would benefit from further study, both experimental and theoretical, and could usefully be extended to include the potential for formation of other oxidants (see, for example, ref. 138).

4.4 CHEMICAL MECHANISMS USED IN AIR POLLUTION MODELLING

Simulation of air quality requires the solution of a set of simultaneous differential equations describing the sources, sinks, transport and chemical transformation of each species. The time required to compute this set of equations will increase rapidly as each phenomena is described in greater detail. For this reason chemical mechanisms which preserve the behaviour of a fully explicit scheme, but which use fewer variables and equations, will reduce the cost of air quality computation. A good many of such schemes have now been put forward.

In general, inorganic chemistry, having fewer species and reactions, remains explicit whilst hydrocarbon chemistry is parameterised in some way[141, 142]. Certain chemical species and reactions may be assumed to be unimportant, in terms of the goal of the modelling exercise, and so neglected (*i.e.* incomplete chemistry); a particular class of compound can be represented by a single member, or generic representation, of the group ("surrogate molecule" and "generic lumping" chemistry); the total VOC loading can be expressed as the sum of various types of carbon-carbon bond ("lumped structure" chemistry); or the chemical mechanism can be translated into a set of equations in notional, or idealised, variables (parameterised chemistry).

Incomplete chemistry assumes prior knowledge of the hydrocarbon chemistry to be modelled and is, for this reason, lacking in generality. The "surrogate molecule" approach appears similar, but differs in the assignment of emissions. Commonly in the surrogate molecule approach, examples of reactive and less-reactive compounds from each class of hydrocarbons are chosen: Selby[143], for example, uses ethane and n-butane, ethene and propene, and toluene and m-xylene to represent alkanes, alkenes and aromatic compounds, respectively. Emissions of species which have not been included are assigned to the compounds which are in the model according to structure and reactivity. In assigning such emissions, either all the carbon mass is allocated to the surrogate, or only that fraction which would give the substituted hydrocarbon the same carbon mass (per mole) as the surrogate. This latter method of addition on a molar basis appears now to be the preferred approach[144].

A second "lumped molecule" approach replaces a group of hydrocarbons with a single, non-specific, variable. For example, the degradation of alkanes can be represented by[145]

$$RH + OH + O_2 \rightarrow RO_2 + H_2O \qquad [4.134]$$
$$RO_2 + NO + O_2 \rightarrow HO_2 + RCHO + NO_2 \qquad [4.135]$$
$$RCHO + h\nu + O_2 \rightarrow HO_2 + CO + 0.5 (CH_3O_2 + RO_2) \qquad [4.136]$$
$$RCHO + OH + O_2 \rightarrow RCO_3 + H_2O \qquad [4.137]$$
$$RCO_3 + NO + O_2 \rightarrow NO_2 + CO_2 + 0.5 (CH_3O_2 + RO_2) \qquad [4.138]$$

The methyl peroxy radical appears in reactions [4.136] and [4.138] to prevent the perpetual regeneration of RO_2 radicals. In addition to the apportionment of emissions, the above scheme must be assigned rate constants that reflect the speciation of emissions in a given scenario. In effect, the specific mechanistic details of the "lumped surrogate molecule" approach is replaced by a more general, "generic" scheme. This increase in generality need not imply increased accuracy.

Rather than lump hydrocarbon compounds by class, the total number of reactive carbon atoms available can be sub-divided by bond type. For example Whitten et al.[146]

define four types of carbon bonds: single bonds (*i.e.* alkanes and alkyl side-chains); reactive double bonds (alkene double bonds, except that in ethene); less reactive double bonds (aromatic compounds and ethene); and carbonyl groups (aldehydes and ketones). The carbon atoms within each emitted compound are then assigned to these classes. Propene, for instance, consists of one (reactive) carbon-carbon double bond and one single bond. The injection of 1 mole of propene is equivalent, therefore, to the injection of 1 mole of reactive double bonds and 1 mole of alkyl bonds. The reactivity of these bond-variables is then estimated to be the average reactivity of the bonds of the individual species from which the variable is made up.

There is a conceptual elegance about this approach. The carbon atoms within any hydrocarbon molecule (except quaternary atoms) will react independently of each other according to their individual reactivity and the dynamics of any collision with a reaction partner. It would be prohibitively expensive to model all these parallel paths, and most "explicit" mechanisms follow the degradation via the one or two most possible paths. The conceptual elegance is lost, however, when we consider physical processes. Should the products of the reactions of one type of carbon bond produce intermediates which are quickly removed from the system by deposition or condensation (the production of organic aerosol from the degradation of aromatic compounds, for example) then some fraction of the carbon atoms in the other classes must also be removed. The quantitative effects of these merits and demerits have yet to be established, although thorough updates and improvements have been made[147]. Generally these modifications consist of re-evaluations of kinetic parameters and the inclusion of explicit or surrogate molecule mechanisms for the reactions of compounds which are of special importance, or not amenable to treatment using the carbon bond method.

Although the lumped parameter and carbon bond mechanisms contain fewer variables than explicit mechanisms, they remain in the form of simultaneous differential equations which must be solved at every time step and grid point in a simulation of the atmosphere. This is expensive in terms of computer usage, particularly when transport processes are also simulated in detail. To reduce a model's computer processing requirements, a set of algebraic equations which approximate the behaviour of the simultaneous differential equations can be used. These parameterising algebraic expressions are derived from the full mechanism using a combination of numerical analysis, insight into the physical chemistry occurring, and analysis of model output. The combination of these techniques is an art, requiring rigorous understanding of the processes involved, and so is sometimes regarded, wrongly in our opinion, as an attempt to second-guess the solution to the problem.

Parameterised chemistry schemes are generally derived from sensitivity studies of more explicit mechanisms. That is, for a kinetic mechanism of the form

$$\frac{dC}{dt} = f(C,P,V,t) \tag{4.14}$$

with initial conditions

$$C_o = C(V) \tag{4.15}$$

where t is the time of reaction; V a vector of controllable variables such as temperature and pressure; C a vector of concentrations; P a vector of parameters such as rate constants; and f a vector of functions. There will be a solution, g, at time, t

$$C(t) = g(P,V) \tag{4.16}$$

which can be approximated by an empirical model, E, that is a function of initial concentrations, C_o, only[148]

$$g(P,V) \approx E(C_o) \tag{4.17}$$

The parameterising function is some fit to the output of a series of runs of the full mechanism in which input variables C, P and V have been systematically changed. The exact mathematical nature of the fit can be determined by the experimental design or fixed as some simple analogy to the overall process: for example, Marsden *et al.*[148] use a log-linear function; MacKenzie *et al.*[139] use a linear function; Dunker[149] uses a second-order Taylor expansion. Cariolle and Déqué[150], in contrast, obtain a fit to the *rate of change* of ozone with time as a function of the sensitivity of ozone production to ozone concentration, temperature and ozone column. With careful choice of model variables and experimental design, the derived empirical fits can give results within 5-10 % of the full scheme, and in a fraction of the time (though some computing effort is required to generate the data from which the empirical fit is derived).

In principle, the idea of parameterised chemistry is appealing, particularly because regional-scale and global-scale grid models separate the chemical and transport operators in the full differential equations (see, *e.g.* Refs. 69, 148, 151 and 152). This implies that the chemistry is simulated in a perfectly mixed isolated box around each grid-point. Under many circumstances the conditions around any one grid-point will be similar to those at another at some point in the model run. It may be the case, then, that an explicit chemistry operator repeatedly solves for closely similar conditions. By efficiently sampling the response surface of the chemical mechanism over a wide range of conditions, the parameterisation procedure removes this redundancy. Of course, any conditions which are outside of the limits of the empirical fit must be solved using the explicit chemistry operator and there will be an optimum range for each variable that maintains the accuracy of the empirical function whilst covering a large range of expected values. It should also be borne in mind that neither explicit, nor parameterised, chemistry schemes can take account of sub-grid scale processes that may have a profound effect on the chemical evolution of airmasses.

In this chapter we have reviewed the chemistry that is important in the formation of photochemical oxidants on global, regional, and local scales. The fundamentals of tropospheric oxidant chemistry have been known for some time. Our understanding of these fundamentals has been improved by detailed kinetic studies which identify the products and measure the rates of each reaction. Some aspects of atmospheric chemistry are not so well advanced, however, and their importance cannot yet be gauged. On the global scale, the effect of biogenic emissions of hydrocarbons and alkyl halides are not fully quantified. The importance of heterogeneous chemistry, within clouds and possibly on the surface of aerosol particles, is only now receiving attention. Unfortunately for those who attempt to simulate tropospheric oxidant behaviour, it appears that a great deal of chemistry may have to be included in the most accurate models.

On the regional and local scales, the speciation and reactivity of VOC emissions are the greatest uncertainty. Little is known quantitatively about the degradation of aromatic compounds even though these can be amongst the most reactive of the hydrocarbons released in urban areas. The ability of organic compounds to affect air quality, on a *per molecule* basis, has begun to be assessed using the concepts of relative reactivity and ozone forming potential but more remains to be done. In particular, the potential for the formation of other photochemical oxidants (such as H_2O_2 and PAN) has not received a great deal of attention.

We have also outlined the most common kinetic schemes used in atmospheric chemistry modelling. Each have their own merits and demerits: their continued development in parallel should provide the means for useful ratification by intercomparison. In our opinion, too few sensitivity studies have been published. It is obvious from our discussions of relative reactivity and model parameterisation that sensitivity studies, when carefully devised, can furnish us with a great deal of chemically and legislatively important information.

1. Wayne, R.P. (1988) *Chemistry of Atmospheres*, Clarendon Press, Oxford, UK, pp 447.

2. Colbeck, I. (1988) Photochemical ozone pollution in Britain, Sci. Prog. Oxf., 72, 207-226.

3. Baulch D.L., Cox, R.A., Crutzen, P.J., Hampson, R.F., Kerr, J.A., Troe, J., and Watson, R.T. (1982) Kinetic and photochemical data for atmospheric chemistry, J. Phys. Chem. Ref. Data, 11, 327.

4. Baulch, D.L., Cox, R.A., Hampson, R.F., Kerr, J.A., Troe, J., and Watson, R.T. (1984) Evaluated kinetic and photochemical data for atmospheric chemistry, J. Phys. Chem. Ref. Data, 13, 1259-1375

5. Atkinson , R., D.L. Baulch, R.A. Cox, R.F. Hampson, J.A. Kerr, and J. Troe (1989) Evaluated kinetic and photochemical data for atmospheric chemistry. Supplement III. IUPAC subcommittee on gas kinetic data evaluation for atmospheric chemistry, J. Phys. Chem. Ref. Data, 18, 881-1097

6. Atkinson , R., D.L. Baulch, R.A. Cox, R.F. Hampson, J.A. Kerr, and J. Troe (1992) Evaluated kinetic and photochemical data for atmospheric chemistry. Supplement IV. IUPAC subcommittee on gas kinetic data evaluation for atmospheric chemistry, J. Phys. Chem. Ref. Data, 21, 1125-1569.

7. Atkinson, R. (1989) Kinetics and mechanisms of the gas phase reactions of the hydroxyl radical with organic compounds, J. Phys. Chem. Ref. Data, 18, monograph 1.

8. Atkinson, R. (1990) Gas-phase tropospheric chemistry of organic compounds: a review, Atmos. Environ., 24A, 1-41.

9. DeMore, W.B., Sander, S.P., Golden, D.M., Hampson, R.F., Kurylo, M.J., Howard, C.J., Ravishankara, A.R., Kolb, C .E., and Molina, M.J. (1992) Chemical kinetics and photochemical data for use in stratospheric modelling, NASA JPL-Publication 92-90, California Inst. Technol., Pasadena, USA, 239pp.

10. Wayne, R.P. (1987). The Photochemistry of Ozone, Atmos. Environ., 21, 1683-1694.

11. Atkins, P.W. (1986) *Physical Chemistry*, 3rd ed., OUP, UK.

12. Wayne, R.P (1988) *Principles and applications of Photochemistry*, OUP, UK.

13. Finlayson-Pitts B. J. and Pitts J. N. (1986) *Atmospheric Chemistry: Fundamentals and Experimental Technique*, John Wiley and Sons, New York.

14. Cox, R.A., (1988) Atmospheric Chemistry of NO_x and hydrocarbons influencing tropospheric ozone, in *Tropospheric Ozone*, I.S.A. Isaksen (ed), Reidel, Dordrecht, pp 263-292.

15. Fairchild C.E., Stone E.J., and Lawrence G.M. (1978) Photofragment Spectroscopy of Ozone in the *u.v.* Region 270-310 nm and at 600 nm, J. Chem Phys., 69, 3632-3638.

16. Sparks R.K., Carlson L.R., Shobatake K., Kowalzyk M.L. and Lee Y.T. (1980) Ozone Photolysis: A Determination of the Electronic and Vibrational State Distributions of Primary Products, J. Chem. Phys., 72, 1401-1402.

17. Ball, S.M., Hancock, G., Murphy, I.J., and Rayner, S.P. (1993) The relative quantum yields of $O_2(a^1\Delta g)$ from the photolysis of ozone in the wavelength range 270 nm $\leq \lambda \leq$ 329 nm, Geophys. Res. Lett., 20, 2063-2066.

18. Crutzen, P.J., I.T.N. Jones, and R.P. Wayne (1971) Calculation of $[O_2 (^1\Delta g)]$ in the atmosphere using new laboratory data, J. Geophys. Res., 76, 1490-1497.

19. Demerjian, K.L., Kerr, J.A., Calvert, J.G. (1974) The mechanisms of photochemicasl smog formation, Adv. Environ. Sci. Technol., 4, 1-262.

20. Davidson J.A., Cantrell C.A., McDaniel A.H., Shetter R.E., Madronich S. and J.G. Calvert (1988) Visible - ultraviolet absorption cross sections for NO_2 as a function of temperature, J. Geophys. Res., 93, 7105-7112.

21. Hicks, E., B. Leroy, P. Rigaud, J. Jourdain and G. Le Bras, 1979, Spectres d'absorption dans le proche ultraviolet et le visible des composes minoritaires atmospheriques NO_2 et SO_2 entre 200 et 300 K, J. Chem. Phys., 76, 693-698.

22. Shetter R.E., Davidson J.A., Cantrell C.A., Burzynski N.J. and Calvert J.G. (1988) Temperature Dependence of the Atmospheric Photolysis Rate Coefficient for NO_2, J. Geophys. Res., 93, 7113-7118.

23. Derwent R.G. and Hov Ø. (1979) Computer Modelling Studies of Photochemical Air Pollution Formation in N.W. Europe, AERE Report No. 9434, HMSO, UK.

24. Fehsenfeld F.C., Parrish D.D., Fahey D.W. (1988) The Measurement of NO_x in the non-urban troposphere in *Troposheric Ozone*, I.S.A. Isaksen (ed.), NATO ASI series C, 227, Reidel, Dordrecht, pp 185-216.

25. Singh, H.B. (1987) Reactive Nitrogen in the Troposphere, Environ. Sci. Technol., 21, 320-327.

26. Roberts, J.M. (1990) The Atmospheric Chemistry of organic nitrates, Atmos. Environ., 24a, 243-287.

27. Carter, W.P.L. and Atkinson, R. (1985) The atmospheric chemistry of alkanes, J. Atmos. Chem., 3, 377-405.

28. Luke W.T., Dickerson R.R., and Nunnermacker L.J. (1989) Direct Measurements of the Photolysis rate Coefficients and Henry's Law Constants of Several Alkyl Nitrates, J. Geophys. Res., 94, 14905-14921.

29. Becker K.H. and Wirtz K. (1989) Gas Phase Reactions of Alkyl Nitrates with Hydroxyl Radicals under Tropospheric Conditions in Comparison with Photolysis, J. Geophys. Res., 9, 419-433.

30. Taylor, W.D., Allston, T.D., Mascato, M.J., Fazekas, G.B., Kowslowski, R., and Takacs, G.A. (1980) Atmospheric photodissociation lifetimes for nitromethane, methyl nitrite and methyl nitrate, Int. J. Chem. Kinetics, 12, 231-240.

31. Roberts, J.M, and Fajer, R.W. (1989) UV absorption cross sections of organic nitrates of potential atmospheric importance and estimation of atmospheric lifetimes, Environ. Sci. Technol., 23, 945-951.

164

32. Luke, W.T., Dickerson, R.R., and Nunnermacker, L.J. (1989) Direct measurements of the photolysis rate coefficients and Henry's Law constants of several alkyl nitrates, J. Geophys. Res., 94, 14905-14921.

33. Molina, M.T., and Molina, M.J. (1981) UV absorption cross sections of HO_2NO_2 vapor, J. Photochem., 15, 97-108.

34. Senum, G.I., Lee, Y.-N., and Gaffney, J.S. (1984) The ultraviolet absorption spectrum of peroxyacetyl nitrate, J. Phys. Chem., 88, 1269-1271.

35. Jenkin M.E., Cox R.A., Hayman G.D. and Whyte L.J. (1988) Kinetic Study of the reactions of $CH_3O_2+CH_3O_2$ and $CH_3O_2+HO_2$ using molecular modulation spectroscopy, J. Chem Soc., Faraday Trans. 2, 84, 913-930.

36. Atkinson R, and Lloyd, A.C. (1984) Evaluation of kinetic and mechanistic data for modelling of photochemical smog, J. Phys. Chem., Ref. Data, 13, 315.

37. Killus J.P. and Whitten G.Z. (1984) Isoprene: A Photochemical Kinetic Mechanism, Environ. Sci., Technol, 18, 142-148.

38. Lloyd A.C., Atkinson R., Lurmann F.R. and Nitta B. (1983) Modelling Potential Ozone Impacts from Natural Hydrocarbons-I, Atmos. Environ., 17, 1931-1950.

39. Brewer D.A., Ogliarvso M.A., Augustison T.R., Levine J.S. (1984) The Oxidation of Isoprene in the Troposphere: Mechanism and Model Calculations, Atmos Environ, 18, 2723-2744.

40. Barrie L.A., Bottenheim J.W, Schnell R.C., Crutzen P.J. and Rasmussen R.A. (1988) Ozone destruction and photochemical reactions at polar sunrise in the lower Arctic atmosphere, Nature, 334, 138-141.

41. Madronich, S. (1987) Photodissociation in the atmosphere. 1: Actinic flux and the effect of ground reflections and clouds, J. Geophys. Res., 92, 9740-9752.

42. Lelieveld J., (1990) *The role of clouds in tropospheric photochemistry*, Ph.D. thesis, Rijkuniversiteit Utrecht. pp 131.

43. Lelieveld J. and Crutzen P.J. (1989) Influences of Cloud Photochemical Processes on Tropospheric Ozone, Nature, 343, 227-233.

44. World Meteorological Organisation (1991) *Global Ozone Research and Monitoring Project*, 25, WMO, Geneva, Switzerland.

45. Zhou X.L. and White J.M. (1990) Surface Photochemistry: Monolayer versus Multilayer for Alkyl halides on Ag {lll}, Chem. Phys. Lett., 167, 205-208.

46. Adamson, A.W. (1990) *Physical Chemistry of Surfaces*, 5th ed., Wiley and Sons, New York, 775 pp.

47. Somorjai, G.A. (1972) *Principles of Surface Chemistry*, Prentice-Hall, New Jersey

48. Schiavello M., Augvgliaro V., Coluccia S., Palmisano L. and Sclafani A. (1989) Photolytic and Redox Mechanisms for the Photodecomposition of Ethanoic Acid Adsorbed over Pure and Mixed Oxides, in *Photochemistry on Solid Surfaces*, M. Anpo and T. Matsnuraz (eds), *Studies in Surface Sci. and Catalysis*, 47, Elsevier, Amsterdam (pp 149-167).

49. Costello S.A., Roop B., Liv, Z.H. and White J.M. (1988) Photochemistry of CH_3Br absorbed on PE {lll} J. Phys Chem, 92, 1019-1020.

50. Grassian V.H. and Pimentel G.C. (1988) Photochemical reactions of cis and trans 1,2 -dichloroethene absorbed on Pd{lll} and Pt {lll}, J. Chem. Phys., 88, 4484-4491.

51. Anpo M. (1989) Photochemistry of Alkyl Ketones in the Absorbed State: Effects of solid Surfaces upon the Photolysis, in *Photochemistry on Solid Surfaces*, M. Anpo and T. Matsnuraz (eds), *Studies in Surface Sci. and Catalysis*, 47, Elsevier, Amsterdam (pp 149-167).

52. Harrison R.M. and Pio C.A. (1983) Kinetics of SO_2 oxidation over carbonaceous particles in the presence of H_2O, NO_2, NH_3, and O_3, Atmos. Environ., 17, 1261-1275.

53. Gupta G., Sabaratnam S. and Dadson R. (1987) A Modified Nitrogen-oxide Ozone Cycle, Water, Air and Soil Pollution, 33, 117-124.

54. Allen A.G., Harrison R.M. and Erisman J.W. (1989) Field Measurements of the Dissociation of Ammonium Nitrate and Ammonium Chloride Aerosols, Atmos Environ., 23, 1591-1599.

55. Dollard G.J., Davies T.J. and Lindstrom J.P.C. (1987) Measurements of the dry deposition rates of some trace gas species, in *Physico-chemical behaviour of atmospheric pollutants*, G. Angeletti and G. Restelli (eds), D. Reidel, Dordrecht. pp 470-479.

56. Huebert B.J. and Robert C.H. (1985) The dry deposition of nitric acid of grass, J. Geophys. Res., 90, 2085-2090.

57. Boodaghians R.B., Canosa-Mas C.E., Carpenter P.J. and Wayne R.P. (1988) The Reactions of NO_3 with OH and H, J. Chem. Soc. Faraday Trans. 2, 84, 931-948.

58. Schwartz, S.E. (1984) Gas- and aqueous phase chemistry of HO_2 in liquid water clouds, J. Geophys. Res., 89, 11,589-11,598.

59. Lind, J.A., and Kok, G.L. (1986) Henry's Law determinations for aqueous solutions of hydrogen peroxide, methylhydroperoxide, and peroxyacetic acid, J. Geophys. Res., 91, 7889-7895.

60. Kozak-Channing, L.F., and Heltz, G.R. (1983) Solubility of ozone in aqueous solutions of 0-0.6 M ionic strength at 5-30°C, Environ. Sci. Technol., 17, 145-149.

61. Ledbury, W., and Blair, E.W. (1925) The partial formaldehyde vapour pressure of aqueous solutions of formaldehyde, II, J. Chem. Soc., 127, 2832-2839.

62. CRC (1987) *Handbook of Chemistry and Physics*, 67th edition, CRC Press, Boca Raton, Florida.

63. Chameides, W.L. (1986) Possible role of NO_3 in the nighttime chemistry of a cloud, J. Geophys. Res., 91, 5331-5337; *and* Reply, J. Geophys. Res., 91, 14,571-14,572.

64. Schwartz, S.E., and White, W.H. (1981) Solubility equilibria of the nitrogen oxides and oxiacids in dilute aqueous solutions, Adv. Environ. Sci. Eng., 4, 1-45.

65. Lee, Y.N., and Schwartz, S.E. (1981) Reaction kinetics of nitrogen dioxide with liquid water at low partial pressures, J. Phys. Chem., 85, 848.

66. Perrin, D.D. (1981) *Ionisation constants of inorganic acids and bases in aqueous solution*, Pergamon Press, New York.

67. Smith, R.M., and Martell, A.E. (1976) *Critical stability constants*, Vol. 4, inorganic complexes, Plenum, New York.

68. Martell, A.E., and Smith, R.M. (1977) *Critical Stability Constants*, vol. 3, *Other Organic Ligands*, Plenum, New York.

69. Seinfeld J.H. (1986) *Atmospheric Chemistry and Physics of Air Pollution*, John Wiley & sons, New York, USA.

70. Calvert, J.G. (ed) (1984) *SO₂, NO and NO₂ Oxidation Mechanisms: Atmospheric Considerations,* Acid Precipitation Series, 3, Butterworth, London, UK.

71. Mozurkewich, M., and Calvert, J.G. (1988) Reaction probability of N_2O_5 on aqueous aerosols, J. Geophys. Res., 93, 15,889-15,896.

72. Dentener, F.J., and Crutzen, P.J. (1993) Reaction of N_2O_5 on tropospheric aerosols: impact on the global distributions of NO_x, O_3, and OH, J. Geophys. Res., 98, 7149-7163.

73. Livingston, F.E., and Finlayson-Pitts, B.J. (1991) The reaction of gaseous N_2O_5 with solid NaCl at 298 K: estimated lower limit to the reaction probability and its potential role in tropospheric and stratospheric chemistry, Geophys. Res. Lett., 18, 17-20.

74. Ravishankara A.R. (1988) Kinetics of Radical Reactions in the Atmospheric Oxidation of CH_4, Ann. Rev. Phys. Chem., 39, 367-394.

75. Khalil, M.A.K., and Rasmussen, R.A. (1985) Causes of increasing atmospheric methane, Atmos. Environ., 19, 397-407.

76. Crutzen, P.J. (1988) Tropospheric ozone: an overview, in *Tropospheric Ozone,* I.S.A. Isaksen (ed.), NATO ASI Ser., D. Reidel, Dordrecht, pp 3-32.

77. Kurylo, M.J., Dagaut, P., Wallington, T.J., Neumann, D.M. (1987) Kinetic measurements of the gas-phase HO_2 & CH_3O_2 cross-disproportionation reaction at 298 K, Chem. Phys. Lett. 139, 513-518.

78. Horie, O., Crowley, J.N., and Moortgat, G.K. (1990) Methyl peroxy self reaction: production and branching ratio between 223 and 333 K, J. Phys. Chem., 94, 8198-8203.

79. Law, K.S., and Pyle, J.A. (1993) Modelling trace gas budgets in the troposphere, 1, Ozone and odd nitrogen, J. Geophys. Res., 98, 18,377-18,400.

80. Vaghjiani, G.L., and Ravishankara, A.R. (1990) Photodissociation of H_2O_2 and CH_3OOH at 248 nm and 298 K, J. Chem. Phys., 92, 996-1003.

81. Wayne, R.P. (1991) *Chemistry of Atmospheres,* 2nd ed, OUP, Oxford, UK.

82. Chameides W.L., Lindsay R.W., Richardson J., Kiang C.S. (1988) The Role of Biogenic Hydrocarbons in Urban Photochemical Smog. Science, 241, 1473-1475.

83. Hough A.M., Derwent R.G. (1990) Changes in the global concentration of tropospheric ozone due to human activities, Nature, 344, 645-648.

84. MacKenzie, A.R., Harrison, R.M., Colbeck, I., and Hewitt, C.N. (1991) The role of biogenic hydrocarbons in the production of ozone in urban plumes in SE England, Atmos. Environ., 25A, 351-359.

85. Atkinson R., Aschmann S.M., Tuazon, E.C., Arey, J. and Zielinska, B. (1989) Formation of 3-methylfuran from the gas phase reaction of OH radicals isoprene and the rate constant for its reaction with the OH radical, Int. J. Chem Kinetics, 21, 593-604.

86. Pierotti D., Wofsy S.C., Jacob D., and Rasmussen R.A. (1990) Isoprene and its oxidation products: methacrolein and methyl vinyl ketone, J. Geophys. Res., 95, 1871-1881.

87. Atkinson R., and Carter W.P.L. (1984) Kinetics and Mechanisms of the Gas-Phase Reactions of Ozone with Organic Compounds under Atmospheric Conditions, Chem. Rev., 84, 437-470.

88. Lopez A., Barthomeuf M.O., and Huertas M.L. (1989) Simulation of Chemical Processes Occuring in an Atmospheric Boundary layer: influence of light and Biogenic Hydrocarbons on the Formation of Oxidants, Atmos. Environ., 23, 1465-1478.

89. Tuazon, E.C., and Atkinson, R. (1990) A product study of the gas-phase reaction of methacrolein with the OH radical in the presence of NO_x, Int. J. Chem. Kinetics, 22, 591-602.

90. Tuazon E.C., and Atkinson R. (1989) A Product Study of the Gas Phase Reaction of Methyl Vinyl Ketone with the OH Radical in the presence of NO_x, Int. J. Chem. Kinetics, 21, 1141-1152.

91. Gery M.W., Whitten G.Z., Killus J.P., and Dodge M.C. (1989) A Photochemical Kinetics Mechanism for Urban and Regional Scale Computer Modelling, J. Geophys. Res., 94, 12,925-12,956.

92. Hatakeyama S., Izumi K., Fukuyama T., and Akimoto H. (1989) Reactions of Ozone with α-Pinene and β-pinene in Air: Yields of Gaseous and Particulate Products, J. Geophys. Res., 94, 13,013-13,024.

93. Nolting F., Behnke W., and Zetsch C. (1988) A Smog Chamber for Studies of the Reactions of Terpenes and Alkanes with Ozone and OH, J. Atmos. Chem., 6, 47-59.

94. Kotzias, D., Fytianos, K., Geiss, F. (1990) Reaction of monoterpenes with ozone, sulphur dioxide and nitrogen dioxide - gas phase oxidation of SO_2 and formation of sulphuric acid, Atmos. Environ., 24A, 2127-2132.

95. Pandis, S.N., Paulson, S.E., Seinfeld, J.H., and Flagan, R.C. (1991) Aerosol formation in the photoxidation of isoprene and β-pinene, Atmos. Environ., 25A, 997-1008.

96. Hampson R.F., Kurylo, M.J., and Sander, S.P. (1989) Chapter 3 of *Scientific Assessment of Stratospheric Ozone: 1989* - reaction rate constants, WMO Global Ozone Res. and Monitoring Project, 20, vol 2, 41-104.

97. Atkinson, R. (1989) Tropospheric reactions of the haloalkyl radicals formed from hydroxyl radical reaction with a series of alternative fluorocarbons, in *Scientific Assessment of Stratospheric Ozone: 1989*, WMO Global Ozone Res. and Monitoring Project, 20, vol 2, 162-205.

98. Niki, H. (1989) An assessment of potential impact of alternative fluorocarbons on tropospheric ozone, in *Scientific Assessment of Stratospheric Ozone: 1989*, WMO Global Ozone Res. and Monitoring Project, 20, vol 2, 409-428.

99. Hinds, W.C. (1982) *Aerosol Technology*, John Wiley, New York, USA, pp423.

100. Finlayson-Pitts, B.J., Livingston, F.E., and Berko, H.N. (1990) Ozone destruction and bromine photochemistry at ground level in the Arctic spring, Nature, 343, 622-624.

101. Niki, H., and Becker, K.H. (1993) *The Tropospheric Chemistry of Ozone in the Polar Regions*, NATO ASI Series, 17, Springer-Verlag, Berlin, Germany.

102. Sturges, W.T., Cota, G.F., and Buckley, P.T (1992) Bromoform emission from Arctic ice algae, Nature, 358, 660-662.

103. Sturges, W.T., Schnell, R.C., Dutton, G.S., Garcia, S.R., and Lind, J.A. (1993) Spring measurements of tropospheric bromine at Barrow, Alaska, Geophys. Res. Lett., 20, 201-204.

104. Sturges, W.T., Sullivan, C.W., Schnell, R.C., Heidt, L.E. and Pollock, W.H. (1993) Bromoalkane production by Antarctic ice algae, Tellus, 45B, 120-126.

105. Sturges W.T. (1990) Excess Particulate and Gaseous Bromine at a Remote Coastal Location, Atmos. Environ., 24A, 167-171.

168

106. Saltzman E.S., and Cooper D.J. (1989) Dimethyl Sulphide and Hydrogen Sulphide in Marine Air, in *Biogenic Sulfur in the Environment*, Saltzman and Cooper (eds), ACS Symposium Series 393, Amer. Chem. Soc.,Washington, D.C., pp 330-351.

107. Fletcher, (1989) in *Biogenic Sulfur in the Environment*, Saltzman and Cooper (eds), ACS Symposium Series 393, Amer. Chem. Soc.,Washington, D.C., pp 489-501.

108. Lovelock, J.E. (1988) *Ages of Gaia*, Oxford University Press, Oxford, UK, pp 252.

109. Plane, (1989) in *Biogenic Sulfur in the Environment*, E.S. Saltzman and D.J. Cooper (eds), ACS Symposium Series 393, Amer. Chem. Soc.,Washington, D.C., pp 404-423.

110. Hewitt. C. N., and Harrison, R. M. (1985) Tropospheric concentrations of the hydroxyl radical - a review, Atmos. Environ., 19, 545-554.

111. Altshuller A.P. (1989) Ambient Air Hydroxyl Radical Concentrations: Measurements and Model Predictions, J. Air. Pollut. Control Assoc., 39, 704-708.

112. Mount, G.H. (1992) The measurement of tropospheric OH by long path absorption I: instrumentation, J. Geophys. Res., 97, 2417-2444.

113. Barnes *et al.*, (1989) in *Biogenic Sulfur in the Environment*, E.S. Saltzman and D.J. Cooper (eds), ACS Symposium Series 393, Amer. Chem. Soc., Washington, D.C., pp 464-475.

114. Chameides W.L., and Davis D.D. (1980) Iodine: Its possible role in tropospheric chemistry, J. Geophys. Res., 85, C12, 7383-7398.

115. Chatfield, R.B., and Crutzen, P.J. (1990) Are there interactions of iodine and sulphur species in marine air photochemistry, J. Geophys. Res., 95, 22,319-22,342.

116. Maguin, F., Mellouki, A., Laverdet, G., Poulet, G., and Le Bras, G. (1991) Kinetics of the reactions of the IO radical with dimethyl sulphide, methanethiol, ethylene and propylene, Int. J. Chem. Kinetics, 23, 237-245.

117. Barnes, I., Bastian, V., Becker, K.H., and Overath, R.D. (1991) Kinetic studies of the reactions of IO, BrO and ClO with dimethylsulphide, Int. J. Chem. Kinetics, 23, 579-591.

118. Seinfeld, J.H. (1989) Urban Air Pollution: State of the Science, Science, 243, 745-752.

119. Kelly, N.A., and Gunst, R.F. (1990) Response of ozone to changes in hydrocarbon and nitrogen oxide concentrations in outdoor smog chambers filled with Los Angeles air, Atmos. Envrion., 24A, 2991-3005.

120. Seila, R.L., and Lonneman, W.A. (1988) paper 88-150.8, presented at the 81st Annual meeting of the Air Pollution Control Assoc., Dallas, TX. Cited in Ref. 118.

121. Tuazon, E.C., Carter, W.P.L., and Atkinson, R. (1991) Thermal decomposition of peroxyacetyl nitrate and reactions of acetyl peroxy radicals with NO and NO_2 over the temperature range 283-313 K, J. Phys. Chem., 95, 2434-2437.

122. Bruckmann, P.W., and Willner, H. (1983) Infrared spectrascopic study of peroxyacetylnitrate (PAN) and its decomposition products, Environ. Sci. Technol., 17, 352-357.

123. Bridier, I., Caralp, F., Loirat, H., Lesclaux, R., Veyret, B., Becker, K.H., Reimer, A., and Zabel, F. (1991) Kinetic and theoretical studies of the reactions $CH_3C(O)O_2 + NO_2 + M \leftrightarrow CH_3C(O)O_2NO_2 + M$ between 248 K and 393 K and between 30 and 760 Torr, J. Phys. Chem., 95, 3594-3600.

124. Lurmann F.W., Lloyd A.C., and Atkinson R. (1986) A Chemical Mechanism for Use in Long-Range Transport/Acid Deposition Computer Modeling, J. Geophys. Res., 91, 10,905-10,936.

125. Atkinson, R., Aschmann, S.M., and Arey, J. (1991) Formation of ring retaining products from the OH radical initiated reactions of o-, m-, and p-xylene, Int. J. Chem. Kin., 23, 77-97.

126. Zellner R., Fritz B., and Preidel M. (1985) A cw UV Laser absorption study of the reactions of the hydroxycyclodienyl radical with NO_2 and NO, Chem. Phys. Lett., 121, 412-416.

127. Atkinson, R., Aschmann, S.M., Arey, J., and Carter, W.P.L. (1989) Formation of ring retaining products from the OH radical initiated reactions of Benzene and toluene, Int. J. Chem. Kinetics, 21, 801-827.

128. Japar, S.M., Wallington, T.J., Richert, J.F.O., and Ball, J.C. (1990) The atmospheric chemistry of oxygenated fuel additives: t-butyl alcohol, dimethyl ether, and methyl t-butyl ether, Int. J. Chem. Kinetics, 22, 1257-1269.

129. Wallington, T.J., and Japar, S.M. (1991) Atmospheric chemistry of diethyl ether and ethyl tert-butyl ether, Environ. Sci. Technol., 25, 410-415.

130. Smith, D.F., Kleindiest, T.E., Hudgens, E.E, McIver, C.D., Bufalini, J.J. (1991) The photooxidation of methyl tertiary butyl ether, Int. J. Chem. Kinetics, 23, 907-924.

131. Wayne, R.P, Barnes, I., Biggs, P., Burrows, J.P., Canosa-Mas, C.E., Hjorth, J., Le Bras, G., Moortgat, G.K., Perner, D., Poulet, G., Restelli, G., and Sidebottom, H. (1991) The nitrate radical, Atmos. Environ., 25A, 1-203.

132. Platt, U., Le Bras, G., Poulet, G., Burrows, J.P., and Moortgat, G.K. (1990) Peroxy radicals from night-time reactions of NO_3 with organic compounds, Nature, 348, 147-149.

133. Carter W.P.L., and Atkinson R. (1987) An experimental study of incremental hydrocarbon reactivity, Environ. Sci., Technol., 21, 670-679.

134. Carter W.P.L., and Atkinson R. (1989) Computer Modelling Study of Incremental Hydrocarbon Reactivity, Environ. Sci. Technol., 23, 864-880.

135. Hough A.M., and Derwent R.G. (1987) Computer Modelling Studies for the Distribution of Photochemical Ozone Production Between different Hydrocarbons, Atmos. Environ., 21, 2015-2033.

136. Chang T.Y., Rudy S.J., Kuntasal G., and Gorse R.A. (1989) Impact of Methanol vehicles on ozone air quality, Atmos. Environ., 23, 1629-1644.

137. Dunker A. (1990) Relative Reactivity of emissions from methanol-fuelled and gasoline-fuelled vehicles in forming ozone, Environ. Sci. Technol., 24, 853-862.

138. Derwent R.G., and Hov O. (1988) Application of sensitivity and Uncertainty Analysis Techniques to a Photochemical Ozone Model, J. Geophys. Res., 93, 5185-5199.

139. MacKenzie A.R., Harrison R.M., and Colbeck I. (1990) The impact of Local Emissions on the Formation of Secondary Pollutants in Urban Plumes, Sci. Tot. Environ., 93, 245-254.

140. Cocks A.T., and Fletcher I.S. (1989) Major Factors Influencing Gas-Phase Chemistry in Power Plant Plumes -II Release Time and Dispersion Rate for Dispersion into an "Urban" Ambient Atmosphere, Atmos. Environ., 23, 2801-2812.

141. UK Photochemical Oxidants Review Group (1987) Ozone in the United Kingdom, HMSO, London, 112pp.

170

142. Hough, A.M. (1988) An intercomparison of mechanisms for the production of photochemical oxidants, J. Geophys. Res., 93, 3789-3812.

143. Selby K. (1987) A Modelling Study of Atmospheric Transport and Photochemistry in the Mixed Layer During Anticyclonic Episodes in Europe, J. Climate and Appl. Meteorol., 26, 1317-1338.

144. Dodge M.C. (1989) A Comparison of Three Photochemical Oxidant Mechanisms, J. Geophys. Res., 94, 5121-5136.

145. Cocks A.T. and Fletcher I.S. (1988) Major Factors Influencing Gas-Phase Chemistry in Powers Plant Plumes, Atmos. Environ., 22, 663-676.

146. Whitten G.Z., Hogo H., and Killus J.P. (1980) The Carbon-Bond Mechanism: A Condensed Kinetic Mechanism for Photochemical Smog, Environ. Sci. Technol., 14, 690-700.

147. Gery, M.W., Whitten, G.Z., Killus, J.P., and Dodge, M.C. (1989) A photochemical kinetics mechanism for urban and regional computer modeling, J. Geophys. Res., 94, 12,925-12,956.

148. Marsden A.R., Frenklach M., and Reible D.D. (1987) Increasing the Computational Feasibility of Urban Air Quality Models that Employ Complex Chemical Mechanisms, J. Air Pollut. Control Assoc., 37, 370-376.

149. Dunker A.M. (1986) The reduction and parameterization of chemical mechanisms for inclusion in atmospheric reaction-transport models, Atmos. Environ., 20, 479-486.

150. Cariolle, D., and Déqué, M. (1986) Southern hemisphere medium scale waves and total ozone disturbances in a spectral general circulation model, J. Geophys. Res., 91, 825-846.

151. Jacob D.J., Sillman, S., Logan, J.A., and Wofsy, S.C. (1989) Least independent variables method for simulation of tropospheric ozone, J. Geophys. Res., 94, 8497-8509.

152. McRae G.J., Goodin W.R., and Seinfield J.H. (1982) Development of a 2nd generation Mathematical Model for Urban Air pollution -I. Model Formulation, Atmos. Environ., 16, 679-696.

CHAPTER 5

THE MEASUREMENT OF PHOTOCHEMICAL OXIDANTS
AND THEIR PRECURSORS

In this chapter we review the analytical techniques which are used, or have been tested for use, in the monitoring of oxidants and their precursors in the atmosphere. Techniques which have been designed to monitor output from industrial sources, and which have detection limits well above typical atmospheric concentrations, will usually be omitted. For the sake of brevity, the analysis of sulphur compounds and particulate phases is also left aside; details can be found in Harrison and Perry[1], Finlayson-Pitts and Pitts[2], and Allegrini and De Santis[3]. A useful short introduction to measurement techniques in atmospheric chemistry is given by Ayers and Gillett[4]. In keeping with the layout of the rest of the book, the techniques for measuring each oxidant or oxidant precursor are discussed together, detailed descriptions of a general technique being given where an example is first cited. As an aid to cross referencing, Table 5.1 gives a summary of the techniques discussed for each of the compounds or compound categories. The plan in subsequent sections will be to discuss optical methods first and then deal with techniques based on chemical analysis.

Table 5.1: Techniques for the determination of oxidant and precursor concentrations in the troposphere. Infrared techniques include TDLAS and FTIR; *uv*-vis techniques include DOAS and DIAL (see text for definitions of instruments).

Compound	Infrared absorption	*uv*-vis. absorption	LIF	ESR	Chemi-luminescence	Chemical	Chroma-tography
Ozone		+			+	+	
Organic Peroxides	+				+	+	
PAN compounds	+					+	+
OH and other radicals		+	+	+		+	
Nitrogen oxides	+	+	+	+	+	+	
Non-methane hydrocarbons					+		+
Oxygenated hydrocarbons	+	+				+	

5.1 OZONE

Today surface ozone is routinely monitored using an ultra-violet (*uv*) photometric technique, which, because the technique does not require calibrations (although it does require intercomparisons), has largely superseded chemiluminescence as the preferred continuous measurement technique. However, the fast response of the chemiluminescence monitors make them more suitable for deposition studies, such as those based on eddy correlation, which require a high sampling frequency. Historically, wet chemical techniques have been used and they remain the subject of research aimed at providing a consistent ozone record as well as low-cost techniques for the use in global networks.

Ultraviolet photometric detectors determine the absolute amount of ozone in a sample cell using the Beer-Lambert law which relates monochromatic light absorption to sample concentration:

$$\log\left(\frac{I}{I_o}\right) = -\epsilon_\lambda c l \tag{5.1}$$

where
I_o = intensity of light passing through ozone-free air

I = intensity of light passing through sample

ϵ_λ = molar absorption coefficient for ozone at, *e.g.* 253.7 nm

i = path length

c = concentration of ozone in sample

and the light source is of wavelength, λ. A schematic diagram of a *uv* photometer is shown

Figure 5.1

Schematic of an ozone uv photometer, based on the Monitor Labs model 8810, showing the sample intake, scrubber loop, 254 nm ultraviolet source, glass optical chamber, detector and signal processing electronics.

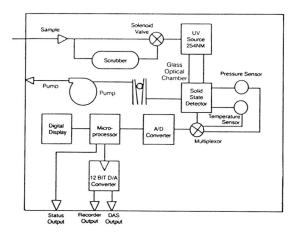

in Figure 5.1. In this single cell configuration, inlet air is sampled directly and through an

173

ozone filter, in turn, to yield values for I and I_o, respectively. The measurement is dependent, therefore, only on fundamental properties of the vapour and as such constitutes a primary calibration standard for ozone subject to the following assumptions[5];

i)   the ozone scrubber is specific for ozone and does not add or remove any other optically active component of the sample;

ii)   there are negligible changes to the absorption at any other wavelength emitted by the source and measured by the detector, during the measurement;

iii)   practically all the light intensity measured is of wavelength λ;

iv)   the temperature and pressure do not change within the cell during the measurement;

v)   ozone destruction occurs only within the ozone scrubber and not on any other surface within the instrument; and

vi)   variations in lamp intensity between sample and "zero" measurements are taken into account.

The three latter sources of error can be minimised by the use of carefully constructed housing, inert materials, and electronic data processing, respectively, resulting in detection limits of 1-2 ppbv (Ref. 1).

The sequential nature of the measurement of zero and sample results in a time resolution of 15-30 s. This, together with the relatively high sampling rate (1-4 l min^{-1}), makes the technique unsuitable for studies in which ozone concentrations vary rapidly, or the experimental volume is small. Other limitations are due chiefly to interference from other compounds absorbing in the same region of the spectrum and which are removed (to some extent) by the ozone scrubber. For example, mercury vapour is very efficient at absorbing the light produced, at its resonant frequency, by the mercury vapour lamp source. Other compounds causing interference include cresols, nitrocresols, other aromatics and the aliphatic keto-acid, pyruvic acid[6]. Concentrations of most of these organic species (and mercury vapour) will generally be very low in ambient air, so their effects will be negligible as long as their absorption strengths are comparable to that of ozone. However, some care must be exercised when operating or calibrating photometric analysers in laboratory conditions where the concentrations of interfering compounds may be significant.

A second type of photometric device, the <u>d</u>ifferential <u>a</u>bsorption <u>li</u>dar (DIAL), is used for detailed, spatially-resolved, measurements. Lidar (<u>li</u>ght <u>d</u>etection <u>an</u>d <u>r</u>anging) systems probe the atmosphere by firing short pulses of laser light into it and measuring the backscattered radiation as a function of time[7]. The amount of light scattered at time, t, is then a measure of the efficiency of backscattering in the volume of air at a distance of ct

from the instrument, where c is the velocity of light. In DIAL measurements, the laser is tuned to a frequency at which there is strong absorption in the ozone spectrum and then retuned to an off-resonance frequency for consecutive laser "shots". The signal returning to the detector is then due to the backscatter of light from air molecules and aerosol particles, modulated by the amount of ozone absorption at each wavelength. Backscatter from air molecules and spherical aerosol particles can be calculated using Rayleigh and Mie theories of scattering, respectively, and each type of scattering is generally known by the name of the theory used to describe it. The ratio of backscatter signals gives the variation in the concentration of ozone in a volume of air remote from the instrument[8]:

$$[O_3]_{av} = \frac{1}{2(R_2 - R_1)[\sigma(\lambda_{on}) - \sigma(\lambda_{off})]} \ln\left(\frac{P_{on}(R_1) P_{off}(R_2)}{P_{off}(R_1) P_{on}(R_2)}\right) \qquad (5.2)$$

where $[O_3]_{av}$ is the average concentration of ozone (molec cm^{-3}) over the range R_1 and R_2; σ is the absorption cross-section given at the on- and off-resonance frequencies, λ_{on} and λ_{off}; and P is the power of the backscattered signal, again at the on- and off-resonance frequencies. Different values for the ranges R_1 and R_2 are selected by varying the time interval between light emission and detection. Equation (5.2) assumes that the optical characteristics of the atmosphere do not change over the time and wavelength interval between on- and off-resonance shots. Temporal variability can be minimised by using a high rate of firing: 1-10 Hz is typical[9]. *Backscatter* and *extinction* differences over the wavelength interval (λ_{on} - λ_{off}) are not so easily dealt with.

A detailed discussion of these and other possible sources of error is given by Fredriksson (chapter 4 in ref. 7). Briefly, *extinction* errors arise from interference by a second gas (usually SO_2 in the case of O_3) and from the wavelength dependence of extinction by aerosol particles. The first of these can be overcome by a careful choice of wavelengths and the use of a calibration cell containing the interfering gas. The calibration cell is placed in the path of the radiation and the on- and off-resonance frequencies adjusted to give zero difference. Extinction errors due to a wavelength dependence of the attenuation by aerosol particles are reduced by choosing wavelengths close to each other. Unfortunately the absorption spectrum of ozone in the region of interest contains no fine structure (Figure 5.2) and the wavelengths used (λ_{on} = 286 nm, λ_{off} = 300 nm; or λ_{off} = 287 nm) are typical values[8, 10]) are a compromise between the need for a substantial difference in absorption cross-sections and the desire to minimise errors which will be difficult to

Figure 5.2

Absorption spectra of ozone and sulphur dioxide around 250 nm. Note the lack of fine structure in the ozone spectrum compared to that of sulphur dioxide: This makes choosing "on-line" and "off-line" wavelengths for ozone troublesome. The absorption of sulphur dioxide in this region makes it an interferent in the measurement of ozone.

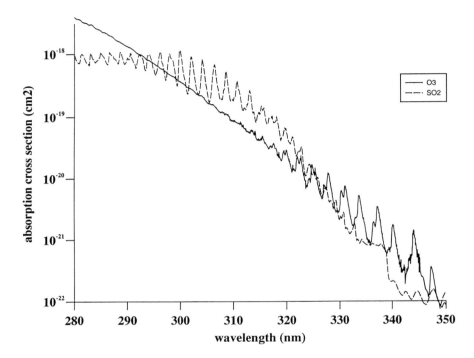

quantify. If a quantification of the aerosol extinction is required, one simple method is to measure a second off-resonance wavelength and compensate for the differential aerosol extinction and backscatter by linear interpolation[11].

Minimising $\Delta\lambda$ will reduce the *backscatter* error that is caused by the wavelength dependence of aerosol backscatter, and by changes in the wavelength dependence of aerosol backscatter, as well as reducing the *extinction* error that is caused by the wavelength dependence of aerosol extinction. In the free troposphere the backscatter error is generally negligible, but when measuring in regions of changing aerosol loading, such as across the top of the planetary boundary layer or through a plume of pollution, some attempt must be made to quantify it. Browell *et al.*[12] have shown that the backscatter ratio, S_R, that is the ratio of Mie backscatter to Rayleigh backscatter, can be found by re-casting the lidar equation at the off-resonance frequency as an ordinary differential equation in S_R with respect to range, R. The aerosol phase function, which describes the angular distribution of scattering by aerosol particles, must be assumed in this method. So too must the wavelength dependence of aerosol scattering, and S_R at some reference range. Alternatively, Papayannis et al.[13] suggest adding a third wavelength measurement to obtain the Raman signal that is due to inelastic scattering from molecular nitrogen in the atmosphere. The ratio of the Raman signal to the Mie-and-Rayleigh signal can then be used to calculate both the aerosol backscatter and extinction. In general, the errors and uncertainties discussed above result in a sensitivity of about 10 ppbv in the first few hundred metres, and of $\sim$ 20 ppbv at around a kilometre[9]. Although these errors may seem large compared to those associated with point measurements by *uv* photometry, the large amount of temporally - and spatially - resolved information available from a DIAL system make it a useful complement to a network of *in situ* monitors.

Figure 5.3

Schematic of the Central Electricity Research Laboratory DIAL system. From Varey *et al.*[10]. Heavy solid lines show the light paths into and out of the device. The system is mobile and so the position of each backscatter return must be logged.

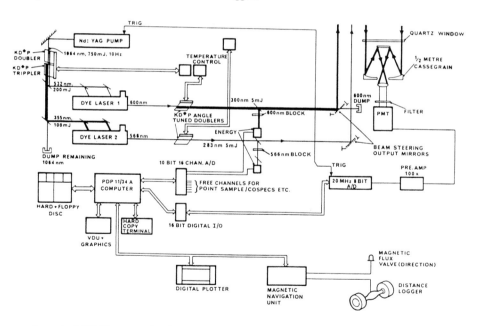

A typical DIAL set-up is shown in Figure 5.3. The essential features of the system are a laser transmitter, a receiver, electronic controls and data reduction equipment. The laser used is required to have high energy to produce a detectable backscatter signal, narrow bandwidth to probe the ozone absorption spectrum and to allow the use of a filter to exclude skylight, and small beam divergence to maintain spatial resolution. Typically Nd-YAG lasers are used to pump dye lasers[7] (Figure 5.3) although some measurements have been made using CO_2 lasers in the infrared[7, 14] which has the advantage of being easily tuned over a wide frequency range to provide measurements of other trace gases in addition to ozone. Two lasers are shown in Figure 5.3 which allows the system to make simultaneous measurements of two species, or to measure ozone at two wavelength pairs thus reducing the effect of interference errors.

Receiver telescopes may be coaxial with the laser source, as shown in Figure 5.3, or sited beside the light source. Steering mirrors direct the emitted beam towards the atmospheric volume under investigation, and direct the backscattered light to the receiver. Zenith and nadir measurements may also be made from airborne instruments. Control and data capture electronics then feed the results onto a monitor for real-time display, or computer for subsequent analysis. The rate of data capture is limited chiefly by the digitiser sampling rate which is typically between 10 and 100 MHz. With pulse repetition rate of 1-10 Hz and the need to average over several shots to get a meaningful result, this gives a temporal resolution for the determination of the range-resolved ozone profile of the order of seconds. The vertical resolution of a DIAL system operating in the planetary boundary layer is typically around 50 m.

Figure 5.4

Time series of vertical profiles of ozone concentration as measured by DIAL. From Varey *et al.*[10]. Late at night ozone concentrations increase with height because ozone in the nocturnal boundary layer is being deposited at the surface. Ozone production near the surface during the day results in a negative gradient of ozone with height.

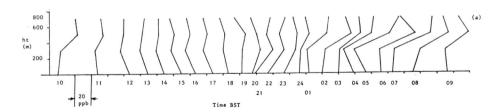

Figure 5.4 and Figure 5.5 give examples of the type of data available via DIAL studies. In Figure 5.4, a time-series of vertical profiles shows a diurnal cycle of ozone concentrations at the surface as we would expect from surface measurements. The negative gradients in the lowest 200m of the day-time profiles are indicative of photochemical ozone

Figure 5.5

Evening vertical cross-sections of O_3 observations using DIAL: (a) east-west, (b) south-north transects. From Ching et al.[15]. O_3 concentrations greater than 10ppbv above the local background value are indicated by black shaded areas. The spatial inhomogeneity of ozone levels is striking.

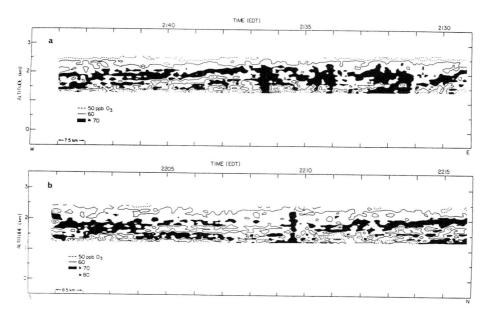

production at the surface. The positive gradients over this same range at night show the dominance of deposition processes in the stable nocturnal boundary layer. The nocturnal measurements document the appearance of a "reservoir" of ozone aloft which, after sunrise, mixes down into the layer depleted of ozone[10].

An example of the spatially-resolved data recorded via DIAL is given in Figure 5.5. This shows regions of elevated ozone concentrations in a part of the free troposphere above the planetary boundary layer (PBL) which had recently been the site of active cumulus convection[15]. The black areas represent regions where ozone levels are more than 10 ppbv above the background. These regions do show some degree of organisation and can be correlated with aerosol backscatter data collected by the same instrument. The position of regions of ozone excess are coincident with regions containing cloud debris as indicated by increased aerosol backscatter. This coincidence of excess ozone and cloud debris demonstrates the importance of convective cloud activity in ventilating the stagnant and polluted PBL during periods of anticyclonic weather (see chapter 6).

Other photometric methods have been employed to measure average ozone

Figure 5.6

Schematic of a differential optical absorption spectrometer (DOAS). From Plane and Nien[18]. Light from a xenon arc lamp is transmitted to a retro-reflector (corner-cube) situated some distance away, and returned for collection and dispersion onto a diode array.

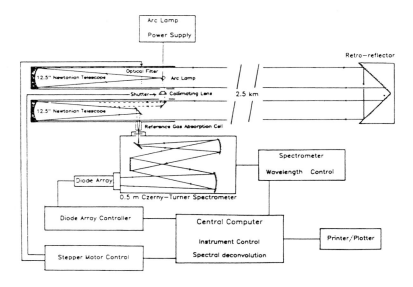

concentrations over a long path. The differential optical absorption spectrometer (DOAS) developed by Platt and Perner[16] has provided data on many species simultaneously, thus allowing a comparison with the output from a photochemical model[17]. A schematic of one such long-path system[18] is given in Figure 5.6. DOAS differs from DIAL in that: it uses a retro-reflector (which may be a topographic feature such as a mountain), which greatly increases the backscatter signal but removes any information on the spatial distribution of ozone along the path; and it uses many wavelengths, which can permit the identification of a molecule by its spectroscopic "fingerprint". A similar method based on atmospheric transmittance of laser light in the solar-blind *uv*, *i.e.*, that part of the solar spectrum which does not reach the ground (220-280nm), has been developed by Trakhovsky[19]. Trakhovsky's method foregoes a complex spectral fitting analysis, such as is used with DOAS, for an analysis based on the simple ratio of the atmospheric transmittance at two wavelengths. This greatly simplifies the analysis, but requires the aerosol extinction to be independent of wavelength over the wavelength interval used.

Remote sensing of the atmosphere, that is, measurement of species' concentrations far from the instrument, need not be earth-bound. Airborne platforms are possible, as we have seen in the case of lidar, and satellite platforms may also be used. Since 1970 the solar backscattered ultraviolet (SBUV) technique has been used to monitor ozone throughout the atmosphere and from 1978 - 1993 almost global coverage has been obtained using the Total Ozone Mapping Spectrometer (TOMS).

TOMS measures the amount of solar radiation scattered vertically upwards to the orbiting satellite. For the simple case of an overhead sun, a vertically downwards view, an atmosphere infinitely deep, and when attenuation of solar radiation by absorption due to ozone greatly exceeds that due to scattering, the flux of solar radiation at pressure level p is

$$S_\lambda(p) = S_\lambda(0) \exp(-k_\lambda \int_0^p [O_3]_{p'} \, dp') \qquad (5.3)$$

where $S_\lambda(0)$ is the solar flux at the top of the atmosphere at a given wavelength, λ, k_λ is the absorption coefficient at the wavelength and $[O_3]_p$ is the concentration of molecules per unit volume at pressure level p'. Then the total scattered radiance, L_λ, leaving the top of the atmosphere is

$$L_\lambda = a \, S_\lambda(0) \int_0^p \exp(-2k_\lambda \int_0^p [O_3]_p \, dp) \, dp \qquad (5.4)$$

where a is a constant of proportionality and the factor 2 is due to the equal attenuations of the solar radiation on the inward and outward paths[20]. The measured backscattered radiance, L_λ can therefore be related to the total column density of absorbing molecules. Satellite measurements of total ozone amounts using this technique have been shown to give data of comparable accuracy and precision to that of a carefully administered Dobson (column amount) measurement station at the surface[21].

Most atmospheric ozone exists in the stratosphere, however, and the TOMS data are primarily of interest for studying the climatology of that ozone layer. Nevertheless, when used in conjunction with vertically resolved datasets (such as the Stratospheric Aerosol and Gas Experiment (SAGE) satellite data) tropospheric residual ozone values can be calculated by subtraction[22]. Figure 5.7 shows the calculated average tropospheric residuals for low latitudes. The contours are given in Dobson units (where 1 DU = 2.69 x 10^{16} molec. O_3 cm^{-2}). A significant maximum occurs off the West coast of Africa in the dry season. This has been attributed to photochemical generation of ozone as a result of biomass burning[23, 24]. Indeed, in some cases the effect of photochemical oxidant generation in the troposphere is apparent in the raw TOMS data[25]. Two caveats are necessary, however. Firstly, the

Figure 5.7

Distribution of tropospheric ozone, calculated as the residual of two satellite data sets. From Fishman *et al.*[23]. Two seasons are shown: note the maximum during the dry season in the tropics. The contours are in Dobson Units (DU), with a contour interval of 4 DU.

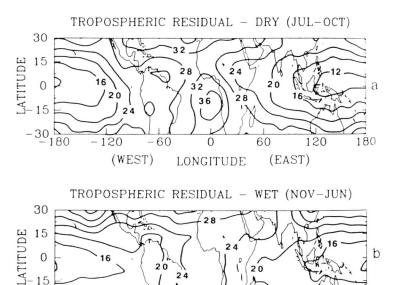

technique relies upon there being a reasonably well understood climatology of stratospheric ozone. This limits the region which can usefully be studied to the tropics, where the tropopause height is relatively constant. Secondly, the TOMS instrument cannot detect ozone below clouds. In the case of cloud cover the data retrieval algorithm adds an amount to the column values based on a tropospheric ozone climatology and an estimate of the cloud height. This can lead to erroneously high or low values when compared to *in situ* profile measurements[25]. But, since regional scale episodes of photochemical pollution occur in warm, clear anticyclonic conditions, this latter problem is not as significant as it might at first appear.

Prior to the adoption of *uv* photometry as the standard method for the routine monitoring of ozone in ground-level networks, gas-phase chemiluminescence was commonly used. This measures the light emitted from the products of the reaction of ozone with ethylene (see, e.g., ref. 5). A schematic diagram of a continuous chemiluminescence

Figure 5.8

Schematic diagram of a chemiluminescence ozone analyser. Air is drawn into the reaction chamber and mixed with excess ethylene (C_2H_4). The ethylene reacts with ozone and light is emitted which is detected by the photomultiplier (P/M) tube.

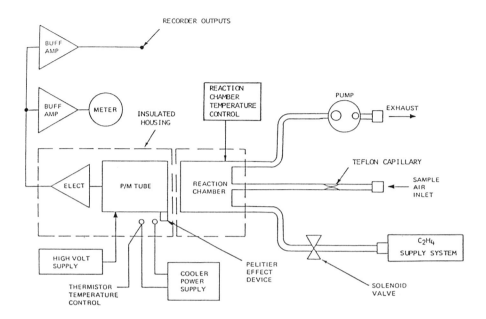

ozone analyser is given in Figure 5.8. Ambient air is drawn into the instrument through a teflon filter and mixed with excess ethylene in a reaction chamber. The light emitted by the reaction is measured using a photomultiplier tube[1]. The instrument is not a primary standard and regular calibration is required when it is in use. This is usually carried out using a *uv* photometric analyser as a transfer standard which is carried to the chemiluminescence analyser on site, and laboratory-based titration of ozone by NO as the absolute standard. A common source of signal-drift is contamination of the air inlet by ambient aerosol particles which can act as a sink for ozone[1, 5]. Variations of the method above usually involve the replacement of ethylene with another compound which will give rise to chemiluminescence when it reacts (specifically) with ozone. Early work used the dye rhodamine B dissolved on silica gel[1]. More recently, Ray *et al.*[26] suggested the use of eosin Y supported on a fibre pad and continuously replenished from a reservoir. Takeuchi *et al.*[27] used indigo - 5,5'- disulphonate in a similar system. The technique, in reverse, may also be used to measure alkenes (see § 5.6). Analysis of ozone by chemiluminescence has two advantages over the photometric technique: fast response and low sample flow-rate. The fast response of chemiluminescence monitors (typically around 1 Hz, about an order of magnitude faster than uv photometer monitors, makes them suitable for use in deposition studies where the fluctuations in ozone concentration are correlated to those of other meterological parameters (*e.g.* refs. 28, 29 and 30). The low sample flow-rate of chemiluminescence monitors also make them particularly useful in laboratory studies where measurements from a small volume (a leaf chamber, for example) are required.

Wet chemical methods are also available. These techniques are important historically because they were used in the earliest quantitative measurements of ambient oxidant levels (see ref. 31 and chapter 6). Such techniques may also have a role to play today in establishing low-cost monitoring networks around the world. In the nineteenth century, "Schönbéin" paper was used extensively to give semi-quantitative measurements of the atmospheric oxidant concentration. This method used the oxidation of iodide to iodine by ozone:

$$O_3(aq) + 2\ I^-(aq) + 2\ H^+(aq)\ \rightarrow I_2(aq) + O_2(aq) + H_2O(l) \hspace{2cm} [5.1]$$
$$I^-(aq) + I_2(aq)\ \rightarrow\ I_3^-(aq) \hspace{4cm} [5.2]$$

the tri-iodide ion formed complexes with starch to give a blue colour[31]. Unfortunately, a strong, non-linear dependence on relative humidity, together with an observed loss of colour for exposure times longer than about 1000 minutes (17h), make this technique unreliable.

A more accurate method, employed at Montsouris on the outskirts of Paris in the nineteenth century, used iodine as a catalysis for the oxidation of arsenite by ozone[31]

$$O_3(aq) + I^-(aq)\ \rightarrow\ IO^- + O_2(aq) \hspace{3cm} [5.3]$$
$$AsO_3^{3-}(aq) + IO^-(aq)\ \rightarrow\ AsO_4^{3-}(aq) + I^-(aq) \hspace{2cm} [5.4]$$

Although not subject to the humidity and ageing effects noted above for Schönbein paper, there are positive interferences from other atmospheric oxidants, such as hydrogen peroxide, and significant negative interferences from formaldehyde and sulphur dioxide. Reasonable estimates of the size of these interferences can be made, however, thus allowing historical ozone levels to be inferred from records of the time (see chapter 6).

The neutral KI technique[1] for ozone analysis has been in use for some time now and is the standard wet chemical method. The stoichiometry of the reaction is assumed to be:

$$O_3(aq) + 3\ KI(aq) + H_2O\ \rightarrow\ KI_3(aq) + 2\ KOH(aq) + O_2(aq) \hspace{1.5cm} [5.5]$$

and the amount of tri-iodide produced is measured by colorimetry. In the presence of large amounts of SO_2, which interferes with the measurement, a chromic acid filter is required.

The neutral KI technique has been used in several long-term measurement programs which have yielded important information on the apparent increase in annual-mean surface ozone concentrations over the last century[32, 33].

Several other reactions have been put forward as suitable for use in the determination of ambient ozone and/or total oxidants. Cohen *et al.*[34] suggested the oxidation of ferrous ammonium sulphate followed by colorimetric measurement of the $Fe(CN)_6^{3-}$ ion formed by the addition of ammonium thiocyanate. The technique is sensitive for ozone and most peroxy compounds and so gives a measure of the total oxidant loading. A similar total oxidants method was developed by Haagen-Smit and Brunelle[35] who used the oxidation of phenolphthalein to highly-coloured phenolphthalin. The reagents 1,2-di(4-pyridyl)ethylene or diacetyl-dihydro-lutidine are more selective for ozone[36, 37]. Chemical reactions which produce colour changes also form the basis of techniques designed to be used as passive sampling devices. Surgi and Hodgeson[38] demonstrated the sensitivity and selectivity of a sampler which used 10,10'-dimethyl-9,9'-biacridylidene. The reagent, phenoxazine, suggested by Lambert *et al.*[39] does also react with NO_2, but yields a different colour on reaction with ozone. Lastly, Williams and Grosjean[40] have employed both KI and phenoxazine coatings on annular denuders for the quantitative collection of ozone. Other oxidants could also be collected with varying degrees of selectivity depending on the coating used. Whilst the denuder technique is probably too labour-intensive to be useful as a method for ozone monitoring, it may prove important as a way of preventing on-filter reaction of collected particulate matter.

With such a diversity of measurement techniques - photometric, gas-phase chemical and wet chemical - measuring ozone under tropospheric conditions is not a problem. The techniques discussed above continue to be refined, of course, so that our confidence in the data will continue to improve. This improvement is particularly important where a rapid response is required, and where interferents may be present.

5.2 HYDROGEN PEROXIDE AND ORGANIC PEROXIDES

In one sense, hydrogen peroxide, H_2O_2, in the ambient air has been measured since the end of the last century: it is one of the species which contributes to the "total oxidants" signal as measured using the neutral KI technique. If these measurements of total oxidants were a true measure of the smog-forming potential and toxicity of secondary air pollution, they would be sufficient. As it is, the individual oxidants have such different chemical lifetimes and toxic effects that a measure of total oxidants is of limited value. Only recently have techniques become available for selective determination of gaseous hydrogen peroxide, and their accuracy is still being determined[41, 42, 43]. Because of its solubility, and its importance in the aqueous-phase oxidation of sulphur compounds, H_2O_2 must be measured in cloud and rain droplets as well as in the gas phase. In keeping with our topic, we discuss here only those techniques for the measurement of H_2O_2 in the gas phase: in fact, many of the gas-phase measurements rely on partitioning the vapour into an aqueous medium before analysis. A summary of the available techniques is given in Table 5.2[44, 45, 46, 47, 48, 49, 50, 51]. Some of the wet chemical methods outlined are also suitable for the determination of organic peroxides using a dual-channel inlet.

Currently the only technique for the analysis of H_2O_2 without prior dissolution is

Figure 5.9

Tunable diode laser absorption spectrometer (TDLAS) system suitable for mounting on an aircraft[52, 104]. Key M_{1-5}, beam directing mirrors; OAP_{1-5}, off-axis parabolic mirrors; M_F, in-focus white cell mirror; M_{d1-2}, out-of-focus white cell mirrors; M_c, corner reflector; BS, 90/10 beam splitter; RC1, reference line-locking cell; RC2, HNO_3 optical calibration cell; D_r, infrared detector (reference side); D_s, infrared detector (sample side); V, absolute pressure valve; IP, ion pump; MV, motorised throttle valve; He-Ne, Helium-Neon alignment laser.

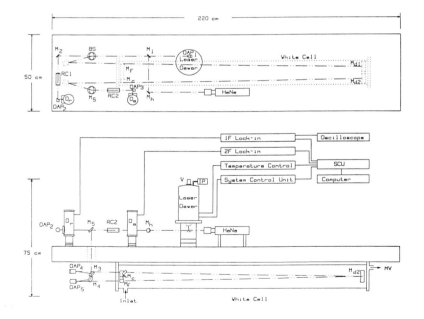

measurement by tunable diode laser absorption spectroscopy[52] (TDLAS; Figure 5.9). An atmospheric sample is pumped into an long-path optical cell under reduced pressure and a single rotational-vibrational line measured. With a sampling time of the order of minutes, a path length of about 40 m, and a cell pressure of about 25 torr, a detection limit of less than 1 ppbv can be achieved[53]. In a recent measurement campaign, Mackay *et al.*[54] found that they could measure gas-phase H_2O_2 with a detection limit of 0.15 ppbv, determined as the concentration from a calibration source that produced a signal to noise ratio of unity. An accuracy of ±20% was found for TDLAS when compared to the standard colorimetric technique.

Several wet-chemical techniques exist for the measurement of H_2O_2 and organic

Table 5.2: Techniques for the measurement of H_2O_2 in air. NR = not reported.

Method	Detection limit	Linear Range	Interferences	References
luminol chemiluminescence	< 1 ppb (aq)	≤ 1 ppm	PAN, SO_2, Fe, Mn, O_3	41, 56
peroxidase mediated depletion of scopoletin fluorescence	2 nM (aq)	NR	None reported	62, 44
peroxidase catalysed dimerisation of POPHA with fluorescence	12 nM (aq) 0.1 ppbv (g)	≤ 0.15 mM ≤ 128 ppbv	Hydroxylamine, ROOR', O_3, SO_2	58, 59
POPHA/MnO_2/nested loop	3 nM (aq) 30 pptv (g)	≤ 3 mM ≤ 30 ppbv	ROOR'	45
POPHA HPLC	5 nM (aq)	≤ 6 mM	None reported	43
peroxyoxalate chemiluminescence	8 nM (aq)	≤ 880 mM	Ca^{2+}, CO_3^{2-}, ethanol	46
fibre optic sensor	2.5 nM (aq)	≤ 80 mM	None reported	47
immobilised fluorophores, TCPO chemiluminescence	10 nM (aq)	≤ 100 mM	None reported	48
enzyme catalysed oxidation of DPD	10 nM (aq)	≤ 100 mM	SO_3^{2-}, ROOR'	49
spectrophotometry using pyridine-2,6-dicarboxylic acid and VO_3^-	1.5 nM (aq) 1 ppbv (g)	≤ 350 mM	O_3 > 2 ppm, Ti or Fe > 4 mg l^{-1}	55
iodometric titration	20 nM (aq)	NR	ROOR'	44
photometry using sodium N-ethyl-N-(sulfopropyl)-aniline salt	0.14 mM (aq)	≤ 40 mM	Transition metal ions > 3 mM	50
TDLAS	0.3 ppbv (g)	≤ 8 ppbv	None reported	53, 54
amperometric detection	0.15 mM (aq)	≤ 147 mM	NO_2^-, HCHO, HSO_3^- > 10 mg l^{-1}	51

peroxides (Table 5.2). These techniques operate by dissolving the gaseous sample and then reacting it with an indicator compound to produce a coloured[55], chemiluminescent[56, 57], or fluorescent[58, 59, 60, 61] product. Peroxides are stripped from the sample air using a cold trap, an impinger, or a diffusion scrubber[62, 63]. A cold trap consists of a clean glass tube which is cooled by immersion in an acetone-dry ice slush. An impinger consists of a vessel containing collection solution through which the gaseous sample is passed by means of a fritted tube. A diffusion scrubber consists of a long tube containing collection solution into which bubbles of sample are introduced. None of the stripping methods prevent interference by S(IV) compounds; impingers and diffusion scrubbers are also affected by ambient ozone which can produce artifact H_2O_2. The S(IV) interference can be eliminated by adding formaldehyde to the conditioning reagent[64]. Ozone may be removed from the system by the addition of excess NO upstream of the scrubbing device[59].

Addition of the enzyme catalase to one channel of a dual-channel analyser results in the destruction of H_2O_2 in that channel. This channel will then give a signal from the organic peroxides that are present, and the H_2O_2 concentration can be found by difference[58]. Only methyl hydroperoxide of the other hydroperoxides reacts with catalase, and this only slightly. The aliasing problem raised by methyl peroxide's reactivity can be minimised by adding only enough catalase to destroy 97 % of the hydrogen peroxide present. Such a system has been used in aircraft studies of the lower troposphere over the eastern United States[65, 66, 67]. Gas-phase H_2O_2 levels were found to be below 1 ppbv in general, with maximum values occurring above a cloud-topped boundary layer emphasising, as did the airborne ozone measurements reported above, the importance of convection through clouds in determining the chemical make-up of the lower free troposphere.

5.3 PAN COMPOUNDS AND OTHER ORGANIC NITRATES

Peroxyacetyl nitric anhydride (PAN), the most abundant of the atmospheric organic compounds containing nitrogen, has been measured in polluted urban air since the mid-fifties[68]. Monthly average values are generally between 1 and 10 ppbv in urban air[69], as discussed in chapter 6. Over the last decade, however, interest has focused on the role of PAN as a source of NO_x in the remote troposphere. Here the concentrations of PAN are generally less than 1 ppbv, perhaps as low as 10 pptv over the remote ocean[70]. New techniques have been developed to enable measurement of such low levels of the gas. These new methods generally employ chromatography and so cannot give a continuous signal. Nevertheless, some progress has been made towards automation[71].

The only photometric technique used to detect PAN is Fourier transform infrared spectroscopy (FTIR). This was the technique used in the early experiments and it continues to be of use in urban areas[72]. Several species (O_3, NO, NO_2, HCHO, HNO_3 and PAN, for example) can be measured simultaneously and *in situ* with a frequency of about 1 measurement per hour. Concentrations are calculated using the Beer-Lambert law (equation 5.1) though some care must be taken to ensure that appropriate values for the absorption coefficients are chosen[72]. A short, critical, review of current integrated band intensities from which absorption coefficients are calculated for PAN, has been given by Rogers[73].

A variant of the spectroscopic technique combines matrix isolation with FTIR[74]. A sample of air is frozen onto a cooled substrate in the presence of an inert matrix (which may, in fact, be the CO_2 already present in the sample). This prevents further reaction within the spectrometer and interactions between absorbing species. Because the trapped compounds cannot rotate within the very cold matrix, rotational structure is not present and vibrational transitions are seen as single lines. This simplifies the spectrum and hence increases the sensitivity of the FTIR spectroscopy. The ambient CO_2 which forms the isolation matrix can also be used as an internal standard for the spectroscopy. Typical detection limits are given in Table 5.3. FTIR is also suitable for the measurement of organic nitrates (that is, $RONO_2$, where R is a hydrocarbon moeity) but, even in moderately polluted air, organic nitrates occur in concentrations at, or below, the detection limit[72].

Table 5.3: Species which have been identified by matrix isolation Fourier transform infrared (MI-FTIR) spectroscopy, and their detection limits. From Griffith and Schuster[74]

Species	Detection Limits (pptv)	Species	Detection Limits (pptv)
N_2^*	30	CH_3CN	-
F-11	10	PAN	50
F-12	20	CH_2O	30
F-22	20	CH_3CHO	50
H_2O	30	HCOOH	20
OCS	5	CH_3COOH	30
SO_2	10	$(CH_3)_2CO$	50
CS_2	5	CH_3OH	-
NO	60	CH_3OOH^a	-
NO_2	5	$N_2O_5{}^a$	20
HNO_2	10	$ClONO_2{}^a$	10
HNO_3	10		

^{a} From laboratory spectra only.

Physico-chemical speciation of organic compounds containing nitrogen is possible using gas chromatography equipped with an electron capture detector[1, 75] (GC-ECD).

Chromatographic columns are usually short, packed, columns containing a polyethylene glycol stationary phase, although laboratory studies show that capillary columns may also be used[76]. Samples may be pre-concentrated using activated charcoal with benzene extraction[77] or a cryogenic trap. Inert materials are used to prevent sample loss or the formation of artifact PAN in the sampling system. Loss of PAN on the column, and detector sensitivity to relative humidity may both be significant, as discussed by Roberts[75]. An alternative procedure determines the PAN concentration by employing alkali hydrolysis in dilute KOH impingers. Ion chromatography is then used to determine acetate or nitrite[78, 79].

Commercial $NO-O_3$ chemiluminescence NO_x analysers give a quantitative response to PAN and so can be used as calibration devices for GC-ECD systems[80, 81], or as a measurement technique in their own right[6, 78]. Commercial NO_x analysers contain a molybdenum catalyst which is heated to about 310 °C. PAN compounds passing through

Figure 5.10

Calibration of a gas chromatographic PAN detector using a commercial NO_x analyzer. From Joos et al.[80].

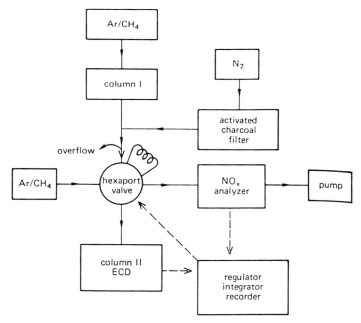

this catalyst are quantitatively decomposed to NO. Figure 5.10 shows the apparatus used by Joos et al.[80]: a sample of PAN is purified by passing through column 1, and detected using the NO_x analyser. When the PAN signal measured by the NO_x analyser is at a maximum the sample is diverted to the GC-ECD which is then calibrated by comparison with the output from the NO_x analyser. This system has the advantage that no storage of a PAN standard is required, but is rather cumbersome for use in the field. Blanchard et al.[81] use a simpler system in which the gas phase PAN standard is produced by bubbling N_2 through a solution of PAN in dodecane. More commonly, calibration has been carried out with stored gas phase PAN samples[75]. These stored samples are subject to degradation and extrapolation must be employed to determine the putative GC-ECD response to the PAN concentration in the bag at the time of its preparation[82].

Detection of PAN by NO-O_3 chemiluminescence[6, 78] is not sufficiently sensitive for routine monitoring, but a similar approach using a commercially available NO_2-luminol chemiluminescence detector[81] offers better prospects. The modified NO_2-luminol system is

Figure 5.11

Schematic diagram of PAN detection using a commercial NO_2-luminol detector. From Blanchard et al.[81]. Air is drawn past an NO_2 diffusion source and through a $FeSO_4$ catalyst to convert the NO_2 into NO. The sample is then passed onto a column to separate out PAN from other organic compounds. Once through the column, the PAN sample is converted to NO_2 by thermal decomposition. Some additional and constant quantity of NO_2 is then added immediately prior to detection.

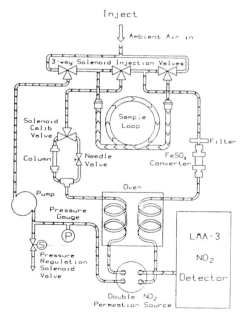

shown schematically in Figure 5.11. Conversion of PAN to NO_2 is carried out by pyrolysis. NO is added to the gas stream at a constant concentration to ensure quantitative conversion of PAN to NO_2 in the pyrolyzer and a constant source of NO_2 is added immediately upstream of the NO_2 detector to ensure a linear response. The PAN present in the air sample is isolated using a column packed with Carbowax 400 on Chromosorb G-AW just as in the GC-ECD technique. The system is calibrated using standard NO_2 sources although at some point the PAN/NO_2 conversion factor for the instrument must also be found and this will involve the use of gas-phase PAN standards.

The NO_2-luminol method described above has been compared to two GC-ECD systems[81], one of which was calibrated using stored gas-phase PAN standards, the other being calibrated by the $NO-O_3$ chemiluminescence technique. All three systems were found to be accurate to around $\pm$ 25%. The $NO-O_3$ chemiluminescence calibration technique performed well against the standard method, and measurements by the NO_2-luminol instrument were close to those of the two GC-ECDs. All three instruments had reported detection limits of 25 pptv. This suggests that a variety of techniques are now available with which to measure ambient PAN concentrations reliably down to the sub-ppbv levels found in all but the remotest parts of the troposphere[70]. Routine monitoring of other organic compounds containing nitrogen is not yet at such an advanced stage.

5.4 OH AND OTHER RADICAL SPECIES

The hydroxyl radical plays a central role in the chemistry of the polluted troposphere. It is commonly the first species with which primary pollutants react. RO_2 radicals are produced which can oxidise NO and shift the O_3-NO_x photostationary state away from values indicative of the clean troposphere (see chapter 4). However, because the rate of photolysis of ozone is relatively slow in the lower troposphere, and the rate of removal of OH by reaction with primary pollutants is fast, the ambient concentration of OH is very low. From modelling studies and measurements we expect OH levels to be of the order of 10^6 cm^{-3} ($\approx$0.1 pptv). Measurement techniques must, therefore, combine sensitivity, selectivity, and careful sampling to avoid loss of the short-lived radical. Instruments for the direct measurement of OH and which satisfy these criteria are generally expensive and difficult to operate. Results are then inevitably limited in terms of spatial and temporal coverage. Nevertheless, a growing body of data is appearing[83, 84]. If auxiliary measurements of other pollutants are also made, measured OH concentrations can become a stringent test of photochemical reaction mechanisms[17].

Two photometric techniques have been employed to measure tropospheric OH: fluorescence and absorption. The first of these is exemplified in the two-wavelength laser-induced fluorescence (LIF) analyser of Rodgers et al.[85], which is shown schematically in Figure 5.12. Two dye lasers are pumped by Nd:YAG lasers: one is set to an "on-line"

Figure 5.12

Two-photon laser induced fluorescence (LIF) instrument for the measurement of OH. From Rodgers et al.[85]. On- and off-line frequencies are sampled within 500 μs of each other to minimise noise due to fluctuations in the background atmospheric spectrum. Key: AMP, dye laser amplifier; AP, aperture; AT, laser beam attenuator; BC, beam combining mirror; BS, beam splitting mirror; CS computer; DPS, dye laser pressure scan systems; EMD, energy monitor diode; FP, collection optics/filter pack; GCI, grated charge integrator; GLP, glan laser polarizer; HVP, high voltage power supply; L, focusing lens; LP, printer; LSS, laser synchronization system; M, dichroic mirror; P, turning prism; PB, Pellin-Broca prism; PC, pocket cell; PMT, photomultiplier tube; PDL, pulsed dye laser; RC, reference cell; SHG, second harmonic generation crystal; T, beam expansion telescope; TD, trigger diode. The inlet, optical cell and PMT detectors are not shown.

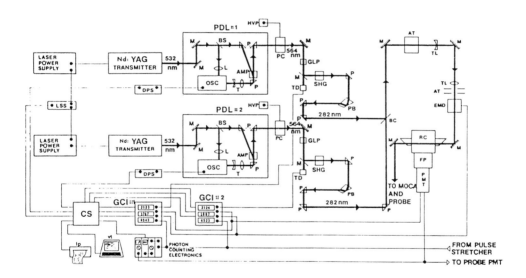

frequency in the OH spectrum: the other is set to an nearby "off-line" frequency. The fluorescence due to OH is then given by the difference between the total fluorescence, OH + background non-resonant fluorescence, at the on-line and off-line frequencies. The lasers and the detection electronics are computer controlled. Pulse duration and energy, together with the laser firing sequence, are adjusted to minimise interferences. Air is drawn into a cell (not shown) where the fluorescence is measured by means of photomultiplier tubes. Gaseous reagents can be injected into the cell to enable calibration in the field.

The LIF technique attempts to measure the weak resonant fluorescence of OH on top of a large and fluctuating background signal, caused by non-resonant fluorescence from the chamber walls, organic vapours and atmospheric aerosol particles. Errors arising from uncertainties in the background signal can be minimised by using a very short time interval between the on- and off-line measurements. The LIF technique is also subject to an inherent interference from artifact OH production by the laser source. Artifact OH production occurs because the electronic transitions of OH in its vibrational groundstate are at frequencies shorter than those required to photolyse O_3 and produce $O(^1D)$:

$$O_3 + h\nu(\lambda = 282 \text{ nm}) \rightarrow O(^1D) + O_2 \qquad\qquad [5.6]$$

Most of the $O(^1D)$ produced is quenched, but some will react with ambient water vapour to produce artifact OH.

$$O(^1D) + H_2O \rightarrow 2 \text{ OH} \qquad\qquad [5.7]$$

$$O(^1D) + M \rightarrow O(^3P) + M \qquad\qquad [5.8]$$

This interference is, therefore, unavoidable in the application of the LIF technique to atmospheric monitoring, although it can be minimised by the use of laser pulses of low energy density and short duration. Interference by both background signal and artifact OH production may also be lessened by reducing the pressure of the sample. This is the basis of fluorescence assay with gas expansion (FAGE) as outlined by Hard et al.[86]. The background fluoresecence is relatively insensitive to pressure. It therefore decreases in importance, at low pressure, relative to the easily quenched OH fluorescence. Artifact OH production is reduced because the rate of the bimolecular reaction [5.7] is reduced at low pressure. However, photochemical modelling of the FAGE system[87] has demonstrated that there can also be significant OH formation in this system. The modellers go on to suggest improvements which could be made to the technique which would reduce artifact production. These are: reduction of the laser energy; realignment of the optical cell; reduction of the laser bandwidth; and the use of deuterated isobutane to remove ambient OH for background measurements. These points have now been addressed[88] and the use of a laser which can deliver a higher frequency of pulses (which can therefore be of lower energy density) has been shown to substantially reduce in situ OH formation.

As with PAN analysers, the calibration of OH detectors is complicated by the fact that stable standards cannot be synthesised and stored. In the laboratory, controlled irradiation of H_2O_2 or an O_3/H_2O mixture can produce known amounts of OH provided that the absorption cross sections and quantum yields are known: i.e.

$$H_2O_2 + h\nu(\lambda = 266 \text{ nm}) \rightarrow 2 \text{ OH} \qquad [5.9]$$

or reactions [5.6] - [5.8] can be used. Both methods have been shown to agree closely, and the O_3/H_2O makes a convenient field calibration technique[85]. Alternatively, the rate of removal of a hydrocarbon species which reacts exclusively with OH from an air sample containing the radical, can be used for calibration[86].

The second photometric technique used to monitor hydroxyl radicals, laser long-path absorption spectroscopy (LPA), measures the absorption of laser light by OH at 307.995nm

Figure 5.13

Laser long-path absorption spectrometer for the measurement of OH. From Dorn *et al.*[91]. The technique is similar to that of the differential optical absorption spectrometer (DOAS) shown in Figure 5.6.

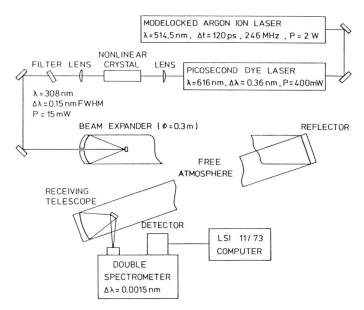

over a path of several kilometres[89, 90] (Figure 5.13). In the instrument shown, a laser light source is expanded to a diameter of 30 cm (thus ensuring a negligible production of OH from the laser itself), reflected off a mirror or retro-reflector 2.9 km away, and received by a photodiode array detector adjacent to the laser source[91]. Figure 5.14 shows a typical atmospheric spectrum, reference spectra for OH and SO_2, and the residual spectrum after the removal of the SO_2 signal. As is obvious from the figure, the SO_2 spectrum must be

Figure 5.14

Measurement of tropospheric OH by laser long-path absorption spectroscopy. From Dorn *et al.*[91]. Top-trace: air spectrum recorded in Julich, Germany, integrated over 100 measurements during 30 minutes; lower trace: residual air spectrum after subtraction of SO_2 absorption (the absorption of pure SO_2 is shown in the upper central spectrum). The absorption due to OH in the residual air spectrum can be seen by comparing the spectrum with that of OH alone (lower central spectrum).

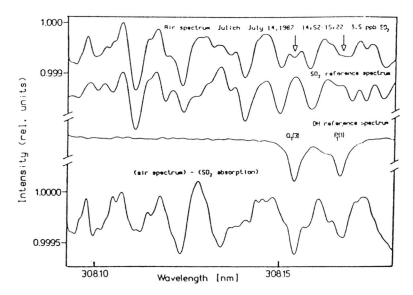

removed in order to obtain a reasonable value for the absorbence due to OH. Carbon disulphide and formaldehyde can also interfere with the OH signal but are not usually present in concentrations high enough to be significant. Calibration comes directly from the Beer-Lambert law. This gives a detection limit of approx 2.4×10^6 molecules cm^{-3} over a 5.8 km path with an integration time of 30 minutes, a sufficiently low value for the measurement of OH in the middle of the day when its photochemical production is maximum. Such measurements have now been performed and compared to model calculations[17]. The model tended to over-predict OH concentrations, particularly when measured OH concentrations were themselves high. This may suggest that there are important sinks for OH which have yet to be included in model calculations.

The decay of a compound by reaction with OH has been used to calibrate LIF systems[86] but may also be used as a measurement technique in its own right. Radioactive ^{14}CO is injected into the sample where it is oxidised by the available OH:

$$^{14}CO + OH + O_2 \rightarrow {}^{14}CO_2 + HO_2 \qquad [5.10]$$

The ^{14}CO is then quantified using proportional counting (refs. 83, 92 and references

Figure 5.15

Schematic of a quartz tube flow reactor for the measurement of OH[92]. The reaction vessel is transparent in the uv so that the photostationary state of ambient air, and hence the OH concentration, is maintained.

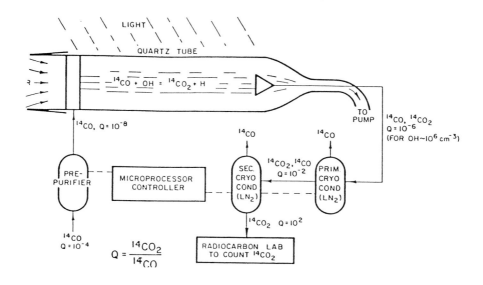

therein). Figure 5.15 shows a flow reactor for OH detection by this method. Ambient air enters a quartz tube and sampled downstream of an array of ^{14}CO injecting nozzles. The quartz tube is transparent to photolytic radiation, thus ensuring that the ambient photostationary state is maintained. After sampling from the centre of the reactor, ^{14}CO$_2$ is collected and concentrated in a series of cryogenic traps. Typically about 30 ppbv of ^{14}CO is added to the flow; the air takes about 12 s to flow downstream to the sampler; and the sample is collected for 100 s or so. The detection limit is determined by uncertainties in the ^{14}CO tracer purity, the separation of the product ^{14}CO$_2$ after sampling, the ^{14}CO$_2$ activity determination, and the ambient ^{222}Rn background. It has been claimed that, with removal of ^{222}Rn, a detection limit of $\approx 10^4$ molecules cm^{-3} can be attained[93]. Such a limit would allow measurement of nocturnal concentrations and will contribute greatly to our understanding of tropospheric chemistry should it be achieved.

The ^{14}CO technique assumes that OH remains in photochemical steady state within the reaction vessel. The technique of Eisele and Tanner[94] differs from the ^{14}CO technique in that it rapidly quenches the OH radical concentration within the sample volume by adding a large excess of $^{34}SO_2$:

$$SO_2 + OH + M \rightarrow HSO_3 + M \qquad\qquad [5.11]$$

The HSO_3 formed is then rapidly converted to H_2SO_4:

$$HSO_3 + O_2 \rightarrow SO_3 + HO_2 \qquad\qquad [5.12]$$

$$SO_3 + H_2O + M \rightarrow H_2SO_4 + M \qquad\qquad [5.13]$$

and the H_2SO_4 converted to a negative ion

$$H_2SO_4 + NO_3^- \rightarrow H_2SO_4^- + NO_3 \qquad\qquad [5.14]$$

prior to detection by ion mass spectrometry. The ratio of $H_2SO_4^-$ to NO_3^-, together with a knowledge of the rate of reaction [5.14], gives the H_2SO_4 concentration. It is assumed that all this (isotopically labelled) H_2SO_4 is due to the reaction of SO_2 with OH, and so the OH concentration can be calculated. Ion mass spectrometry is an extremely sensitive detection technique, allowing the measurement of concentrations of the order of 1 ion cm^{-3}. The uncertainty in measurements of OH by this technique lies not in the sensitivity of the detector, but in the specificity of the complex chain of reactions leading from the species which is of interest (OH) to the species which is measured ($H_2SO_4^-$). The first of a series of gas injectors ensures that the OH initially present in the sample reacts exclusively with SO_2, whilst the second injector adds propane to quench any OH produced from HO_2 within the reactor. NO_3^- ions are generated in a separate air flow and then fed into the sample: this avoids any other ions reacting with the H_2SO_4 in the sample. The system as a whole has a detection limit of the order of 10^5 molecules cm^{-3}, has a rapid response (mass spectrometric signals are integrated for 5 s in the raw output), and so can see fluctuations in the ambient OH concentration due to transient incidents such as clouds passing overhead.

OH concentrations can be implied from measurements of other, more abundant or more easily measured, species. Such indirect measurements have been reviewed by Hewitt and Harrison[83] and by Altshuller[84], and will not be reiterated here. The decay of hydrocarbon species in the plume of the 1991 Kuwaiti oil fires is a good example of the use of indirect measurements to imply OH concentrations[95]. Scaling tropospheric OH fields to observed methyl chloroform levels has proved to be a convenient way of parameterising oxidant chemistry in global three-dimensional models[96].

The measurement of radicals other than OH has not received much attention. The important night-time oxidant NO_3 can be measured by DOAS[97, 98], using an apparatus like that in Figure 5.6. The differential absorption of NO_3 is measured using the absorption maximum at 662 nm, and the absorption minima at 650 nm and 674 nm. The concentration of NO_3 present then comes directly from the Beer-Lambert Law.

Matrix isolation techniques have been used in the measurement of radical species as well as for organic nitrates as discussed earlier (§ 5.3). Mihelcic et al.[99, 100] describe a technique which traps NO_2 and RO_2 (where R = H or an organic group) in an ice matrix for subsequent analysis by electron spin resonance (ESR). A typical measurement of an atmospheric sample is shown in Figure 5.16: the RO_2 signal becomes apparent after removal of the spectrum associated with the cold finger itself and that of NO_2. Using D_2O to form

Figure 5.16

Deconvolution of the electron spin resonance spectrum of an atmospheric sample[100]. The composite spectrum (A) is an average of 49 scans. It was recorded in a D_2O matrix at 77 K and a modulation amplitude of 2 Gauss. (B) is the intrinsic signal from the sample holder. (C) = (A-B). (D) NO_2 reference spectrum. (E) = (C-D)x5. (F) H_2O reference spectrum (x5). (G) residual (x10).

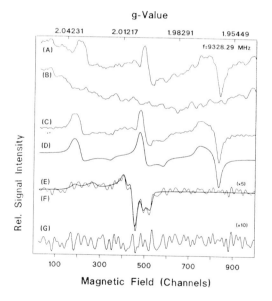

the ice matrix, the spectra of HO_2, NO_3 and CH_3COO can be distinguished from other (principally alkyl) peroxy radicals. An alternative technique for radical measurement using ESR has been put forward by Watanabe et al.[101], in which OH radicals are trapped by reaction with a spin-trapping reagent. The reported detection limit of about 2 x 10^5

molecules cm^{-3} is quite adequate for tropospheric measurements, but the method does suffer from rather large imprecision which may limit its usefulness[83].

By the very nature of radical species, their detection is never likely to become routine. However, the position of radical species as reactive intermediates in photochemical oxidation means that their measured concentrations, and particularly measured concentrations of OH, will always be the ultimate test of photochemical theories. No completely satisfactory techniques yet exist for the measurement of radicals at sub-pptv concentrations in the atmosphere. That said, the progress that has been made on several fronts, ranging from purely optical techniques to sophisticated chemical-transformation-and-detection techniques, suggests that those relatively sporadic measurements which are performed will become increasingly reliable.

5.5 OXIDES OF NITROGEN

The distribution of oxides of nitrogen, particularly in relation to the distribution of hydrocarbons, determines whether ozone is produced or destroyed in the troposphere (see Fig. 4.7). The speciation of nitrogen compounds in an air sample is also important: total odd-nitrogen (NO_y) measurements must be sub-divided into measurements of reactive oxides which take part in the ozone photostationary state ($NO_x = NO + NO_2$), reactive intermediates (e.g. NO_3 and HONO), and reservoir species (e.g. HNO_3, PAN, organic nitrates and N_2O_5). Aspects of the detection of each kind of nitrogen compound are discussed below, but we will concentrate principally on the generic species NO_x and NO_y. Organic compounds containing nitrogen have been discussed in § 5.3 and that discussion will not be reiterated here. Note that the simultaneous measurement of NO_x and NO_y enables the concentration of reservoir compounds to be calculated by simple molar subtraction, since the concentrations of reactive intermediates are, by definition, small.

Photometric detection of NO_x is possible using absorption spectroscopy. Hanst et al.[72] describe the use of long-path FTIR to measure several components of photochemical smog simultaneously (see § 5.3). Combining FTIR spectroscopy with matrix isolation techniques gives greater sensitivity: NO and NO_2 may then be measured down to detection limits of 60 pptv and 5 pptv, respectively[74] (Table 5.3). Reactions within the matrix are virtually absent so that the NO_2/NO ratio in the sample will be maintained in the matrix. The principal difficulty with the method is designing a quantitative sampling system. Sampling is usually cryogenic: air is dried by passing through a tube held at about 250 K and then trapped in a vessel held at 77 K. Both NO and O_3 have appreciable vapour pressures at 77 K, so they will not be entirely trapped.

Biermann *et al.*[102] report the use of FTIR for the measurement of nitric acid at concentrations above $\approx$ 5 ppbv (*i.e.* urban air). Diode lasers and low pressure, multi-pass optical cells may also be used: the TDLAS technique (Figure 5.9) measures NO_x compounds using rotational-vibrational lines in the 1900 cm^{-1}(NO) and 1600 cm^{-1}(NO_2) regions[103, 104, 105]. Detection limits are about 40 pptv and 25 pptv, respectively, for a five minute measurement cycle. Accuracies for both species are estimated to be ±15% (ref. 103). The instrument is calibrated for NO using a standard gas bottle. It is calibrated for NO_2 using a permeation tube, that is, a tube made of a permeable material and containing condensed NO_2. The rate of NO_2 permeation through the tube wall is a known function of temperature, so a calibration gas stream can be generated by passing a known flow of clean air over the permeation tube as it is maintained at a known temperature in an oven. Individual measurements of each gas can be carried out very rapidly: typically measurements are made with a frequency of 10 Hz or so, and averaged over the residence time of air in the cell. The outputs of the diode lasers are sensitive to temperature and so care must be taken to ensure that they remain in a thermally stable environment at between 20 and 80 K. Nevertheless, if this can be achieved, the system is sufficiently robust and reliable to be left unattended for 2-3 days at a time[105], and sufficiently light to be mounted on an aircraft[52, 104]. Nitric acid vapour may also be measured by TDLAS, although losses in sampling can be much larger than for the NO_x compounds[105, 106] since nitric acid deposits very efficiently onto surfaces.

Differential absorption lidar (DIAL; see § 5.1), using dye, CO or CO_2 lasers, can be used for both NO_x species. For example, Galle *et al.*[107] studied the ventilation of an urban area by using DIAL to map the distribution of NO_2. DOAS spectroscopy has been used to study several compounds including NO_2, HONO and NO_3 (refs. 16 and 102).

Two proton LIF has also been used to detect NO, NO_2 can also be detected if the instrument is fitted with a photolytic converter[108]. The NO detection system, which is similar to that used for OH and shown schematically in Figure 5.12, produces laser light at wavelengths of 226 nm and 1.1 µm to carry out rotational-vibrational transitions in the sample NO. The excited NO then relaxes back to the ground-state by fluorescence. Nitrogen dioxide is analysed by photofragmentation, using laser light, and subsequent detection of the NO produced by LIF. Detection limits are in the low pptv range for an integration time of 2 minutes. No interferences are reported for NO measurement: NO_2 detection is subject to small interferences due principally to the photolysis of compounds other than NO_2 in the photolytic converter, and due to the reaction of the product NO with ambient ozone which has been drawn into the instrument. A similar technique for the measurement of nitric acid by fluorescence has been reported by Papenbrock and Stuhl[109]: HNO_3 is photolysed to excited OH and the subsequent fluorescence monitored. The detection limit is around 100 pptv for a 10 minute sampling cycle, and no major interferences have been reported to date.

At long-term monitoring sites nitrogen oxides are usually measured by chemiluminescence. A proportion of the reaction of NO with O_3 produces excited NO_2 molecules

$$NO + O_3 \rightarrow NO_2 + O_2 \qquad\qquad [5.15]$$
$$NO + O_3 \rightarrow NO_2^* + O_2 \qquad\qquad [5.16]$$

which may be quenched collisionally, or emit light:

$$NO_2^* + M \rightarrow NO_2 + M + heat \qquad\qquad [5.17]$$
$$NO_2^* \rightarrow NO_2 + hv \qquad\qquad [5.18]$$

A schematic diagram of a chemiluminescence monitor based on this principle is given in

Figure 5.17

Schematic of an ozone-NO chemiluminescence monitor for the detection of oxides of nitrogen[1] based on a Thermo electron instrument.

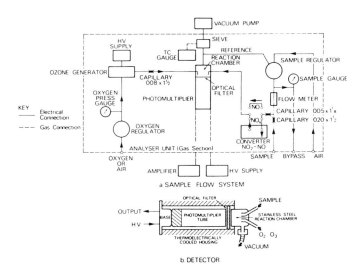

Figure 5.17. Ambient air flowing into the reaction chamber is mixed with excess ozone and

the resulting emission measured using a photomultiplier tube. The signal detected is a function of the intensity of emitted light (I), the efficiency of light collection (G) and the sensitivity of the collector (A): $S = GAI$ (ref. 110). The efficiency, G, is maximised by using a polished, gold-coated, reaction chamber to reflect the luminescence. The sensitivity, A, is maximised by using a low-noise IR-sensitive photomultiplier tube. The intensity, I, is maximised by using a fast flow of air at reduced pressure and containing a large amount of ozone.

Oxides of nitrogen other than NO can be measured by inserting a catalytic converter upstream of the reaction cell. The type of converter used depends on the measurement required[111]. Specific conversion of NO_2 only, could it be achieved, results in a NO_x measurement and the concentration of NO_2 is found by subtraction of the NO-only signal. A general catalyst converts all the oxides of nitrogen present to NO and hence gives a measure of NO_y. Molybdenum catalysts are generally used as NO_2 converters in commercial instruments, but have been shown to convert other nitrogen-containing compounds with varying efficiencies[6, 78]. Photolysis at wavelengths longer than 320 nm can be made to be highly specific for NO_2 but is not 100 % efficient and care must be taken to correct the detector signals for artifact NO_2 formation in the photolysis cell[112]. Ferrous sulphate crystals will quantitatively convert NO_2 to NO but are also efficient converters of PAN and other nitrogen compounds. Gold catalysts quantitatively convert all NO_y compounds to NO without interference from NH_3 and N_2O (Ref. 113). In an effort to circumvent the non-specificity of catalytic converters, Braman et al.[114] have carried out an extensive survey of denuder tube coatings and deduced a sequence that can give selective preconcentration of HNO_3, HONO, NO_2 and NO. These can then be measured individually at the sub-ppbv level using thermal desorption and a chemiluminescence detector fitted with a gold catalyst. Some nitrogen-free species can also interfere with chemiluminescent measurements of NO and NO_x (refs. 115 and 116), probably by the generation of chemiluminescence on reaction with ozone in the reaction chamber, and/or by the quenching of electronically excited NO_2. This quenching effect will be particularly significant when measuring combustion products at source[116].

Without modification a commercial NO_x analyser manufactured in the past 10 years will have a detection limit of about 2 ppbv for NO and NO_x. Customised instruments, however, can perform significantly better: Tanner[110] reports a limit of 0.2 ppbv for NO and NO_x; Ridley et al.[112] report limits of below 0.01 ppbv (10 pptv), also for both gases. Detection limits of this magnitude are required particularly for measurements in the remote troposphere.

Analysis of NO$_2$ by the ozone chemiluminescence technique is indirect because the NO$_2$ must be converted into NO prior to detection. An alternative chemiluminescent reaction, that of NO$_2$ with luminol, is specific for NO$_2$. A schematic of the detector is given in Figure 5.18[117]. Reagent solution continuously flows across a filter paper which is exposed

Figure 5.18

Schematic of a luminol detector for the measurement of NO$_2$. From Wendel *et al.*[117]. Light from the reaction between NO$_2$ and luminol is detected by a photomultiplier tube (RCA 1P28).

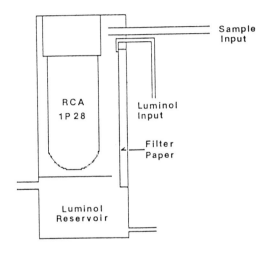

to a flow of ambient air. Chemiluminescence occurs at the gas-liquid interface and is detected by a photomultiplier tube. Reagent solution draining from the filter paper is recycled. The reagent contains luminol (5-amino-2,3-dihydro-1,4-phthalazinedione), sodium sulphite, sodium hydroxide and methanol in aqueous solution at concentrations which maximise the detector's sensitivity to NO$_2$ and minimise interference by O$_3$ (ref. 117). This solution and the filter paper across which it flows, must be replaced regularly to preserve sensitivity. The time constant for the detector is around 0.5 s (for a 20 ppbv modulation of the NO$_2$ signal), and the detection limit around 30 pptv. Just as for the ozone-NO chemiluminescence instrument, the addition of catalytic converters upstream enables the measurement of other oxides of nitrogen. The performance of the systems as a monitor for NO depends on the efficiency of the oxidising catalyst employed to convert NO to NO$_2$ prior to detection. A modification to the instrument which enables it to detect

PAN quantitatively, has been described in § 5.3.

Intercomparisons of the spectroscopic and chemiluminescent techniques[105, 118, 119, 120] have shown that:

(i) at high NO_x levels (> 2 ppbv, say) TDLAS, $NO-O_3$ chemiluminescence with photolytic or catalytic convertor, and luminol-chemiluminescence all agree well[105, 118]. At lower NO_x concentrations the luminol technique must be corrected for interference by O_3 and PAN (refs. 118 and 121).

(ii) $FeSO_4$, when used as a converter in the ozone-chemiluminescence method, converts PAN, as well as NO_2, to NO and so the measurement is not specific for NO_2 (ref. 119).

(iii) no interferences are reported for the ozone-chemiluminescence technique with photolytic converter. A small adjustment to the signal is required to compensate for artifact NO_2 produced within the instrument[118].

(iv) At low NO_x levels (≤ 50 pptv), the TDLAS, LIF and chemiluminescence methods are less well correlated for NO_2 (ref. 119). For NO concentrations below 80 pptv, measurements by LIF and chemiluminescence agree to within the estimated precision and accuracy of the instruments.

(v) The TDLAS system appears to overestimate NO_2 concentrations near its detection limit, probably as a result of finding spurious correlations between the background noise and the reference spectrum.

We can conclude that, whilst further work is needed to improve the performance of all systems in the low pptv range, it is already evident that several well-characterised techniques are available for the measurement of NO_x in the troposphere. These can be used in routine monitoring networks or airborne measurement campaigns, thus allowing a complete picture of the anthropogenic and biogenic influences on global photochemistry to be built up.

The techniques discussed above appear to be the most promising for the accumulation of data with the temporal resolution that is required for the constraint of detailed numerical models. Other techniques are available for the measurement of oxides of nitrogen over longer time periods. The selective preconcentration and chemiluminescence technique of Braman et al.[114] described earlier, is an example of these integrated measurements although it, and the matrix-isolation and electron spin resonance spectroscopy technique[100, 101] (see § 5.4), still provide data averaged over only an hour or less. The spectrophotometric techniques of Nash[122], Baveja et al.[123], and Raman[124] also fall into this category of measurements with resolution of the order of hours. Filter packs and denuder and diffusion tubes, on the other hand, typically provide measurements integrated over a day or more.

Filter packs fitted with nylon filters are the current standard method for nitric acid determination[125, 126, 127]. Air is drawn through the filter at a regulated rate and the sample returned to the lab for analysis by wet chemical or chromatographic procedures. The technique is simple, robust and inexpensive, but not free from technical difficulties. As for any method that involves air sampling, losses of HNO_3 in the inlet and on the walls of the filter pack can be significant. In addition, the equilibrium

$$HNO_3(g) + NH_3(g) \; (+ \; H_2O(l)) \; \rightleftarrows \; NH_4NO_3(s,aq) \tag{5.19}$$

may be disturbed during sampling, causing artifact HNO_3 to be produced.

Nitrous acid (HONO) may also be measured indirectly using two KOH filters in series[128]: a nitrite signal in the analysis of the first filter is produced by HONO, NO and NO_2; the signal from the second filter will be due to NO and NO_2 only. The level of HONO is given therefore, by subtraction. Triethanolamine (TEA) impregnated filters may also be employed[129]. This allows simultaneous determination of sulphur dioxide and nitrogen dioxide. PAN has been discovered to interfere strongly with the measurement of NO_2 using this reagent, however[130]. HONO may also be measured by dissolving it, and feeding the scrubbing solution, containing dissolved NO_2^- from HONO, into an ion chromatograph[131]. Dissolution takes place in a diffusion scrubber to maximise the contact between the scrubbing solution and the ambient air. Calibration is via the well characterised thermal decomposition of NH_4NO_2 into NH_3 and HONO.

Base-coated denuder tubes are also used extensively for the measurement of nitric and nitrous acids[126, 132, 133]. Air is drawn through a tube, or the annular space between two tubes. The tube walls are coated with a sink for the compound to be measured. The compound is stripped from the airstream by molecular diffusion to the walls, and subsequently analysed in the same way as a filter sample. The concentration of the compound escaping from the end of the denuder, [A], decreases rapidly as the flow rate, F, inside the denuder is decreased, and as the length of the tube, L, is increased[134]:

$$\frac{[A]}{[A]_o} = 0.819 \exp\left(-14.6272 \frac{DL}{4F}\right) \tag{5.5}$$

where $[A]_o$ is the concentration of the compound entering the tube; and D is the gas phase diffusion coefficient of the compound.

Denuder tubes have the advantage that particulates and other pollutant gases pass through unperturbed and can be collected separately down-stream. This avoids the possibility of artifact nitric acid production from evaporating NH_4NO_3 aerosol particles because evaporation of the aerosol particles occurs over a much longer timescale than transport through the denuder[135, 136, 137]. Again, as with filters, denuders can be connected in series for the collection of different compounds. The denuder-chemiluminescence techniques of Braman et al.[114] and Roberts et al.[138] work on this principle. Detection limits of better than 1 pptv can be achieved by both filters and denuders for 24 hour samples analysed using ion chromatography[125, 139]. Some automation of both methods is possible using a system of electrically-actuated valves or, in the case of denuder tubes, by continuously supplying fresh absorbent to the surface of the tube[139].

Average ambient NO_2 concentrations can be measured cheaply and accurately using Palmes' passive diffusion tubes[140, 141]. These devices consist of acrylic tubes (7 cm in length and 1 cm internal diameter) open to the air at one end and containing TEA-coated stainless steel discs at the other. When in operation, either indoors[142, 143] or outdoors[144], NO_2 diffuses into the tube and is collected by the coated disc. Subsequent analysis for nitrite in the coating is by spectrophotometry[140] or ion chromatography[145]. The technique's accuracy is determined by blank measurements of around 100 ppbv h and detection limits of approximately 300 ppbv h (ref. 146). The blank values appear to be due to desorption of NO_2 from the acrylic tube and diffusion of ambient NO_2 through the tube and cap during storage. The high detection limit restricts the use of diffusion tubes to long-term monitoring at polluted sites, and to determining the exposure of individuals to pollution in the workplace (e.g. Refs. 147, 148, 149).

It appears, then, that oxides of nitrogen, and in particular the NO_x compounds, can be measured over a great variety of spatial and temporal ranges. The standard ozone-chemiluminescence technique is fast, robust and, if not exactly lightweight, sufficiently compact to fit conveniently alongside other pollutant monitors in mobile laboratories. It is specific for NO, but care must be taken to fully characterise the specificity of catalytic converters when these are employed to measure other oxides of nitrogen. The best technique for the catalytic conversion of NO_2 remains a matter of some dispute.

Laser spectrophotometric techniques for the measurement of NO_2, such as FTIR and DOAS, are inherently more specific than the ozone-chemiluminescence measurement, but they are expensive, bulky, and require expert operators. At the other extreme stripping methods, such as filters, denuders and passive diffusion tubes are cheap, highly compact and straightforward to use. Automatic valve control can be used to enable the stripping methods to give information on a useful timescale, especially with regard to the validation of chemical mechanisms (which are, after all, simply a concise description of our understanding of atmospheric chemistry). Furthermore, it is becoming apparent from toxicological studies that the effects of exposure are determined to some extent by the time over which the dose is administered (chapter 7). Should this prove to be the case then the diversity of measuring techniques described above will prove very useful. Indeed, with the advent of new techniques - thin film sensors, for example[150] - and the continuing refinements of current methods, comprehensive measurements of NO_x compounds on all scales of interest should become possible. The techniques for other nitrogen oxides are less well developed, with wall losses in sample lines continuing to be a major problem.

5.6 NON-METHANE HYDROCARBONS

If the techniques employed to measure oxides of nitrogen are remarkable in their diversity, the techniques employed for the other major group of photo-oxidant precursors, the hydrocarbons, are singularly uniform. In almost all cases hydrocarbons are collected, separated by gas chromatography (gc), and then detected by flame-[151, 152, 153, 154] or photo-ionisation[155, 156]. Total hydrocarbon amounts can be monitored continuously using a flame ionisation detector (FID) without prior separation on a gc column[1] but, because of the wide range of hydrocarbon reactivities, such data are of limited use in the discussion of photochemical oxidant production.

The collection of hydrocarbons has been carried out using three approaches: whole air samples may be collected using metal canisters (*e.g.* s. 155 and 157) or a variety of bags, aspirators and syringes made from inert materials such as glass or PTFE (ref. 1); hydrocarbons may be adsorbed onto porous adsorbents (*e.g.* ref. 151) or activated carbon (*e.g.* refs. 153 and 154); or cryogenic traps can be employed (*e.g.* ref. 152). None of these methods is ideal[156]: canisters and other containers may not collect a sufficiently large volume to enable the gases with lowest concentrations to be measured; adsorption of hydrocarbons onto solid traps may not be completely reversible or, if reversible, may require temperatures that will bring about decomposition or rearrangement of less stable gases; and cryogenic traps are inconvenient because of the need for a supply of coolant such as liquid nitrogen or Argon. Perhaps, though, the overriding problem for all the methods of collection is that of contamination. All collecting devices, of whatever kind, must have a very low *blank* (*i.e.*, hydrocarbon concentration when empty) and be

Table 5.4: Relative change in the composition of air collected in containers equipped with a metal bellows sealed valve with a Kel-F tip compared to containers with all stainless steel valves. Average and variance from five pairs of measurements.

Compound	Relative Change (%)	Variance (%)
C_2H_6	+0.5	0.7
C_2H_2	-0.2	2.0
C_3H_8	-1.0	1.0
C_3H_6	+1.0	4.0
$n\text{-}C_5H_{12}$	+2.0	3.0
$n\text{-}C_6H_{14}$	+88.0	30.0
$n\text{-}C_7H_{16}$	+28.0	15.0
C_6H_6	+7.0	9.0
F-12	+2.0	1.0
F-22	+80.0	26.0
F-113	+6.0	1.0
CH_3Cl	+12.0	7.0
CH_2Cl_2	+6.0	1.0
$CHCl_3$	+9.0	8.0
CCl_4	+0.6	1.0
C_2HCl_3	+113.0	27.0
C_2Cl_4	+92.0	70.0
CH_3CCl_3	+11.0	9.0

demonstrably free from *memory effects* (*i.e.* a sample is not influenced by those previously collected in the same device). The primary cause of contamination is contact of the air sample with organic polymer in the form of seals, gaskets, tubing, etc.[156]. Table 5.4 demonstrates this effect. Two - and three-carbon compounds appear to be affected less strongly, but very significant artifact signals can be seen for the longer chain compounds.

The inherently small sampling volume of canisters can be increased by filling them to a higher pressure, using metal-bellows pumps to avoid contamination. Other problems remain, however. In the time between sampling and analysis there may be loss of analyte to the walls of the vessel, or reaction of analyte with another constituent of the whole air sample. Passivated stainless steel canisters perform well on both counts particularly if treated with hydrocarbon-free water. This water is believed to compete for the active sites which trap hydrocarbon molecules on the surface of the steel. Passivated aluminium canisters have also been tested[158] but with no greater success. Interference with the FID response is another potential problem. This was observed during the measurement of toluene and some chlorinated organics in a sample containing a large amount of HCl (such as might be found in waste incinerator effluent[159]).

Figure 5.19

Automated drying system for use in the detection of non-methane hydrocarbons by gas chromatography. From Pliel *et al.*[161]. The flow of zero air and the heater are controlled by separate relays. Air entering the analytical system passes through the un-heated drier, removing water vapour from the sample. Prior to taking the next sample, the drier is heated and purged with dry air.

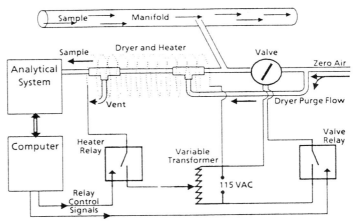

When using cryogenic traps, or hydrophilic adsorbents such as activated charcoal, ambient water vapour will be trapped and care must be taken to prevent this water from clogging the trap, blocking the gc column, or extinguishing the FID hydrogen flame. With very small samples, such as those collected in a section of cooled capillary column[160] saturation of the column is not a problem. Using such small samples reduces the sensitivity of the technique, however. Alternatively, a drying procedure may be adopted. Passing the sample through a bed of dessicant, such as $Mg(ClO_4)_2$ or K_2CO_3, can lead to large changes in sample composition[156] and the dessicant itself can become less efficient over time unless they are periodically reconditioned or replaced. Permeable membranes overcome both the changing composition and variable efficiency problems. Sample air is separated from a sheath of dry air by a membrane that is permeable to water and other light, polar, compounds. Water vapour in the sample diffuses through this membrane and is vented. Figure 5.18 shows such a drying device that has been improved by application of heat and dry zero air prior to sample injection[161]. This procedure also reduces memory effects due to the dryer. However, some doubt has been cast on the integrity of the sample once passed through the dryer[162].

It is also possible to remove water vapour prior to sample collection. An application of this technique to the detection of monoterpenes in forest air[163] is shown in Figure 5.20.

Figure 5.20

Cryotrap device used in the analysis of monoterpenes in humid and ozone rich air. 1 = Whatman glass-fibre filter; 2 = ozone scrubber; 3 = glass coil for the condensation of water; 4 = cartridge provided with ground glass joint and Luer lock; 5 = tenax bed of cartridge for collection of monoterpenes; 6 = insulated box filled with crushed ice; 7 = silicone rubber tube; 8 = diaphragm pump; 9 = flow meter; 10 = gas meter, H = outlet. From Juttner[163].

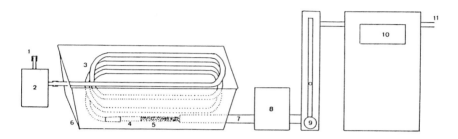

The absorbent bed and a long inlet tube are immersed in crushed ice. Ambient water vapour will then condense out onto the walls of the inlet tube leaving the absorbent sufficiently dry to prevent clogging of the trap, etc. Figure 5.20 also illustrates the use of an ozone scrubber in the sampling line. This will prevent the degradation of absorbed hydrocarbons by reaction with ozone, although some loss of polar compounds in the scrubber has been reported[166].

Immediately prior to separation on the gc column, the sample is preconcentrated, or cryofocussed as it is often termed. Preconcentration involves condensing the hydrocarbons from a large volume of sample air to a much smaller volume in a cold trap. This can be carried out externally to the gc column[154] or on the top of the column if it can be cooled to sub-ambient temperatures[151, 160]. External preconcentration loops may make use of porous glass beads, porous silica, alumina or graphite as adsorbents[156]. After preconcentration, the sample is injected into the gc proper by withdrawing the cryogenic liquid and flushing the warmed preconcentration loop. Placing a movable dewar around the preconcentration loop is one way to facilitate this procedure but it is not convenient to use in an automated system. An alternative device is described by Rudolph et al.[162] and is shown in Figure 5.21.

Figure 5.21

Cryogenic preconcentration device for use in the measurement of non-methane hydrocarbons by automated gas chromatography. The level of liquid nitrogen in the flask is controlled by the vent valve. From Rudolph *et al.*[162].

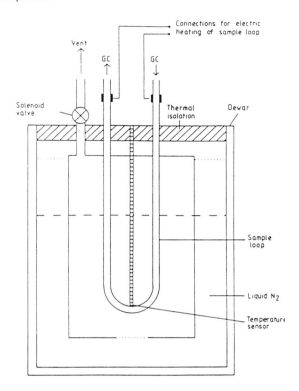

When the vent shown is open, liquid nitrogen bathes the preconcentration loop condensing the sample. Closing the vent increases the vapour pressure over the liquid and reduces its volume. The exposed tube is then heated electrically to evaporate the preconcentrated sample. Vapour released through the vent can be recondensed and recycled thus enabling unattended operation. A similar device has been described by McClenny *et al.*[164].

Once on the column, gas chromatographic separation can take place. However, the large range of retention times of the hydrocarbons in an ambient air sample makes the use of a single column impractical. In order to achieve optional separation of light (C_2-C_5) components, the heavier fractions must be removed. Often a "backflush" technique is employed; a short pre-column retains the heavier hydrocarbons whilst allowing the lighter fraction to elute onto the main column (*e.g.* refs. 154, 156 and 165). The heavier components are then removed from the pre-column which is cleansed using a reverse flow

Figure 5.22

Complete system for the analysis of non-methane hydrocarbons by gas chromatography [156]. I-IV are multiport valve; 1-6 are shut-off valve; R = pressure regulator; and A, B, C are needle valves. a) sample preconcentration in a cold loop (T ≈ 80 K). b) sample injection and pre-separation on the first column. c) separation of light fraction on second column and transfer of heavier fraction from the first column to the capillary column (3) d). separation of heavier fraction on the capillary column and re-conditioning of the preconcentration loop.

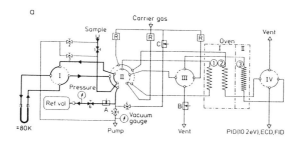

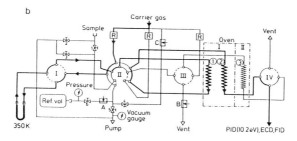

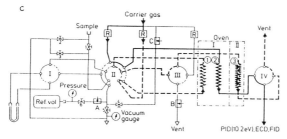

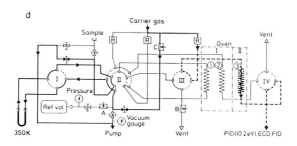

213

of eluent. This fraction may then be fed onto a third column which is optimised to separate the heavier hydrocarbon compounds[156], or discarded. Columns packed with porous silica, modified porous silica or porous polymers, such as porapak Q or QS; are generally used for the separation of lighter hydrocarbons[154, 155, 156]; capillary columns appear to be more appropriate for the analysis of the heavier components[151, 155, 156]. A fully integrated system is illustrated in Figure 5.22. At the left of the diagram, the sample preconcentration loop can be seen (Figure 5.22a). The sample is then passed on to a pre-column (Figure 5.22b) where it is divided into two fractions using a "backflush" technique. The light hydrocarbon fraction elutes first from column 2 (Figure 5.22c), and is followed by the output from the third column (Figure 5.22d) as the preconcentration equipment is reconditioned.

The light alkanes (C_2-C_5) which are important in the formation of photochemical oxidants, are best detected by flame ionisation[152, 154, 160]. This detector may also be used to measure unsaturated compounds such as monoterpenes[151] although Rudolph and Jebsen[165] have found that photoionisation detectors (PID) are between 3 and 20 times more sensitive for the detection of alkenes and aromatics. Electron capture detectors (ECD) are many orders of magnitude more sensitive to halogenated compounds than FID or PID. Maximum information can be extracted from a sample, therefore, by connecting two, or more, of these detectors in parallel[153], or in series[156, 165]. Figure 5.23 gives an example of output from FI-, PI- and EC-detectors connected in series[165] (the FID must be placed last in the series, of course). The superior sensitivity that PID has over FID for aromatics is exemplified in the signals for benzene in Figure 5.23a and Figure 5.23d. Comparing Figure 5.23a and Figure 5.23b similarly illustrates the enhanced sensitivity of ECD for the detection of halogenated organic compounds.

Combining all the above steps, from sampling, through preconcentration, dehydration, pre- and main column separation, to detection yields a system capable of determining a wide range of hydrocarbons at levels down to a few pptv[151, 154, 155, 156, 164]. This is an admirable achievement and, with the development of automated systems[154, 162, 164] routine measurements are now beginning to be made. It is likely, then, that gas chromatography will remain the standard technique for the measurement of airborne hydrocarbons. Recently, however, a technique for selective and continuous monitoring gaseous isoprene has been described[166]. It is shown schematically in Figure 5.24. The principle of the technique is the same at that used to measure ozone by chemiluminescence (§ 5.1) but with sample and reactant gases reversed. Air containing isoprene is drawn into a reaction chamber where it is mixed with excess ozone. On reaction,

Figure 5.23

Chromatograms obtained from a 1.7 *l* air sample detected by photoionization (PID), electron capture (ECD) and flame ionization (FID) devices in series[165]. Notice the different sensitivities of each device to individual compounds. a) FID trace; b) ECD trace; c) PID trace with a 11.7 eV lamp.

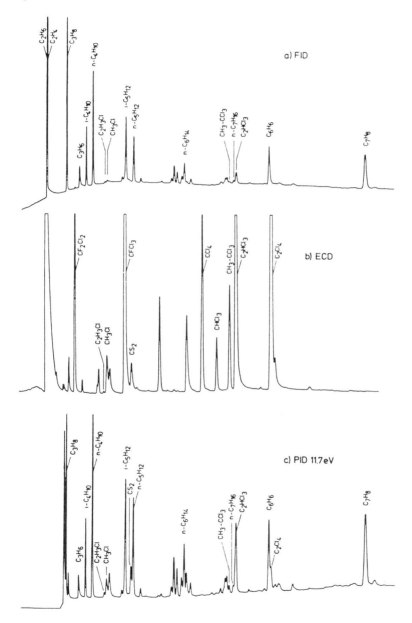

Figure 5.24

Schematic of a chemiluminescence detector for the continuous measurement of isoprene[166]. The technique is the same as the chemiluminescent detection of ozone shown in Figure 5.8, but with sample and reagent gases reversed. FC = flow controller; PMT = photomultiplier tube; P = pump.

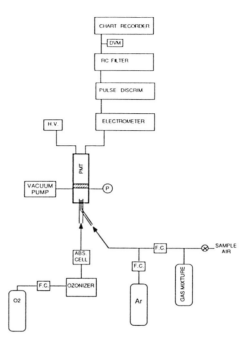

products are formed in excited states and decay to their ground states with emission of a photon. This chemiluminescence may then be monitored using a photomultiplier tube which is sensitive in the blue end of the spectrum (425-550 nm) but not in the red, where $NO+O_3$ chemiluminescence occurs. The instrument is linear over three orders of magnitude, has a fast response, and can detect isoprene down to around 400 pptv: isoprene levels are generally expected to be of this order or less except very close to emitting vegetation. The chemiluminescence technique is surprisingly selective given the general nature of the ozone-alkene reaction. Only ethene and propene react with sufficient rapidity to interfere with the isoprene measurement. If these are present in significant quantities they can be removed from the sample using a cold trap. Isoprene will condense into this trap at a much higher temperature than either ethene or propene. The technique can, of course, also be used to measure other alkenes[167, 168].

To summarise, gas chromatographic determination of non-methane hydrocarbons is at an advanced stage of development. Columns and detectors are well characterised and their advantages and disadvantages for the determination of different hydrocarbons well known. Sample collection remains a problem, however, with no one method being clearly preferred. Sample integrity after the removal of water from the sample is also a matter of some dispute. If and when these technical difficulties are overcome, the detection of hydrocarbons down to very low levels will be made possible. That said, the expense and intricacy of current gc instruments for hydrocarbon measurement means that regional monitoring networks may yet be some way off.

5.7 OXYGENATED HYDROCARBONS

Aldehydes and ketones are important intermediates in the atmospheric photo-oxidation of hydrocarbons (chapter 4) and are released directly by anthropogenic sources (chapter 3). Formaldehyde, in particular, is of interest because it is the most abundant gaseous carbonyl in the atmosphere, is a suspected animal carcinogen (chapter 7), and is an important source of free radicals via photolysis (chapter 4). The proposed introduction of methanol fuelled vehicles would add an extra source of formaldehyde in urban areas because it is present in the fuel and is also formed as an intermediate product of the oxidation of methanol. The discussion below will focus on techniques for measuring formaldehyde but will also touch briefly on the detection of organic acids which are the products of the oxidation of aldehydes and ketones. A recent comparison of the techniques for measuring formaldehyde is reported in detail in a special issue of *Aerosol Science and Technology* devoted to the results of the Carbonaceous Species Methods Comparison Study[169].

There are three photometric methods in use for the determination of formaldehyde concentrations in ambient air: FTIR, TDLAS, and DOAS. Both FTIR (see § 5.3) and TDLAS (see § 5.2) operate in the infrared region of the spectrum, FTIR using an optical path of about 1 km in the ambient atmosphere, and TDLAS using a low pressure White cell. Like FTIR, DOAS uses a long optical path at atmospheric pressure (see § 5.1). Absorbance measurements are taken in the near-*uv* and visible regions of the spectrum (339 nm for HCHO). Overlapping absorption features of NO_2 and HONO spectra must be subtracted from the DOAS measurement to isolate the absorbance due to HCHO alone. This is easily accomplished given the unique spectroscopic "fingerprint" of each compound. Detection limits of the FTIR and DOAS systems are both a few ppbv[169]; the TDLAS

217

Figure 5.25

Comparison of formaldehyde measurement techniques. The plots show a time series of measurements of formaldehyde in the ambient air of Glendora, CA, USA. At the time of the study this location was experiencing the most severe ozone episodes in the South Coast Air Basin of Southern California[169]. a) differential optical absorption spectroscopy (DOAS), Fourier transform infrared spectroscopy (FTIR), and tunable diode laser absorption spectroscopy (TDLAS) measurements of ambient formaldehyde concentrations. b) the average spectroscopic (DOAS/FTIR/TDLAS) measurement of formaldehyde, and that from enzyme-fluorescence method. c) the average spectroscopic measurement of formaldehyde, and that from the diffusion scrubber method.

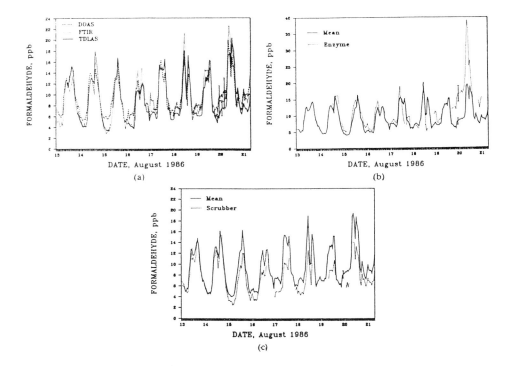

instrument can achieve a detection limit as low as 0.15 ppbv (ref. 54) because of the accuracy with which the formaldehyde absorbence can be measured at low pressure. A comparison of measurements from the three techniques is shown in Figure 5.25a. The formaldehyde concentrations shown are typical of those found in polluted areas which can experience severe photochemical smog episodes. Clearly the spectrophotometric methods all agree well although there is some evidence that the DOAS instrument overestimates nocturnal HCHO levels at or near its detection limit[169]. No statistically significant differences were found between the FTIR and TDLAS systems operating with sampling times of between 3 and 5 minutes. This implies that the sampling of ambient air into the low pressure cell of the TDLAS instrument caused no response time effects on this timescale. Overall these results suggest that it is reasonable to use any or all of these methods as calibration standards for other, non-spectroscopic, techniques in the same way that *uv* photometers are used for ozone (§ 5.1).

The non-spectroscopic methods which require occasional calibration of this kind generally involve stripping formaldehyde from the ambient air and into aqueous solution. This can be achieved using a mixing coil[170], a diffusion scrubber[171], or a absorbent cartridge[172]. One such analytical set-up is shown in Figure 5.26: air containing ambient

Figure 5.26

Schematic of an enzyme-fluorescence instrument for the detection of formaldehyde. From Lazrus et al.[170]. Formaldehyde is partitioned into solution in the stripping coil and then mixed with reagents to produce a fluorescent from of NADH. The measured fluorescence then gives a direct indication of the amount of formaldehyde in solution and hence in the original air sample.

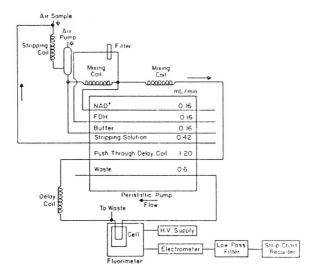

amounts of HCHO is drawn into a mixing coil where it dissolves into aqueous solution. After separation from the gaseous residue, the liquid fraction is mixed with reagents to produce, in this case, a fluorescent product:

$$HCHO(g) + H_2O(l) \rightarrow H_2C(OH)_2(aq) \qquad [5.20]$$

$$H_2C(OH)_2(aq) + FDH(aq) \rightarrow HCOO^-(aq) + H_3O^+(aq) + H_2FDH(aq) \qquad [5.21]$$

219

$$H_2FDH(aq) + NAD^+(aq) \rightarrow NADH(aq) + FDH(aq) \qquad [5.22]$$

where FDH is the enzyme formaldehyde dehydrogenase which catalyses the reduction of nicotinamide adenine dinucleotide (NAD^+) to its fluorescent form, NADH, by formaldehyde[170]. Reaction of formaldehyde with the Nash reagent gives an alternative product which can be detected fluorometrically[171] or electrochemically[173]. A strongly absorbing compound can be produced by further reaction of the NADH:

$$NADH + INT \xrightarrow{\text{diaphorase}} NAD^+ + \text{formazan} \qquad [5.23]$$

where INT is the reagent 2-(4-iodophenyl)-3-(4-nitro-phenyl)-5-phenyltetrazolium[174]. Such systems typically have detection limits of a few tenths of a ppbv, and can be set up to monitor HCHO continuously. The results of an intercomparison between two fluorescence techniques - one employing a diffusion scrubber and the Nash reagent, the other as shown in Figure 5.26 - and the three spectrophotometric methods discussed above, are shown in Figure 5.25b and Figure 5.25c. The two wet chemical techniques do not agree as well as the three spectroscopic techniques and there appears to be a systematic bias in each wet chemical technique: positive in the case of the mixing coil/NADH technique; and negative for the diffusion scrubber/Nash reagent technique[169]. These systematic biases are attributed to difficulties in maintaining an optimal system over a sustained period.

Cartridges filled with absorbent materials can also be used to strip formaldehyde from ambient air[169, 172]. The filling material is coated with phosphoric acid, and 2,4-dinitrophenylhydrazine (DNPH) which is converted to hydrazones on exposure to aldehydes. The DNPH derivatives are then separated and measured by high performance liquid chromatography (HPLC). Measurable amounts of formaldehyde can be collected from a few hundred litres of air so that sampling times are generally of the order of 1 hour. Detection limits are typically around 1 ppbv or less. When compared to spectroscopic measurements, the DNPH cartridge method agreed tolerably well, but appeared to give systematically low values, especially over longer sampling periods[169]. Nonetheless, the simplicity and relative cheapness of the technique make it a useful one.

Formaldehyde may also be absorbed onto sulphite-coated glass-fibre filters[175]. Large amounts of formaldehyde can be collected selectively and without artifact formation allowing the determination of isotope ratios by mass spectrometry. The ratio of ^{13}C to ^{12}C can be used in source apportionment studies since the $^{13}C/^{12}C$ ratio of formaldehyde produced from methane oxidation differs from that due to NMHC oxidation. The technique is sensitive to ambient humidity, however, and only preliminary data have been reported to date.

Organic acids are measured using techniques similar to those used for the aldehydes. FTIR can measure formic acid in ambient air using its Q-branch absorbence at 1105 cm^{-1}. The detection limit is about 2 ppbv over a 1 km path[176]. Alternatively, alkaline traps may be used for routine monitoring across a network. As with the DNPH cartridge technique described above, subsequent analysis is by chromatography, such as liquid chromatography with *uv* detection. Detection limits of a few tenths of a ppbv can be achieved using this technique. Both aldehydes and PAN compounds are potential interferents, however[177]. In the alkaline conditions of the trap aldehydes may disproportionate via the *Cannizzaro* reaction:

$$2 \text{ RCHO} + \text{OH}^- \rightarrow \text{RCH}_2\text{OH} + \text{RCOO}^- \qquad [5.24]$$

and PAN compounds may decompose to release carboxylate:

$$\text{RC(O)OONO}_2 + 2 \text{ OH}^- \rightarrow \text{RCOO}^- + \text{NO}_2^- + \text{O}_2 + \text{H}_2\text{O} \qquad [5.25]$$

These interferences appear to be negligible for filter traps, but may be of some significance when using cartridges. A recent comparison of the alkaline trap and FTIR techniques revealed a good agreement without noticeable bias[178].

In general, it can be said that techniques for measuring ambient oxygenates are becoming increasingly well characterised, so that it is unlikely, for instance, that determining formaldehyde levels will be a problem in future measurement campaigns. Spectroscopic techniques are now sufficiently accurate to be used as calibration standards. The accuracy of cheap and simple techniques, however, may yet have to improve if they are to be used in routine monitoring networks. In this sense the state of the art in formaldehyde detection is typical of that for most of the products and precursors involved in photo-oxidant chemistry (except, perhaps, volatile organic compounds): complex and highly sophisticated spectroscopic techniques can identify absorbances due to tens of pptv of some compounds from the myriad absorbances occurring simultaneously across even a small spectral region. These systems are probably more important as bench-marks than for the measurements they can provide on their own and, as such, allow us to gauge the accuracy and precision of simpler methods which can then be installed over a wider area in monitoring networks.

1. Harrison, R.M. and Perry, R. (eds.) (1986) *Handbook of Air Pollution Analysis*, 2nd Edition, Chapman and Hall, UK.

2. Finalyson-Pitts, B.J. and Pitts, J.N. (1986) *Atmospheric Chemistry: Fundamentals and Experimental Technique*, John Wiley and Sons, New York, USA.

3. Allegrini, I., and F. De Santis (1989) Measurement of atmospheric pollutants relevant to dry acid deposition, Crit. Rev. Anal. Chem., 21, 237-255.

4. Ayers, G.P., and Gillett, R.W. (1990) Tropospheric chemical composition: overview of experimental methods in measurement, Rev. Geophys., 28, 297-314.

5. Elsworth, C.M. and Galbally, I.E. (1984) Accurate Surface Ozone Measurements in Clean Air: Fact or Fiction, in *Proc. 8th International Clean Air Conference*, Clean Air Society of Aus. and N.Z.

6. Grossjean, D. and Harrison, J. (1985) Peroxyacetyl Nitrate: Comparison of Alkaline Hydrolysis and Chemiluminescence Methods, Environ. Sci. Technol., 19, 749-752.

7. Measures, R.M. (1988) *Laser Remote Chemical Analysis*, John Wiley and Sons, N.Y., USA.

8. Browell, E.V. (1989) Differential Absorption Lidar Sensing of O_3, Proc. IEEE, 77, 419-432.

9. Varey, R.H. (1986), in Harrison and Perry[1] (eds.), *Handbook of Air Pollution Analysis*, 2nd Edition, Chapman and Hall, UK.

10. Varey, R.H, Ball, D.J., Crane, A.J., Laxen, D.P.H. and Sandells, F.J. (1988) Ozone Formation in the London Plume, Atmos. Environ., 22, 1335-1346.

11. Ancellet, G., Papayannis, A., Megie, G. and Pelon, J. (1988) Tropospheric Ozone Lidar Measurements, in *Tropospheric Ozone*, I.S.A. Isaksen (ed.), D. Reidel, Dordrecht, Holland.

12. Browell, E.V., Ismail, S. and Shipley, S.T. (1985) Ultraviolet DIAL Measurements of O_3 Profiles in Regions of Spatially Inhomogeneous Aerosols, Applied Optics, 24, 2827-2836.

13. Papayannis, A., G. Ancellet, J. Pelon, and G. Megie (1990) Multiwavelength lidar for ozone measurements in the troposphere and the lower stratosphere, Appl. Opt., 29, 467-476.

14. Itabe, T., Asai, K., Ishizu, M., Aruga, T. and Igarashi, T. (1989) Measurements of the Urban Ozone Vertical Profile with an Airborne CO_2 DIAL, Appl. Opt., 28, 931-934.

15. Ching, J.K.S., Shipley, S.T. and Browell, E.V. (1988) Evidence for Cloud Venting of Mixed Layer Ozone and Aerosols, Atmos. Environ., 22, 225-242.

16. Platt, U. and Perner, D. (1983) in Killinger and Mooradian, A. (eds.) *Optical and Laser Remote Sensing*.

17. Perner, D., Platt, U., Trainer, M., Hübler, G., Drummond, T., Junkermann, W., Rudolph, J., Schubert, B., Volz, A., Ehhalt, D.H., Rumpel, K.J. and Helas, G. (1987) Measurements of Tropospheric OH Concentrations: A Comparison of Field Data with Model Predictions, J. Atmos. Chem., 5, 185-216.

18. Plane, J.M.C., and C.-F. Nien (1992) A differential optical absorption spectrometer for measuring atmospheric trace gases, Rev. Sci. Instrum., 63, 1867-1876.

19. Trakhovsky, E. (1985) Ozone Amount Determined by Transmittance Measurements in the Solar-blind Ultra-violet Spectral Region, Appl. Opt., 24, 3519-3522.

20. Houghton, J.T., F.W. Taylor and C.D. Rodger (1984) *Remote sounding of atmospheres*, Cambridge University Press, Cambridge, U.K.

21. Bharta, P.K., K.F. Klenk, C.K. Wong, D. Gordon, and A.J. Fleig (1984) Intercomparison of the NIMBUS 7 SBUV/TOMS total ozone data sets with Dobson and M83 results, J. Geophys. Res., 89, 5239-5247

22. Fishman, J., and J.C. Larson (1987) Distribution of total ozone and stratospheric ozone in the tropics: implications for the distribution of tropospheric ozone, J. Geophys. Res., 92, 6627-6634.

23. Fishman, J., C.E. Watson, and J.C. Larson (1989) The distribution of total ozone, stratospheric ozone and tropospheric ozone at low latitudes deduced from satellite data sets, *Ozone in the Atmosphere*, (proc. Quad. Ozone Symp. 1988), R.D. Bojkov and P. Fabian (eds.), A. Deepak, Hampton, VA, U.S.A., pp411-414.

24. Fishman, J. (1991) Probing planetary pollution from space, Environ. Sci. Technol., 25, 612-620.

25. Fishman, J., F.M. Vukovich, D.R. Cahoon, M.C. Shipham (1987) The characterisation of an air pollution episode using satellite total ozone measurements, J. Clim. Appl. Meteor., 26, 1638-1653.

26. Ray, J.D., Stedman, D.H. and Wendel, G.J. (1986) Fast Chemiluminescent Method for Measurement of Ambient Ozone, Anal. Chem., 58, 598-600.

27. Takeuchi, K., Kutsuna, S. and Ibusuki, T. (1990) Continuous Measurement of Ozone in Air by Chemiluminescence using Indigo-5,5'-disulphanate, Anal. Chim. Acta, 230, 183-187.

28. Gusten, H., H. Gunther, R.W.H. Schmidt and U. Schurath (1992) A novel ozone sensor for direct eddy flux measurements, J. Atmos. Chem., 14, 73-84.

29. Neumann, H.H. and Den Hartog, G. (1985) Eddy Correlation Measurements of Atmospheric Fluxes of Ozone, Sulfur and Particulates during the Champaign Intercomparison Study, J. Geophys. Res., 90, 2097-2110.

30. Droppo, J.G. (1985) Concurrent Measurements of Ozone Dry Deposition using Eddy Correlation and Profile Flux Methods, J. Geophys. Res., 90, 2111-2118.

31. Kley, D., Volz, A. and Mulheims, F. (1988) Ozone Measurements in Historic Perspective, in *Tropospheric Ozone*, I.S.A. Isaksen (ed.), D. Reidel, Dordrecht, Holland, 63-72.

32. Volz, A. and Kley, D. (1988) Evaluation of the Montsouris Series of Ozone Measurements made in the Nineteenth Century, Nature, 332, 240-242.

33. Feister, U. and Warmbt, W. (1987) Long-term Measurements of Surface Ozone in the GDR, J. Atmos. Chem., 5, 1-21.

34. Cohen, I.R., Purcell, J.C. and Altshuller, A.P. (1967) Environ. Sci. Technol., 1, 247.

35. Haagen-Smit, A.J. and Brunelle, M.F. (1958) The Application of Phenolphthalin Reagent to Atmospheric Oxidant Analysis, Int. J. Air Pollut., 1, 51-59.

36. Hauser, T.R. and Bradley, D.W. (1966) Specific Spectrophotometric Determination of Ozone in the Atmosphere using 1,2-di-(4-pyridyl)ethylene, Anal. Chem., 38, 1529.

37. Nash, T. (1967) Colorometric Determination of Ozone by Diacetyl-dihydro-lutidine, Atmos. Environ., 1, 679-687.

38. Surgi, M.R. and Hodgeson, J.A. (1985) 10,10'-dimethyl-9,9'-biacridylidene Impregnated Film Badge Dosimeters for Passive Ozone Sampling, Anal. Chem., 57, 1737-1740.

39. Lambert, J.L., Liaw, Y-L and Paukstelis, J.N. (1987) Phenoxazine as a Solid Monitoring Reagent for Ozone, Environ. Sci. Technol., 21, 503-505.

40. Willams, E.L. and Grossjean, D. (1990) Removal of Atmospheric Oxidants with Annular Denuders. Environ, Sci. Technol., 24, 811-814.

41. Kleindienst, T.E., Shepson, P.B., Hodges, D.N., Nero, C.M., Arnts, R.R., Dasgupta, P.K., Hwang, H., Kok, G.L., Lind, J.A., Lazrus, A.L., MacKay, G.I., Mayne, L.K. and Schiff, H.I. (1988) Comparison of Techniques for Measurement of Ambient Levels of Hydrogen Peroxide, Environ. Sci. Technol., 22, 53-61.

42. Lawson, D.R., Dasgupta, P.K., Kaplan, I.R., Sakugawa, H., Kok, G.L., Heikes, B.G., MacKay, G.I., Schiff, H.I., Tanner, R.L. (1988) Inter-Laboratory Comparison of Ambient H_2O_2 Measurements in Los Angeles, Abs. Amer. Chem. Soc., 196, 86.

43. Gunz, D.W. and Hoffmann, M.R. (1990) Atmospheric Chemistry of Peroxides: A Review, Atmos. Environ., 24A, 1601-1633.

44. Kieber, R.J. and Helz, G.R. (1986) Two-method Verification of Hydrogen Peroxide Determinations in Natural Waters, Anal. Chem., 58, 2312-2315.

45. Dasgupta, P.K. and Hwang, H. (1985) Application of a Nested Loop System for the Flow Injection Analysis of Trace Aqueous Peroxides, Anal. Chem., 57, 1009-1012.

46. Beltz, N., Jaeschke, W., Kok, G.L., Gitlin, S.N., Lazrus, A.L., McLaren, S.E., Shakespeare, D. and Mohnen, V.A. (1987) A Comparison of the Enzyme Flourometric and the Peroxyoxalate Chemiluminescence Methods for Measuring H_2O_2, J. Atmos. Chem., 5, 311-322.

47. Abdel-latif, M.S. and Guilbault, G.G. (1988) Fibre Optic Sensor for the Determination of Glucose using Micellar Enhanced Chemiluminescence of the Peroxyoxalate Reaction, Anal. Chem., 60, 2671-2674.

48. Gubitz, G., van Zoonen, P., Grooijer, C., Velthorst, N.H. and Frei, R.W. (1985) Immobilized Flourophores in Dynamic Chemiluminescence Detection of Hydrogen Peroxide, Anal. Chem., 57, 2071-2074.

49. Bader, H., Sturzenegger, V. and Hoigné, J. (1988) Photometric Method for the Determination of Low Concentrations of Hydrogen Peroxide by the Peroxidase Catalysed Oxidation of N,N-diethyl-p-phenylenediamine (DPD), Water Res., 22, 1109-1115.

50. Madsen, B.C. and Kromis, M.S. (1984) Flow Injection and Photo-metric Determination of Hydrogen Peroxide in Rainwater with N-Ethyl-N-(sulfopropyl) Aniline Sodium Salt, Anal. Chem., 56, 2849-2850.

51. Kauken, M.P., Schoenebeck, C.A.M., van Wensveen-Louter, A. and Slamina, J. (1988) Simultaneous Sampling of NH_3, HNO_3, HCl, SO_2 and H_2O_2 in Ambient Air by a Wet Annular Denuder System, Atmos. Environ., 22, 2541-2548.

52. Schiff, H.I., D.R. Karecki, G.W. Harris, D.R. Hastie, and G.I. MacKay (1990) A tunable diode laser system for aircraft measurement of trace gases, J. Geophys. Res., 95, 10,147-10,154.

53. Slemr, F., Harris, G.W., Hastie, D.R., MacKay, G.I. and Schiff, H.I. (1986) Measurement of Gas Phase Hydrogen Peroxide in Air by Tunable Diode Laser Absorption Spectroscopy, J. Geophys. Res., 91, 5371-5378.

54. Mackay, G.I., Mayne, L.K. and Schiff, H.I. (1990) Measurement of H_2O_2 and HCHO by TDLAS during 1986 Carbonaceous Species Methods Comparison Study in Glendora, CA, Aer. Sci. Technol., 12, 56-63.

55. Hartkamp, H. and Bachhausen, P. (1987) A Method for the Determination of Hydrogen Peroxide in Air, Atmos. Environ., 21, 2207-2213.

56. Kok, G.L., Holler, T.P., Lopez, M.B., Nachtrieb, H.A. and Yuan, M. (1978) Chemiluminescent Method for Determination of Hydrogen Peroxide in the Ambient Atmosphere, Environ. Sci. Technol., 12, 1072-1076.

57. Jacob, P., Tavares, T.M. and Klockow, D. (1986) Methodology for the Determination of Gaseous Hydrogen Peroxide in Ambient Air, Fres. Z. Anal. Chem., 325, 359-364.

58. Lazrus, A.L., Kok, G.L., Lind, J.A., Gitlin, S.N., Heikes, B.G. and Shetter, R.E. (1986) Automated Fluorometric Method for Hydrogen Peroxide in Air, Anal. Chem., 58, 594-597.

59. Tanner, R.L., Markouits, G.Y., Ferreri, E.M. and Kelly, T.J. (1986) Sampling and Determination of Ozone by Gas-Phase Reaction with Nitric Oxide, Anal. Chem., 58, 1857-1865.

60. Dasgupta, P.K., Dong, S., Hwang, H., Young, H-C and Genfa, Z. (1988) Continuous Liquid-phase Fluorometry Coupled to a Diffusion Scrubber for the Real-time Determination of Atmospheric Formaldehyde, Hydrogen Peroxide and Sulphur Dioxide, Atmos. Environ., 22., 949-963.

61. Tanner, R.L. and Shen, J. (1990) Measurement of Hydrogen Peroxide in Ambient Air by Impinger and Diffusion Scrubber, Aer. Sci. Technol., 12, 86-87.

62. Zika, R., Saltzman, E.S., Chameides, W.L. and Davis, D.D. (1982) H_2O_2 Levels in Rainwater Collected in S. Florida and the Bahama Islands, J. Geophys. Res., 87, 5015-5017.

63. Sakugawa, H. and Kaplan, I.R. (1987) Atmospheric H_2O_2 Measurement: Comparison of Cold-trap Method with Impinger Bubbling Method, Atmos. Environ., 21, 1791-1798.

64. Lazrus, A.L., G.L. Kok, S.N. Gitlin, J.A. Lind, S.E. MacLaren (1985) Automated fluorometric method for hydrogen peroxide in atmospheric precipitation, Anal. Chem., 57, 917-922

65. Heikes, B.G., Kok, G.L., Walega, J.G., Lazrus, A.L. (1987) H_2O_2, O_3 and SO_2 Measurements in the Lower Troposphere over the Eastern United States during Fall, J. Geophys. Res., 92, 915-931.

66. Barth, M.C., Hegg, D.A., Hobbs, P.V., Walega, J.G., Kok, G.L., Heikes, B.G. and Lazrus, A.L. (1989) Measurements of Atmospheric Gas-phase and Aqueous-phase Hydrogen Peroxide Concentrations in Winter on the East Coast of the United States, Tellus (Ser. B.), 41, 61-69.

67. Heikes, B.G., Wale, J.G., Kok, G.L., Lind, J.A. and Lazrus, A.L. (1988) Measurements of H_2O_2 during Watox-86, Global Biogeochemical Cycles, 2, 57-61.

68. Stephens, E.R. (1987) Smog Studies in the 1950's, EOS Trans., 68, 89-92.

69. Grossjean, D. (1984) Worldwide Ambient Measurements of Peroxyacetylnitrate (PAN) and Implications for Plant Injury, Atmos. Environ., 18, 1489-90.

70. Singh, H.B. (1987) Reactive nitrogen in the troposphere, Environ. Sci. Technol., 21, 320-327.

71. Muller, K.P., and J. Rudolph. (1989) An automated technique for the measurement of peroxyacetylnitrate in ambient air at ppb and ppt levels, Int. J. Environ. Anal. Chem., 37, 253-262.

72. Hanst, P.L., Wong, N.W. and Bragin, J. (1982) A Long-path Infra-red Study of Los Angeles Smog, Atmos. Environ., 16, 969-981.

73. Rogers, J.D. (1990) Infrared Absorptivities and Integrated Band Intensities for Gaseous Peroxyacetylnitrate (PAN), Atmos. Environ., 24A, 2891-2892.

74. Griffith, D.W.T. and Schuster, G. (1987) Atmospheric Trace Gas Analysis using Matrix Isolation-Fourier Transform Infrared Spectroscopy, J. Atmos. Chem., 5, 59-81.

75. Roberts, J.M. (1990) The Atmospheric Chemistry of Organic Nitrates, Atmos. Environ., 24A, 243-287.

76. Roberts, J.M., Fajen, R.W. and Springstone, S.R. (1989) Capillary Gas Chromatographic Separation of Alkyl Nitrates and Peroxycarbocylic Nitrate Anhydrides, Anal. Chem., 61, 771-772.

77. Atlas, E. (1988) Evidence for $\geq C_3$ Alkyl Nitrates in Rural and Remote Atmospheres, Nature, 331, 426-428.

78. Grosjean, D. and Harrison, J. (1985) Response of Chemiluminescence NO_x Analysers and Ultraviolet Ozone Analysers to Organic Air Pollutants, Environ. Sci. Technol., 19, 862-865.

79. Helmig, D., Müller, J. and Klein, W. (1989) Improvements in Analysis of Atmospheric Peroxyacetyl Nitrate (PAN), Atmos. Environ., 23, 2187-2192.

80. Joos, L.F., Landolt, W.F. and Lenenberger, H. (1986) Calibration of Peroxyacetylnitrate Measurements with an NO_x Analyser, Environ. Sci. Technol., 20, 1269-1273.

81. Blanchard, P., Shepson, P.B., So, K.W., Schiff, H.I., Bottenheim, J.W., Gallant, A.J., Drummond, J.W. and Wong, P. (1990) A comparison of Calibration and Measurement Techniques for Gas Chromatographic Determination of Atmospheric Peroxyacetylnitrate (PAN), Atmos. Environ., 24A, 2839-2846.

82. Brice, K.A., Bottenheim, J.W., Anlauf, K.G. and Wiebe, H.A. (1988) Long-term Measurements of Atmospheric Peroxyacetylnitrate (PAN) at Rural Sites in Ontario and Nova Scotia: Seasonal Variations and Long-range Transport, Tellus, 40B, 408-425.

83. Hewitt, C.N. and Harrison, R.M. (1985) Tropospheric Concentrations of the Hydroxyl Radical - A Review, Atmos. Environ., 19, 545-554.

84. Altshuller, A.P. (1989) Ambient air hydroxyl radical concentrations: measurements and model predictions, J. Air Pollut. Control Assoc., 39, 704-708.

85. Rodgers, M.O., Bradshaw, J.D., Sandholm, S.T., Kesheng, S. and Davies, D.D. (1985) A 2-λ Laser Induced Fluorescence Field Instrument for Ground-based and Airborne Measurements of Atmospheric OH, J. Geophys. Res., 90, 12819-12834.

86. Hard, T.M., O'Brien, R.J., Chan, C.Y. and Mehrabzadeh, A.A. (1984) Tropospheric Free Radical Determination by FAGE, Environ. Sci. Technol., 18, 768-777.

87. Smith, G.P. and Crosley, D.R. (1990) A Photochemical Model of Ozone Interference Effects in Laser Detection of Tropospheric OH, J. Geophys. Res., 95, 16427-16442.

88. Chan, C.Y., Hard, T.M., Mehrabzaden, A.A., George, L.A., and O'Brien, R.J. (1990) Third-generation FAGE Instrument for Tropospheric Hydroxyl Radical Measurement, J. Geophys. Res, 95, 18569-18576.

89. Hofzumahaus, A., H.-P. Dorn, J. Callies, U. Platt and D.H. Ehhalt (1991) Tropospheric OH concentration measurements by laser long path absorption spectroscopy, Atmos. Environ., 25A, 2017

90. Mount, G.H. (1992) The measurement of tropospheric OH by long path absorption I: instrumentation, J. Geophys. Res., 97, 2427-2444.

91. Dorn, H.P., Callies, J., Platt, U., Ehhalt, D.H. (1988) Measurement of Tropospheric OH Concentrations by Laser Long-path Absorption Spectroscopy, Tellus, 40B, 437-445.

92. Campbell, M.J., Farmer, J.C., Fitzuer, C.A., Henry, M.N., Sheppard, J.C., Hardy, R.J., Hopper, J.F. and Muralidhar, V. (1986) Radiocarbon Tracer Measurements of Atmospheric Hyroxyl Radical Concentrations, J. Atmos. Chem., 4, 413-427.

93. Felton, C.C., Sheppard, J.C. and Campbell, M.J. (1988) Measurements of the Diurnal OH Cycle by a [14]C Tracer Method, Nature, 335, 53-56.

94. Eisele, F.L. and D.J. Tanner (1991) Ion-assisted tropospheric OH measurements, J. Geophys. Res., 96, 9295-9308

95. McKenna, D.S., C.J. Hord, and J.M. Kent (1993) Hydroxyl radical concentrations and Kuwaiti oil fire emission rates for March 1991, submitted to J. Geophys. Res.

96. Spivakovsky, C.M., Wotsy, S.C. and Prather, M.J. (1990) A Numerical Method for Parameterization of Atmospheric Chemistry: Computation of Tropospheric OH, J. Geophys. Res., 95, 18433-18440.

97. Platt, U., D. Perner, A.M. Winer, G.W. Harris and J.N. Pitts (1980) Detection of NO_3 in the polluted troposphere by differential optical absorption, Geophys. Res. Lett., 7, 89-92.

98. Noxon, J.F., R.B. Norton, and E. Marovich (1980) NO_3 in the troposphere, Geophys. res. Lett., 7, 125-128.

99. Mihelcic, D., Musgen, P., Ehhalt, D. (1985) An improved Method of measuring Tropospheric NO_2 and RO_2 by matrix isolation and electron spin resonance, J. Atmos. Chem., 3, 341-361.

100. Mihelic, D., Volz-Thomas, A., Pätz, H.W., Kley, D. and Michelcic, M. (1990) Numerical Analysis of ESR Spectra from Atmospheric Samples, J. Atmos. Chem., 11, 271-279.

101. Watanabe, T., Yoshida, M., Fujiwara, S., Abe, K., Onoe, A., Hirota, M. and Igarashi, S. (1982) Spin Trapping of Hydroxyl Radical in the Troposphere for Determination by Electron Spin Resonance and GC/MS, Anal. Chem., 54, 2470-2474.

102. Biermann, H.W., Tuazon, E.C., Winer, A.M., Wallington, T.J. and Pitts, J.N. (1988) Simultaneous Absolute Measurements of Gaseous Nitrogen Species in Urban Ambient Air by Long Path-length Infrared and Ultraviolet-visible Spectroscopy, Atmos. Environ., 22, 1545-1554.

103. Schiff, H.I., Harris, G.W. and Mackay, G.I. (1987) Measurement of Atmospheric Gases by Laser Absorption Spectroscopy. In Johnson R.W. and Gordon G.E. (eds.) *The Chemistry of Acid Rain*, ACS Symp. Ser. 349, Amer. Chem. Soc., Washington, DC, USA.

104. Schiff, H.I., D.R. Karecki, G.W. Harris, D.R. Hastie and G.I. MacKay (1990) A tunable diode laser system for aircraft measurement of trace gases, J. Geophys. Res., 95, 10,147-10,153.

105. Schmidtke, G., Kohn, W., Klocke, U., Knothe, M., Riedel, W.J. and Wolf, H. (1989) Diode Laser Spectrometer for Monitoring up to Five Atmospheric Trace Gases in Unattended Operation, Applied Optics, 28, 3665-3670.

106. Walega, J.G., Stedman, D.H., Shetter, R.E., MacKay, G.I., Iguchi, T. and Schiff, H.I. (1984) Comparison of a Chemiluminescent and a Tunable Diode Laser Absorption Technique for the Measurement of Nitrogen Oxide, Nitrogen Dioxide and Nitric Acid, Environ. Sci. Technol., 18, 823-826.

107. Galle, B., Sunesson, A. and Wendt, W. (1988) NO_2 - Mapping using Laser-Radar Techniques, Atmos. Environ., 22, 569-573.

108. Sandholm, S.T., Bradshaw, J.D., Dorris, K.S., Rodgers, M.O. and Davies, D.D. (1990) An Airborne Compatible Photofragmentation Two Photon Laser Induced Fluorescence Instrument for Measuring Background Tropospheric Levels of NO, NO_x and NO_2, J. Geophys. Res., 95, 10155-10162.

109. Papenbrock, T. and Stuhl, F. (1990) A Laser-Photolysis Fragment-Fluorescence (LPFF) Method for the Detection of Gaseous Nitric Acid in Ambient Air, J. Atmos. Chem., 10, 451-469.

110. Tanner, R.L. (1987) Chemical Instrumentation of Atmospheric Wet Deposition Processes, in *The chemistry of acid rain: sources and atmospheric processes*, R.W. Johnson, G.E. Gordon, W. Calkins and A.Z. Elzerman (eds.), ACS Symp. Ser., 349, 289-302.

111. Photochemical Oxidants Review Group (PORG), (1990) Oxides of Nitrogen in the United Kingdom. HMSO, London, UK.

112. Ridley, B.A., Carroll, M.A., Gregory, G.L. and Sachse, G.W. (1988) NO and NO_2 in the Troposphere - Technique and Measurements in Regions of a Folded Tropopause, J. Geophys. Res., 93, 15813-15830.

113. Fahey, D.W., Eubank, C.S., Hubler, G., and Fehsenfeld, F.C. (1985) Evaluation of a catalytic reduction technique for the measurement of total reactive odd nitrogen NO in the atmosphere, J. Atmos. Chem., 3, 435-468.

114. Braman, R.S., de la Cantera, M.A. and Han, Q.X. (1986) Sequential, Selective Hollow Tube Preconcentration and Chemiluminescence Analysis System for Nitrogen Oxide Compounds in Air, Anal. Chem., 58, 1537-1541.

115. Zafiriou, O.C. and True, M.B. (1986) Interferences in Environmental Analysis of NO by NO plus O_3 Detectors: A Rapid Screening Technique, Environ. Sci. Technol., 20, 594-596.

116. Tidona, R.J., Nizami, A.A. and Cernansky, N.P. (1988) Reducing Interference Effects in the Chemiluminescent Measurement of Nitric Oxides from Combustion Systems, J. Air Pollut. Control Assoc., 38, 806-811.

117. Wendel, G.J., Stedman, D.H., Cantrell, C.A. and Damrauer, L. (1983) Luminol-based Nitrogen Dioxide Detector, Anal. Chem., 55, 937-940.

118. Fehsenfeld, F.C., Drummond, J.W., Roychowdhury, U.K., Galvin, P.J., Williams, E.J., Buhr, M.P., Parrish, D.D., Hubler, G., Langord, A.O., Calvert, J.G., Ridley, B.A., Grahek, F., Heikes, B.G., Kok, G.L., Shetter, J.D., Walega, J.G., Elsworth, C.M., Norton, R.B., Fahey, D.W., Murphy, P.C., Hovermale, C., Mohnen, V.A., Demerjian, K.L., MacKay, G.I. and Shiff, H.I. (1990) Intercomparison of NO_2 Measurement Techniques, J. Geophys. Res., 95, 3579-3597.

119. Gregory, G.L., Hoell, J.M., Carroll, M.A., Ridley, B.A., Davies, D.D., Bradshaw, J., Rodgers, M.O., Sandholm, S.T., Schiff, H.I., Hastie, D.R., Karecki, D.R., MacKay, G.I., Harris, G.W., Torres, A.L. and Fried, A. (1990) An Intercomparison of Airborne Nitrogen Dioxide Instruments, J. Geophys. Res., 95, 10103-10127.

120. Gregory, G.L., Hoell, J.M., Torres, A.L., Carroll, M.A., Ridley, B.A., Rodgers, M.O., Bradshaw, J., Sandholm, S.T. and Davies, D.D. (1990) An Intercomparison of Airborne Nitric Oxide Measurements, J. Geophys. Res., 95, 10129-10138.

121. Kelly, T.J., Spicer, C.W. and Ward, G.F. (1990) An Assessment of the Luminol Chemiluminescence Technique for Measurement of NO_2 in Ambient Air, Atmos. Environ., 24A, 2397-2403.

122. Nash, T. (1970) An Efficient Absorbing Reagent for Nitrogen Dioxide, Atmos. Environ., 4, 661-665.

123. Baveja, A.K., Chambe, A. and Gupta, V.K. (1984) Extractive Spectrophotometric Method for the Determination of Atmospheric Nitrogen Dioxide, Atmos. Environ., 18, 989-993.

124. Raman, V. (1990) Spectrophotometric Determination of Nitrogen Dioxide (Nitrite), Sci. Tot. Environ., 93, 301-306.

125. Hildemann, L.M., Russell, A.G. and Cass, G.R. (1984) Ammonia and Nitric Acid Concentrations in Equilibrium with Atmospheric Aerosols: Experiment vs. Theory, Atmos. Environ., 18, 1737-1750.

126. Allen, T.M., Rao, S.T. and Sistla, G. (1989) On the Problem of Ozone Compliance in the Northeastern Urban Corridor, Air Waste Management Assoc. Meeting, Calif. 25-30 June 1989, Paper No. 89-3.3.

127. Huebert, B.J., Luke, W.T., Delany, A.C. and Brost, R.A. (1988) Measurements of Concentrations and Dry Surface Fluxes of Atmospheric Nitrates in the Presence of Ammonia, J. Geophys. Res., 93, 7127-7136.

128. Brocco, D., Franco, A., Liberti, A., Tappa, R. (1987) The Use of a Filter Pack for the Monitoring of Organic and Inorganic Acid Species in Air, in Angeletti, G. and Restelli, G. (eds.), *Physico-Chemical Behaviour of Atmospheric Pollutants*, D. Reidel, Dordrecht, Holland, 142-147.

129. Sickles, J.E., Grohse, P.M., Hodson, L.L., Salmons, C.A., Cox, K.W., Turner, A.R. and Estes, E.D. (1990) Development of a Method for the Sampling and Analysis of Sulphur Dioxide and Nitrogen Dioxide from Ambient Air, Anal. Chem., 62, 338-346.

130. Hisham, M.W.M. and Grosjean, D. (1990) Sampling of Atmospheric Nitrogen Dioxide using Triethanolamine: Interference from peroxyacetylnitrate, Atmos. Environ., 24A, 2523-2525.

131. Vecera, Z. and P.K. Dasgupta (1991) Measurement of ambient nitrous acid and a reliable calibration source for gaseous nitrous acid, Environ. Sci. Technol., 25, 255-260.

132. Febo, A., DeSantis, F. and Perrmo, C. (1987) Measurement of Atmospheric Nitrous and Nitric Acid by means of Annular Denuders, in Angeletti, G. and Restelli, G. (eds.), *Physico-Chemical Behaviour of Atmospheric Pollutants*, D. Reidel, Dordrecht, Holland, 121-125.

133. Kitto, A.M.N. and Harrison, R.M. (1992) Nitrous and Nitric Acid Measurements at Sites in Southeast England, Atmos. Environ., 26A, 235-241.

134. Gormley, P. and Kennedy, M. (1949) Proc. Roy. Ir. Acad., Sect. A, 52A, 163-169.

135. Larson, T.V. and Taylor G.S. (1983) On the Evaporation of Ammonium Nitrate Aerosol, Atmos. Environ., 17, 2489-2495.

136. Harrison, R.M. and MacKenzie, A.R. (1990) A Numerical Simulation of Kinetic Constraints upon Achievement of the Ammonium Nitrate Dissociation Equilibrium in the Troposphere, Atmos. Environ., 24A, 91-102.

137. Wexler, A.S. and Seinfeld, J.H. (1990) The Distribution of Ammonium Salts among a Size and Composition Dispersed Aerosol, Atmos. Environ., 24A, 1231-1246.

138. Roberts, J.M., Norton, R.B., Goldam, P.D. and Fehsenfeld, F.C. (1987) Evaluation of the Tungsten Oxide Denuder Tube Technique as a Method for the Measurement of Low Concentrations of Nitric Acid in the Troposphere, J. Atmos. Chem., 5, 217-238.

139. Slanina, J., Van Wensveen, A.M., Schoonebeek, C.A.M. and Voors, P.I. (1987) Automated Denuder Systems, in Angeletti, G. and Restelli, G. (eds.) *Physico-Chemical Behaviour of Atmospheric Pollutants*, D. Reidel, Dordrecht, Holland, 25-32.

140. Palmes, E.D., Gunnison, A.F., DiMattio, J. and Tomczyk, C. (1976) Personal Sampler for Nitrogen Dioxide, Am. Ind. Hyg. Assoc. J., 37, 570-577.

141. Campbell, G.W. (1988) Measurement of nitrogen dioxide concentrations at rural sites in the United Kingdom using diffusion tubes, Environ. Pollut., 55, 251-270.

142. Apling, A.J., Stevenson, K.J., Goldstein, B.D., Melia, R.J.W. and Atkins, D.H.F. (1979). Air Pollution in Homes 2: Validation of Diffusion Tube Measurements. Warren Spring Laboratory Report, LR311 (AP), Stevenage, Herts., U.K.

143. Boleij, J.S.M., Lebert, E., Hoek, F., Noy, D. and Brunekreff, B. (1986) The Use of Palmes' Diffusion Tubes for Measuring NO_2 in Homes, Atmos. Environ., 20, 597-600.

144. Atkins, D.H.F., Sandalls, J., Law, D.V., Hough, A.M. and Stevenson, K.J. (1986) The Measurement of Nitrogen Dioxide in the Outdoor Environment using Passive Diffusion Tube Samplers, AERE Report R12133, Harwell Laboratory, Oxon, UK.

145. Miller, D.P. (1984) Ion Chromatographic Analysis of Palmes Tubes for Nitrite, Atmos. Environ., 18, 891-892.

146. Miller, D.P. (1988) Low-level Determination of Nitrogen Dioxide in Ambient Air using the Palmes Tube, Atmos. Environ., 22, 945-947.

147. Krochmal, D., M. Rzeminski and K. Gorski (1987) Testing of precision and influence of weather conditions on accuracy of Amaya-Sugiura's permeation method for NO_2 measurement in ambient air, Chemia Analityczna, 32, 581-589.

148. Krochmal, D. and L. Gorski (1991) Modification of Amaya-Sagiura passive smapling spectrophotometric method of nitrogen dioxide determination in ambient air, Fres. J. Anal. Chem., 340, 220-222.

149. Krochmal, D., and L. Gorski (1991) Determination of nitrogen dioxide in ambient air by use of a passive sampling technique and triethanolamine as absorbent, Environ. Sci. Technol., 25, 531-535.

150. Sberveglieri, G., Benussi, P., Coccoli, G., Gropelli, S. and Nelli, P. (1990) Reactively Sputtered Iridium Tin Oxide Polycrystalline Thin Films as NO and NO_2 Gas Sensors, Thin Solid Films, 186, 349-360.

151. Roberts, J.M., Fehsenfeld, F.C., Albritten, D.L. and Sievers, R.E. (1983) Measurement of Monoterpene Hydrocarbons at Niwot Ridge, Colorado, J. Geophys. Res., 88, 10667-10678.

152. Lonneman, W.A. (1987) The Measurement of Ozone Precursors and their Relationship to Ozone Formation, Proc. N. Amer. Oxidant Symp., Quebec, 1987, Ministère de l'Environment du Quebec, Canada, 302-321.

153. Muller, J. and Reidel, F. (1987) Measurement of Airborne Hydro-carbons by Headspace-Chromatography, in Angeletti, G. and Restelli, G. (eds.) *Physico-Chemical Behaviour of Atmospheric Pollutants*, D. Reidel, Dordrecht, Holland, 157-165.

154. Mowner, J. and Lindskog, A. (1990) Automatic Unattended Sampling and Analysis of Background Levels of C_2-C_5 Hydrocarbons, Atmos. Environ., 25A, 1971-1979

155. Rudolph, J. and Khedim, A. (1985) Hydrocarbons in the Non-urban Atmosphere: Analysis, Ambient Concentrations and Impact on the Chemistry of the Atmosphere, Int. J. Environ. Anal. Chem., 20, 265-282.

156. Rudolph, J., Johnen, F.J. and Khedim, A. (1986) Problems connected with the Analysis of Halocarbons and Hydrocarbons in the Non-urban Atmosphere, Int. J. Environ. Anal. Chem., 27, 97-122.

157. Rudolph, J., Khedim, A. and Wagenbach, D. (1989) The Seasonal Variation of Light Non-Methane Hydrocarbons in the Antarctic Troposphere, J. Geophys. Res., 94, 13039-13044.

158. Gholson, A.R., Jayanty, R.K.M. and Storm, J.F. (1990) Evaluation of Aluminium Canisters for the Collection and Storage of Air Toxics, Anal. Chem., 62, 1899-1902.

159. Gholson, A.R., Storm, R.F., Jayanty, R.K.M., Fuerst, R.G., Logan, T.J. and Midgett, M.R. (1989) Evaluation of Canisters for Measuring Emissions of Volatile Organic Air Pollutants from Hazardous Waste Incineration, J. Air Pollut. Control Assoc., 39, 1210-1217.

160. Stephens, E.R. (1989) Valveless Sampling of Ambient Air for Analysis by Capillary Gas Chromatography, J. Air Pollut. Control Assoc., 39, 1202-1205.

161. Pleil, J.D., Oliver, K.D. and McClenny, W.A. (1987) Enhanced Performance of Nafion Dryers in Removing Water from Air Samples prior to Gas Chromatographic Analysis, J. Air Pollut. Control Assoc., 37, 244-248.

162. Rudolph, J., Johnen, F.I., Khedim, A. and Pilwat, G. (1990) The Use of Automated "on-line" Gas chromatography for the Monitoring of Organic Trace Gases in the Atmosphere at Low Levels, Int. J. Environ. Anal. Chem., 38, 143-155.

163. Juttner, F. (1988) A cryotrap technique for the quantitation of monoterpenes in humid and ozone-rich forest air, J. Chromatogr., 442, 157-163.

164. McClenny, W.A., Pleil, J.D., Holdren, M.W. and Smith, R.N. (1984) Automated Cryogenic Preconcentration and Gas Chromatographic Determination of Volatile Organic Compounds in Air, Anal. Chem., 56, 2947-2951.

165. Rudolph, J. and Jebsen, C. (1983) The Use of Photoionization, Flameionization and Electron Capture Detectors in Series for the Determination of Low Molecular Weight Trace Components in the Non-urban Atmosphere, Int. J. Anal. Chem., 13, 129-139.

166. Hills, A.J. and Zimmerman, P.R. (1990) Isoprene Measurement by Ozone-induced Chemiluminescence Anal. Chem., 62, 1055-1060.

167. Schurath, U., Wiese, A. and Becker, K.H. (1976) A chemiluminescence analyser for unsaturated hydrocarbons in air, Staub.-Reinhalt. Luft., 9, 379-385.

168. Becker, K.H., Schurath, U. and Wiese, A. (1976) A new chemiluminescent olefin detector for ambient air, US EPA Report no. EPA 600/3 - 77-001a.

169. Lawson, D.R., Biermann, H.W., Tuazon, E.C., Winer, A.M., MacKay, G.I., Schiff, H.I., Kok, G.L., Dasgupta, P.K. and Fung, K. (1990) Formaldehyde Measurement Methods Evaluation and Ambient Concentrations during the Carbonaceous Species Methods Comparison Study, Aerosol Sci. Technol., 12, 64-76.

170. Lazrus, A.L., Fong, K.L. and Lind, J.A. (1988) Automated Fluorometric Determination of Formaldehyde in Air, Anal. Chem., 60, 1074-1078.

171. Dasgupta, P.K., Dong S. and Hwang, H. (1990) Diffusion Scrubber-based Field Measurements of Atmospheric Formaldehyde and Hydrogen Peroxide, Aer. Sci. Technol., 12, 98-104.

172. Ciccioli, P., Draisci, R., Cecinato, A. and Liberti, A. (1987) Sampling of Aldehydes and Carbonyl Compounds in Air and Their Determination by Liquid Chromatographic Techniques, in Angeletti, G. and Restelli, G. (eds.) *Physico-Chemical Behaviour of Atmospheric Pollutants*, D. Reidel, Dordrecht, Holland, 133-141.

173. Facchini, M.C., Chiavan, G. and Fuzzi, S. (1987) Analysis of Carbonyl Compounds in the Atmospheric Liquid Phase, in Angeletti, G. and Restelli, G. (eds.) *Physico-Chemical Behaviour of Atmospheric Pollutants*, D. Reidel, Dordrecht, Holland, 188-197.

174. Ho, M.H. and Richards, R.A. (1990) Enzymatic Method for the Determination of Formaldehyde, Environ. Sci. Technol., 24, 201-204.

175. Johnson, B.J. and Dawson, G.A. (1990) Collection of Formaldehyde from Clean Air for Carbon Isotopic Analysis, Environ. Sci. Technol, 24, 898-902.

176. Tuazon, E.C., A.M. Winer, R.A. Graham, and J.N. Pitts Jr. (1980) Atmospheric measurements of trace pollutants by kilometer pathlength Fourier transform IR spectroscopy, Adv. Environ. Sci. Technol., 10, 259-300

177. Grosjean, D. and Parmar, S.S. (1990) Interferences from Aldehydes and Peroxyacetyl Nitrate when Sampling Urban Air Organic Acids on Alkaline Traps, Environ. Sci. Technol., 24, 1021-1026.

178. Grosjean, D., Tuazon, E.C. and Fujita, E. (1990) Ambient Formic Acid in Southern California: A Comparison of Two Methods, Fourier Transform Infrared Spectroscopy and Alkaline Trap-liquid Chromatography with u.v. Detection, Environ. Sci. Technol., 24, 144-146.

CHAPTER 6

CONCENTRATIONS OF OZONE
AND OTHER PHOTOCHEMICAL OXIDANTS

It has been shown that, averaged over the globe, chemical sources and sinks for ozone are in approximate balance, and are similar in magnitude to the source from the stratosphere and the sink at the ground[1,2]. Seasonal variations in both natural and anthropogenic sources lead to distinct annual ozone cycles according to the location of the site as shown in Figure 6.1 There are hundreds of places around the globe where ozone concentrations near the ground are being measured. Most of these stations are in locations strongly influenced by local/regional pollution and therefore, the information is of limited, or no, use for studies of background levels. Measurements made at remote sites are generally assumed to reflect free tropospheric ozone concentrations. Such sites, far removed from major anthropogenic sources of ozone are few and even these may be influenced by distant sources[3,4]. When comparing many measurements of the same parameter, one must remember that all measurements are different. This may be due to the variety of instruments and techniques used, systematic errors or natural variations.

Most pollutants show significant concentration variations in time and space. Ozone, in particular, exhibits a diurnal variation and hence 24 hourly mean values for each day are distributed about the mean with scatter that varies from day to day. Pollution studies generally require the maximum hourly concentration and a measure of the frequency concentrations exceeding a given level. During each month, daily pollution levels vary, so a mean value for the month and maximum hourly concentration are often quoted. Over a year, the individual hourly or daily mean values may be represented as frequency distributions to reveal the amount of spread of dispersion in concentrations between the highest and lowest.

Ozone concentrations are often high for several consecutive days and occur at several sites separated by hundreds of kilometres. This is often referred to as an ozone episode. The selection criteria for definition of an ozone episode is somewhat arbitrary. Here we will assume an episode to have occurred when the daily maximum hourly ozone concentration has exceeded 80 ppbv simultaneously at several monitoring stations for 2 or more consecutive days.

In this chapter we discuss ozone concentrations at remote, urban and rural sites with regard to seasonal and diurnal variations and the influence of long-range transport. A

Figure 6.1

Mean monthly values of ozone concentration. Data refs 185, 143, 119, 201, 327, 328 and 241.

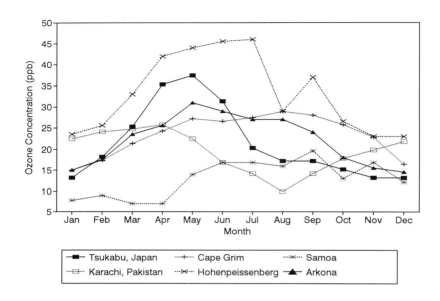

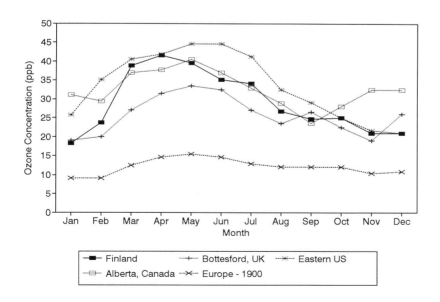

discussion of the meteorological conditions favourable for the occurrence of ozone episodes is postponed until § 6.7 after the discussion of the frequency of episodes in the UK, Europe and the rest of the world (§ 6.4, 6.5, 6.6). Trends in ozone levels will also be examined along with (necessarily) shorter discussions of the distributions of some other oxidants. Measurements considered will be those made at or near to the ground. There is, however, a considerable volume of observational data in tropospheric ozone from balloon and aircraft measurement[1].

Table 6.1: Typical summertime daily maximum ozone concentrations

Region	Ozone (ppbv)
Urban-suburban	100-400
Rural	50-120
Remote-tropical rainforest	20-40
remote marine	20-40

Table 6.1 shows the typical daily maximum ozone concentration at various locations. As discussed later, under certain meteorological conditions rural concentrations can exceed those in suburban areas.

6.1 OZONE CONCENTRATIONS AT REMOTE SITES

Many workers have extracted data from measurements whose primary concern was stratospheric ozone to obtain latitudinal distributions of tropospheric ozone. There are now several sites which monitor surface concentrations continuously, such as those in the US. Geophysical Monitoring for Climatic Change (GMCC) network located at Barrow, Alaska (71°N, 157°W), Mauna Loa, Hawaii (20°N, 155°W), Samoa (14°S, 171°W) and the South Pole (90°S) and those in the WMO background air pollution monitoring network (BAPMoN). It should be noted that the stations at higher altitude locations (*e.g.* Mauna Loa) provide data which is representative of the free troposphere while those at lower altitudes (*e.g.* Barrow) are influenced by boundary layer effects.

The seasonal cycle is a prominent feature of the tropospheric ozone distribution. Both the phasing and variability in this cycle can be useful in explaining its origin at a given location[5]. Figure 6.2 shows the annual cycle at the four GMCC sites. At Mauna Loa, maximum ozone concentrations occur in May or June in the upper troposphere. Therefore the ground level maximum in April is not the result of direct downward mixing from the

Figure 6.2

Mean average surface ozone partial pressures (After Oltmans and Komhyr[5]).

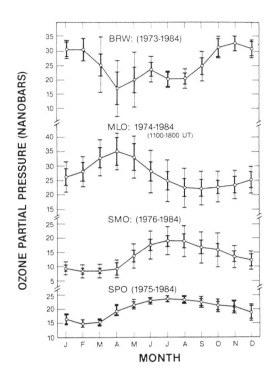

stratosphere but from horizontal advection from higher latitudes of tropospheric air rich in ozone. At some locations the seasonal variations appear to be more closely linked to horizontal transport processes in the troposphere than to downward transport of ozone from the stratosphere. Stratospheric-tropospheric exchange processes are stronger in mid-latitude regions. In the Southern hemisphere, for example, the relatively small phase difference between the sites suggests that ozone is injected into the troposphere from the stratosphere in southern mid-latitudes, then transported in the troposphere to higher and lower latitudes simultaneously . There can also be a chemical signature in the seasonal cycle, however, and recent observations of a large summer minimum in ozone concentration in the unpolluted marine boundary layer of the Southern Hemisphere suggest a clear link between ozone loss and hydrogen peroxide production (see chapter 4). This demonstrates that *in situ* photochemistry, rather than transport, is the major cause of the seasonal cycle[6].

To determine long-term trends, measurements must be made continuously at a site for many years. Data obtained from aerial measurements during meridional flights[7-10], shipboard expeditions[11,12] and short-term surface measurements[12,14] are useful for yielding information on specific phenomena. Several campaigns have recently provided data for the tropics, a region in which data is particularly lacking[14,17]. *In situ* measurements show unexpectedly high ozone concentrations of 60-70 ppbv in the middle troposphere in the southern spring. Satellite data have shown that photochemical ozone production occurs in the tropics[17,18]. The data suggest that emission of CO, hydrocarbons and NO_x from biomass burning lead to photochemical formation of ozone[19].

Analysis of vertical profiles, aircraft data, shipboard expeditions in combination with ground based measurements enable the latitudinal distribution of ozone to be investigated.

Figure 6.3

Meridional distribution of ozone over the Atlantic Ocean measured aboard ships during the period 1977 through 1986 (After Winkler[12]).

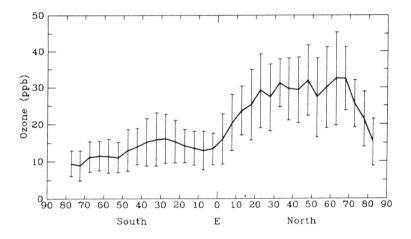

The latitudinal distribution of ozone over the Atlantic ocean is illustrated in Figure 6.3. Concentrations will of course vary from month to month and with longitude[20]. We may conclude that the mean background ozone concentrations range from 10 to 40 ppbv.

Table 6.2: The value of the annual average of the hourly mean concentrations (ppbv) (After Evans et al.[22])

Site	Elevation (m)	1979	1980	1981	1982	1983	1984	1985	1986	1987
South Pole, Antarct.	2835		30	30	30	28	30	38	30	
Bitumount, Alb., Can.	350	26								
Barrow, AK	11		26	27	27	27	24	28	25	29
Theodore Roosevelt, N.P., ND	727						29	32		
Custer National Forest, MT	1006	32	44			35				
Ochoco National Forest, OR	1364		38	31	34	34				
Birch Mtn, Alb., Can.	850	37								
White River Oil Shale Proj., UT, U-4	1600						39	44		
Fortress Mtn., Alb., Can.	2103								41	48
Apache National Forest, AZ	2424		47	35	41	38				
Mauna Loa, HI	3397		38	41	38	42	37	38	36	
Whiteface Mountain, NY	1483	44	41	36	33	36	39	39	40	42
Hohenpeissenberg, FRG	975	32	33	36	39	33	37			
American Samoa	82		14	13	13	13	14	14	14	14

Clean air sites need not necessarily be remote. Even stations in Antarctica are periodically influenced by long-range transport phenomena[21]. As illustrated in Table 6.2, some sites in the US and Canada are as clean as the South Pole and cleaner than Mauna Loa[22]. The identification of clean monitoring sites is fraught with difficulties. The application of land use criteria to monitoring sites does not guarantee that all sites are free from anthropogenic emissions and ozone precursors. It has been shown that some remote national forest sites in the US appear isolated from local sources, whilst other remote sites are influenced by long range transport of ozone and its precursors[23]. Lefohn and Jones[24] reported that several isolated ozone monitoring sites, independent of land use designations, showed a tendency to experience few hourly mean concentrations at or near the minimum detectable level and experience few occurrences of hourly average concentrations above 80 ppbv. Sites in polluted regions tend to experience many hourly average ozone concentrations at or near detectable levels and more occurrences of high ozone concentrations[22]. It is, for all the inherent difficulties, an essential task to identify remote sites so that any trends in background ozone concentrations can be quantified (§ 6.8).

6.2   OZONE CONCENTRATIONS AT URBAN SITES

In chapter 4 we saw how the chemistry of the polluted troposphere differs from that of the clean troposphere; with NO_x acting as a catalyst, ozone production from carbon monoxide and hydrocarbon oxidation is enormous. The precursors of ozone are in greatest quantities in urban-industrial areas. Some of these precursors are also chemical sinks for ozone. Hence the ozone concentration in an urban area will depend on several factors such as location, dispersion of precursors and time-scale from emission.

Air pollution was originally regarded as a sign of industrial activity and economic progress. In the late 1940s and early 1950s, a new phenomenon was observed in Los Angeles. Under specific meteorological conditions, when traffic levels were high, oxidants, mainly ozone, were formed by gas-phase oxidation reactions[25,26]. It was initially assumed that this photochemical smog was specific to the Los Angeles area. However, this form of pollution has been observed in many metropolitan areas throughout the world[27-34,35,36,37].

Figure 6.4

Typical mixing ratio profiles, as a function of time of day, in a photochemical smog cycle.

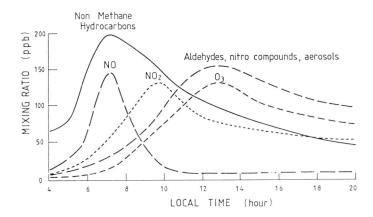

In heavily polluted urban areas, subject to photochemical air pollution, the concentrations of NO, NO_2 and ozone, and other secondary pollutants tend to follow characteristic patterns, as illustrated in Figure 6.4. Emissions of NO and hydrocarbons are high in the early morning, as a result of heavy traffic in the rush hour. These emissions will destroy some of the ozone present, whether it be of natural origin or left from the previous day, during this time. This results in an increase in NO_2 concentrations, which peak when the atmosphere has developed sufficient oxidizing capability to oxidize the NO emissions without total consumption of ozone. In a typical urban cycle, ozone concentrations peak between 12.00 and 15.00 hours when the solar intensity is a maximum and the NO_2/NO ratio has become large. Concentrations fall as the NO_2 photolysis diminishes and stable products such as HNO_3 begin to accumulate. Fresh emissions of NO may also rapidly

Figure 6.5

Mean diurnal variation of ozone, April - September, in Central London.

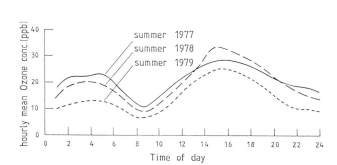

Central London

deplete the ozone (not shown). In the United Kingdom, such diurnal profiles have been observed (Figure 6.5), but peak ozone concentrations are generally rather later. Much of the ozone in the UK is generated during long-range transport of air pollution, so the situation is more complex than that outlined.

Before comparing measurements from various cities around the world, it is important to recognise that the data generally come from only one site in each city and so do not necessarily represent the maximum levels occurring in these urban areas. Frequency distributions or number of days or hours above a given concentration provide a good indication of human exposure levels. Non-kerbside sites give the most realistic indication os exposure. Table 6.3 lists the second highest 1-hour ozone values for 1979 to 1982 and 1988 to 1990 for metropolitan areas with a population in excess of 1 million. Collectively these cities represent approximately 40% of the 1980 US population. The median concentration (not shown) in all years exceeded the US ambient air quality standard for ozone. It is chastening to note here the rather modest improvements that have been achieved according to this statistic over the decade of the 1980's. The average difference between the 1979-82 values and the 1988-90 values is 6%.

Ozone levels at urban, suburban and rural sites for the UK are summarized in Table 6.4. The maximum hourly average concentration per year and the number of hours in

Table 6.3: Second-highest 1-hour ozone concentrations reported for Standard Metropolitan Statistical areas having populations greater than 1 million

Standard Metropolitan Statistical Area	Ozone concentration, ppmv						
	1979	1980	1981	1982	1988	1989	1990
Anaheim - Santa Ana - Garden Grove, CA	0.35	0.29	0.31	0.18	0.24	0.24	0.21
Atlanta, GA	0.16	0.15	0.14	0.14	0.15	0.12	0.15
Baltimore, MD	0.14	0.18	0.17	0.14	0.19	0.13	0.14
Boston, MA	0.22	0.15	0.13	0.16	0.17	0.12	0.11
Buffalo, NY	0.11	0.14	0.12	0.11	0.15	0.11	0.11
Chicago, IL	0.22	0.34	0.14	0.12	0.22	0.12	0.11
Cincinnati, OH - KY - IN	0.13	0.16	0.13	0.13	0.17	0.12	0.15
Cleveland, OH	0.14	0.12	0.12	0.12	0.14	0.12	0.12
Columbus, OH	0.10	0.12	0.11	0.13	0.15	0.11	0.11
Dallas - Fort Worth, TX	0.17	0.18	0.15	0.17	0.13	0.13	0.14
Denver - Boulder, CO	0.16	0.13	0.13	0.14	0.12	0.11	0.11
Detroit, MI	0.12	0.15	0.15	0.16	0.16	0.14	0.12
Fort Lauderdale - Hollywood, FL	0.10	0.12	0.11	0.09	0.15	0.12	0.10
Houston, TX	0.24	0.30	0.23	0.21	0.22	0.23	0.22
Indianapolis, IN	0.12	0.14	0.13	0.12	0.14	0.12	0.11
Kansas City, MO - KS	0.12	0.16	0.12	0.10	0.19	0.11	0.11
Los Angeles - Long Beach, CA	0.44	0.44	0.35	0.32	0.33	0.33	0.27
Miami, FL	0.05	0.15	0.14	0.14	0.13	0.12	0.11
Milwaukee, WI	0.17	0.14	0.17	0.13	0.19	0.15	0.13
Minneapolis - St. Paul, MN - WI	0.10	0.13	0.10	0.10	0.11	0.10	0.10
Nassau - Suffolk, NY	0.18	0.17	0.14	0.13	0.16	0.15	0.14
New Orleans, LA	0.12	0.12	0.11	0.17	0.12	0.11	0.11
New York, NY - NJ	0.19	0.18	0.18	0.17	0.18	0.13	0.16
Newark, NJ	0.15	0.15	0.14	0.17	0.18	0.13	0.13
Philadelphia, PA - NJ	0.18	0.24	0.17	0.18	0.20	0.16	0.14
Phoenix, AZ	0.12	0.15	0.16	0.12	0.12	0.11	0.14
Pittsburg, PA	0.17	0.17	0.16	0.14	0.16	0.13	0.11
Portland, OR - WA	0.11	0.10	0.15	0.12	0.13	0.09	0.15
Riverside - San Bernardino - Ontario, CA	0.42	0.38	0.34	0.32	0.28	0.28	0.30
Sacramento, CA	0.16	0.17	0.17	0.16	0.17	0.14	0.16
San Antonio, TX	0.11	0.12	0.12	0.14	0.12	0.11	0.10
San Jose, CA	0.17	0.19	0.14	0.14	0.12	0.13	0.12
San Juan, PR	ND	ND	0.07	0.02	0.09	0.06	0.07
San Francisco - Oakland, CA	0.14	0.18	0.14	0.14	0.10	0.09	0.06
San Diego, CA	0.36	0.22	0.24	0.21	0.19	0.19	0.17
Seattle - Everett, WA	0.13	0.09	0.12	0.09	0.11	0.11	0.13
St. Louis, MO - IL	0.16	0.18	0.15	0.15	0.15	0.13	0.13
Tampa - St. Petersburg, FL	0.11	0.13	0.11	0.11	0.12	0.10	0.11
Washington, DC - MD - VA	0.18	0.19	0.15	0.15	0.18	0.10	0.13

Table 6.4: Ozone levels at selected sites in the United Kingdom (Calendar year hourly maximun and number of hours in excees of 90ppbv)

	Sibton	Stevenage	Harwell	Central London[1]	Central London[2]	Bottesford
	rural	suburban	rural	urban	urban	rural
1972						
1973	142 (33)				126 (20)	
1974	85 (0)				136 (78)	
1975	161 (66)				164 (21)	
1976	104 (13)	207 (149)	258 (137)		92 (1)	
1977	99 (51)	135 (12)			144 (37)	
1978	127 (37)	100 (4)			99 (3)	
1979	100 (3)	108 (9)			149 (17)	
1980	63 (0)	84 (0)			95 (2)	
1981	207 (141)	68 (0)			76 (0)	79 (0)
1982	124 (18)	164 (87)	142 (41)		128 (12)	111 (21)
1983	133 (19)	168 (25)	114 (32)		78 (0)	99 (7)
1984	124 (11)	174 (81)	103 (6)		77 (0)	111 (26)
1985	88 (0)	123 (13)	137 (13)		68 (0)	109 (18)
1986		103 (14)	107 (2)		149 (28)	103 (5)
1987	100 (3)	85 (0)	81 (0)		105 (5)	104 (7)
1988	106 (4)	70 (0)	117 (29)		67 (0)	68 (0)
1989	122 (35)	101 (5)	132 (37)		78 (0)	83 (0)
1990	145 (35)	136 (42)	91 (2)		108 (5)	138 (27)
1991	86 (0)	84 (0)	116 (31)	63 (0)	88 (0)	49 (0)
1992	116 (21)	89 (0)		76 (0)		81 (0)

Central London sites at (1) Bridge Place, second floor office overlooking a backstreet near Victoria station and (2) situated facing a quiet backstreet located in Victoria.

excess of 90 ppbv (the Department of the Environment poor air quality band) are given. Comparison with the two previous tables reveals that UK concentrations are not atypical of other cities. In the London area in 1976 there was a period of 5 consecutive days or more when ozone levels exceeded 100 ppbv[38]. The industrial threshold limit value was exceeded in outdoor air over much of south-east England for the equivalent of a working week.

Measurements of ozone concentrations have been carried out in London since 1972, when a peak hourly ozone concentration of 112 ppbv was recorded, at that time the highest value to be recorded in Britain[39]. It was suggested that a significant portion of measured ozone in central London was transported there from the continent[40]. The results of ozone measurements in Greater London during the fine summers of 1975 and 1976 (150 ppbv in 1975; 212 ppbv in 1976) have suggested two possible sources for the elevated concentrations[34,41,42]. The first is long-range transport from the European continent and the second is from photochemical reactions involving locally emitted precursors. From an analysis of the ozone concentration profile across Greater London in terms of wind speed and direction, Ball and Barnard[41] concluded that precursor pollutants from London were a significant source of ozone and could produce up to 80 ppbv ozone.

Not all high urban levels are the result of local photochemical processes. In order to estimate the natural ozone component in urban air, measurements are made in rural areas and from these the natural ozone concentrations are estimated. For a typical natural rural concentration of 40 ppbv in the USA, it has been estimated that urban concentrations will have 20 ppbv "natural" ozone[43]. Others have found evidence of high ozone levels in rural areas of New York State being transported into urban areas through advection and vertical mixing[44,45]. Further evidence of long distance transport of ozone in urban areas was provided by Chan et al.[50] who measured ozone aloft and at the surface upwind of Philadelphia. By relating ozone aloft to surface measurements they found ozone transport ranged from a few ppbv to 123 ppbv.

Certain cities are particularly prone to meteorological conditions favouring ozone production (see below) These include : Athens[33,47-50], Los Angeles[51,52], Melbourne[53,54], Sydney[55], Tokyo[56], Harmia in Japan[57], and in Tel Aviv[58] amongst others. In these regions photochemical pollution abatement strategies must take into account horizontal advection and recirculation of pollutants. Over the sea, ozone and its precursors may find conditions such as a greater atmospheric stability, limited vertical exchange and lower destruction processes and hence may be transported over fairly long distances without dilution or transformation. Due to a smaller deposition velocity, ozone has a lifetime 15 times longer over the sea than over dry land.

6.3   OZONE CONCENTRATIONS AT RURAL SITES

Pollutants in the atmosphere are carried to other locations (advection), diluted to lower concentrations (diffusion), modified in their physico-chemical form (conversion) and eventually removed at the Earth's surface (deposition). The overall result of these transport and transformation processes on a pollutant depends on the motions and other natural properties of the atmosphere, the presence of other pollutants and the properties of the substance in question. All these processes affect the lifetime of the pollutant.

Table 6.5: Time and space scales for air pollution transport

Category	Distances	Timescale
Local	0- 10km	1 hour - 1 day
Urban	10- 100km	1 hour - 1 day
Long-range		
Mid range	100- 250km	3 hours - 1 day
Regional	200- 1000km	8 hours - 4 days
Continental	5000- 5000km	1 day - 2 weeks
Global	5000-400000km	2 weeks or more

Pollution transport is affected not only by atmospheric motions, and meteorological variables, which fluctuate over a wide spectrum of time and distance scales, but also by the emission patterns and rates of transformation and deposition of the pollutants. A number of different terms and values may be applied to the typical time and distance scales that characterise air pollution transport as shown in Table 6.5. Pollution transport may be divided into four types based on the distances involved: local, urban, long-range and global. Of particular importance for rural ozone is long-range or regional scale transport.

It was initially thought that photochemical pollution was restricted to urban areas where the emissions of chemical precursors occur. However, the results of monitoring programmes have shown that the situation is not a local, relatively isolated urban problem, but a multi-regional problem. In fact, rural areas generally show a higher frequency of elevated ozone than urban areas as air pollution emissions move more or less freely from one region to another. Over the years, elevated ozone concentrations have been measured in a number of downwind rural locations where local ozone sources were considered insignificant. It has been suggested that long distance atmospheric transport is responsible for many cases of high ozone concentrations found over rural areas. The lifetime of ozone for long-range transport is controlled by dry deposition to the earth's surface. The residence time is typically about 60 hours which gives a mean transport distance in the range 1000-1500 km. Recent measurements have indicated that ozone derived from anthropogenic pollution has a hemisphere wide effect at northern temperate latitudes[356].

Elevated ozone levels in rural areas are almost solely of anthropogenic origin, but the ozone and/or ozone precursors are transported from urban areas, often over distances of hundreds of miles. Other than in stagnant conditions, air is continuously advected from an urban area and ozone formation may continue in the resultant plume if it still contains NO_x and hydrocarbons. It has been shown that the hydrocarbon/NO_x ratio increases downwind of an urban area[59], while Isaksen et al.[60] showed that the higher this ratio the more effective the ozone production becomes, with between 5 and 9 ozone molecules being produced for each NO_x molecule released during the first day of transport. This negative effect of NO_x is due, initially, to large amounts of nitrogen oxides which suppress the build-up of peroxy radicals, and the consequent reduction in the number of NO to NO_2 conversions that can take place before the mixture is chemically deactivated. As the air parcel moves away from an urban area into suburban-rural areas, precursor emissions decline. Ozone concentrations increase, initially the result of the more reactive hydrocarbons and later, possibly 5-10 hours, the results of medium reactivity hydrocarbons. Photochemical production of ozone ceases at the onset of sunset. The ozone-rich air may be trapped aloft at night-time above a surface-based radiation inversion. The ozone in this convectively stable layer is depleted by dry deposition while that aloft is effectively cut off from all sinks. Ozone levels at the surface will rise the following morning as the inversion breaks up and the second day's photochemistry would feed on any remaining hydrocarbons.

In rural areas, the diurnal variation of ozone has been well described during the last 50 years[30,61,62-68]. The form of the diurnal pattern in ozone concentration can provide clues as to the formation, scavenging and transport of ozone at a site. For instance, the shape of the diurnal curve can be used to ascertain if the site is exposed to locally generated or long-range transported ozone. The warm, sunny anticyclonic conditions which bring high daytime ozone levels (see below) are also conducive to the formation of night-time inversions due to radiative cooling at the surface. This establishes a shallow layer near the ground under the influence of mechanical stress[329], which is decoupled from an overlying layer of at least intermittent turbulence, by a zone with little or no turbulence. This severely restricts the vertical mixing of air. Balloon observations[330,331] have shown that above the inversion, the ozone is effectively cut off from all sinks while below the inversion, ozone is removed by dry deposition at the ground. At greater heights, the reduction in concentration is small, depending mostly on the downward mixing of air through the stable layer. Following the disappearance of the inversion in the morning, downward mixing of air from the previous day's reservoir aloft results in a rapid increase in ozone at ground-level. Fehsenfeld et al.[66] found that the mid-morning ozone concentrations were typical of upper-air values, while Harrison and Holman[69] postulated that these concentrations were indicative of those arising from long-range transport of ozone. From data collected at three sites, Kelly et al.[68] concluded that although most of the photochemical ozone production occurred between 1000 and 1400 local time, only very small amounts of ozone were formed (6 ppbv O_3 per ppbv NO_x) and some increase in ozone during this period was due to further vertical mixing.

Figure 6.6

Mean diurnal variations of ozone, July - December 1983, at Stodday, near Lancaster.

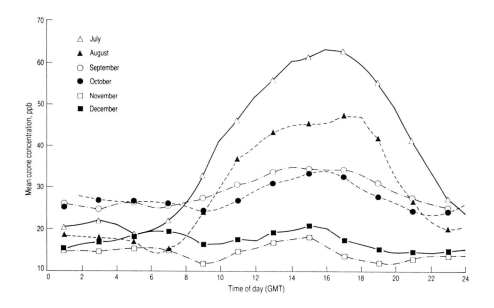

The diurnal variation of ozone in rural areas shows a significant seasonal difference (Figure 6.6). During the summer ozone concentrations show a systematic diurnal variation, while in the winter the concentration remains almost constant. Individual diurnal cycles can be very different to the average, however. In the summer, on days when no nocturnal inversion forms, the night-time levels stay near the day-time values. In winter, winds are generally stronger and the weather overcast, hence the formation of the nocturnal inversion may be inhibited. Alternatively, in clear winter weather this stable layer may persist for several days and ozone concentrations can remain near zero for long periods. Photochemical production of ozone is also limited in winter because of the much shorter daylight period, reduced level of uv radiation and lower temperatures. Bohm et al.[70] concluded that, for areas near coniferous forests in the western US, a given day of hourly data can be placed into one of 17 cases of diurnal curves simply by knowing the 24h mean and coefficient of variation or range. The diurnal curves are shown in Figure 6.7. The curves designated 'A' exhibit little diurnal variation whilst those designated 'E' are characterized by large diurnal variations. The number following the letter relates to the relative magnitude of the 24h

Figure 6.7

Seventeen typical patterns of hourly ozone concentrations as determined by cluster analysis (After Bohm et al.[70]).

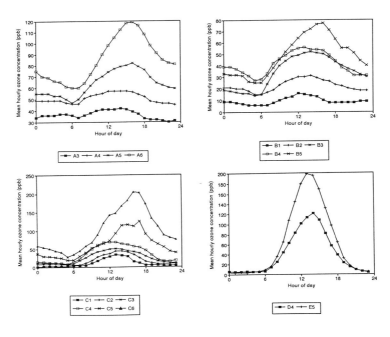

mean (1 has the lowest mean). The variation among the curves may be related to the theory of ozone formation, scavenging and transport. Sites close to large urban areas exhibit a large diurnal variation whilst sites far from urban areas have curves with little variation. Hence sites located close to large urban centres can be characterized by urban patterns, B1, C1, C2 and C3. Those located further away from the source areas exhibit patterns, B3, B4 and B5 and sites on the fringe of large urban settlements experience, in addition to the above, patterns C5 and C6.

Urban plumes

The continuous emission of pollutants into a moving fluid will result in the formation of a pollutant plume which dilutes and disperses with a characteristic time indicative of the flow in the fluid. In the planetary boundary layer, and under stable conditions, such plumes are easily recognisable when emitted from point-sources. However, diffuse source regions, such as highly developed urban complexes, are of sufficient magnitude and structural diversity to make direct human observation impossible. Nevertheless, the analogy with point-source plumes has been made and is seen to yield meaningful results[71]. It is apparent that urban plumes can contribute significantly to the total ozone budget in rural environments[72]. Concentrations across the plume are approximately constant, but they are a complex function of meteorological and kinetic parameters in the mean wind direction[72].

The literature on field studies of ozone in urban plumes is summarised in Table 6.6. To keep the table brief, those studies composed of many individual measurements are cited as giving a range of ozone increments. Thus, instead of in- and out-of-plume measurements, a single parameter, ΔO_3, is reported. ΔO_3 is defined as the ozone concentration in an urban plume minus the ozone concentration of air surrounding the plume. Studies which report ozone concentrations downwind of urban centres without any information on the surrounding non-urban air quality, are not reported (*e.g.* Ref 73). Up to 150 ppbv is attributed to photochemistry in a plume crossing New Jersey, USA[74]; the maximum cited for a European city is 72 ppbv downwind of London, UK. Only Varey *et al.*[75] cite negative values and this is perhaps surprising given the discussion in § 6.2 above. The reason for this lack of plume ozone deficits appears to be the method used to detect the plume. In many cases (*e.g.* Ref 72,69), it is the presence of increased ozone levels that is in fact used to identify the arrival of an urban plume at a receptor site. Only when additional meteorological information is available[75] or, an unrelated chemical indicator, such as $CFCl_3$,

Table 6.6: Ozone increments in urban plumes

REFERENCE	SOURCE AREA	DATE	RECEPTOR AREA	OZONE CONCENTRATION, ppbv WITHIN PLUME	OUTSIDE OF PLUME	DISTANCE DOWNWIND, km	TIME OF DAY
Sexton[33]	Madison, Janesville; Wisconsin; Beloit and Rockford, IL	1977/78	Wisconsin and Illinois	#ΔO_3 = 20-30		50-65	Afternoon
Sexton and Westberg[76]	Chicago NW Indiana	15.8.77	Wisconsin	150	70-80	170	1450-1720CDT
Altshuller[72]	St Louis, MO	July 1975	St Louis Environs	ΔO_3 = 30-120		various	various
White et al.[334]	St Louis, MO	18.7.75	New Springfield, Illinois	130	70	140	1631-1656CDT
Karl[335]	St Louis, MO	19.8.75	N & W of St Louis	180-250	125-250	25-40	1200-1600CDT
Spicer et al.[336]	Springfield, IL	3.8.77	Illinois	70-80	≤50	70	Afternoon
Lindsay and Chameides[337]	Atlanta, Georgia	1983/84	NW of SE of Atlanta	ΔO_3 = 20-40		10-20	various
Chung[338]	Toronto, Canada	Summer 1976	Ajax nr. Toronto	ΔO_3 ≤ 60		20-50	various
Harrison and Holman[69]	Industrialised regions of England	1978	Heysham, Lancs	ΔO_3 = 20-50		80-400	Afternoon
Varey et al.[75]	London, UK	1982-86	London environs	ΔO_3 = (-20)-70		30-90	various
Colbeck and Harrison[81,82]	Midlands of England	July 1983	Stodday, Lancs, UK	ΔO_3 =15-160		30-140	various
Van Valin and Luria[339]	East Coast, USA	26.3.85-5.3.85	W. Atlantic W. Atlantic	5 65	50 30	100 unknown	1255-1345EST 1630-1720EST
Spicer et al.[340]	Boston, MA	18.8.78-23.8.78	Atlantic Ocean Atlantic Ocean	130 154	60 NA*	100 120	1540EDT 1600-1630EDT
Zeller et al.[341]	Boston, MA	10.8.75	West of Boston	100-150	70-100	50-100	1130-1430EDT

Table 6.6(contd): Ozone increments in urban plumes

REFERENCE	SOURCE AREA	DATE	RECEPTOR AREA	OZONE CONCENTRATION, ppbv		DISTANCE DOWNWIND, km	TIME OF DAY
				WITHIN PLUME	OUTSIDE OF PLUME		
Siple et al.[342]	New York St. Long Island	14.8.75	Atlantic Ocean	214	NA	200	1600EDT
Spicer et al.[343]	New York City	18.7.75	NW & W of Boston	150-200	60	400	2100EDT
Spicer et al.[344]	New York City and adjacent areas	23.7.75	S. Connecticut	250-300	70-85	125	1500EST
			NE Connecticut	150-200	80-90	200	2100EST
			NW of Boston	100-150	60-70	300	2400EST
		24.7.75	NE of Boston	130	NA	350	Morning
			Atlantic Ocean	145	NA	450	Afternoon
		9.8.75	Atlantic Ocean	130	NA	160	Evening
Cleveland et al.[74,345,346]	New Jersey	July 1974	New Jersey	150-270	50-120	30-50	Afternoon
Chan et al.[46]	Philadelphia, PA	16.7.79	Pennsylvania	150	60	40	1500-1500LST
Wolff et al.[347]	Philadelphia, PA	10.8.75	New Jersey	194	139[+]	45	Afternoon
Clark and Clarke[348]	Washington, DC & Baltimore, MD	14.8.80	Pennsylvania	140-170	NA	160	1534-1641EST
Possiel et al.[349]	Tulsa	1977	N. of Tulsa	$\Delta O_3 = 60$		34	Afternoon
Blake et al.[350] (1993)	London	12.7.88	East Anglia	$\Delta O_3 = 15$		130	Afternoon

[*] NA Not Available
[+] Upwind of Philadelphia
[#]ΔO_3 Plume Ozone-Background Ozone

249

is used[76] can the plume be positively inferred without necessarily assuming that photochemistry has "matured" the air mass to reveal some or all of its potential. Whilst high levels of ozone in urban plumes has received much attention for the perfectly understandable reason that it is high levels of ozone which are hazardous, a sufficient understanding of photochemical oxidant behaviour requires a rationalisation of the behaviour of the urban plume under all conditions. A further general point merits note. In the great majority of cases cited, ozone concentrations outside the plume are well above ambient background levels, and in some instances exceed the current US NAAQS guideline hourly average of 120 ppbv. Varey et al.[75] provide evidence for a positive correlation between estimated ΔO_3 and 7 hr average ozone outside the plume. That is, photochemical ozone generation within an urban plume often appears to be embedded within a regional-scale episode. At present it is not clear whether this implies that maximum urban ozone effects require the injection of fresh precursors into an already polluted air parcel, or that the correlation is simply an artefact of other inter-relating factors which affect both parameters.

The London plume

In 1982 a number of research institutions initiated the London Ozone Survey. Its primary objectives were to establish an inter-calibrated network of ground-based monitoring sites in southeast England for ozone and other relevant pollutants, and, from this data to investigate the behaviour of ozone throughout the boundary layer, with emphasis on both long-range transport and on ozone formation in the urban plume[77]. Varey et al.[75] extracted data for days between 1982 and 1985 in which ozone concentrations in some part of the region exceeded 80 ppbv. Figure 6.8 shows the ground-level ozone data for one day during one of these episodes together with wind data from the Post Office tower. To distinguish the generation of ozone due solely to the plume from that occurring on a regional scale, mean regional out-of-plume maxima were subtracted from the plume maximum to give an "urban effect" for ozone downwind of London. This urban effect correlates significantly with mean regional ozone, wind speed, maximum daytime temperature and time between the airmass passing the urban centre and being measured. The plume ozone data can also be plotted on an emission/travel time surface, Figure 6.9. Here two areas of elevated maxima can be identified, both of which occur at approximately the same time of day, 14-1800 (BST). The circled area near the x-axis represents the optimum time for local ozone formation in or near the urban centre: air parcels pass over

Figure 6.8

Hourly mean ozone concentrations for each site operating on 11 July 1983, which formed an input to the London Ozone Study. Wind speed and direction from the Post Office Tower.

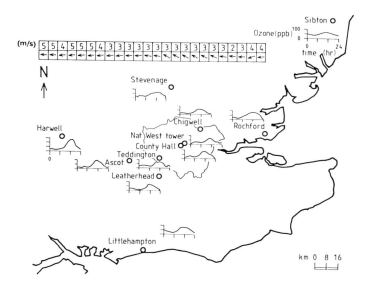

the city at around solar noon and travel for only a few hours before reaching a maximum. The second circled area encompasses the optimum conditions for elevated ozone concentrations in the plume at greater travel times: an air parcel crossing the city earlier in the day requires longer to produce maximum ozone levels than in the previous case. Presumably this is because the precursors emitted then require a greater integrated photon flux to generate ozone than in the previous situation. Interestingly, neither area coincides with the morning or evening rush hours. Modelling studies suggest that this may be due to the strong ozone depleting capacity of urban NO emissions in the short-term[78].

Measurements made from aircraft enable the chemical history of urban emissions to be established as they are transported over hundreds of kilometres. Flights across London have measured ozone, sulphur dioxide and other gases from a Jetstream aircraft[79]. Figure 6.10 shows ozone results from a flight on 12th July 1983. On this day, winds were predominantly north easterly and ozone levels in excess of 100 ppbv were attained at ground level sites downwind of London[34]. The flight details show higher ozone concentrations on the downwind side of Greater London, providing further evidence of local

Figure 6.9

The relation of maximum plume ozone concentrations to time at which the air parcel passed over central London (emission time) and time taken to reach the monitoring site (time downwind). The data covers all ozone episodes in southeast England from 1982-1985.

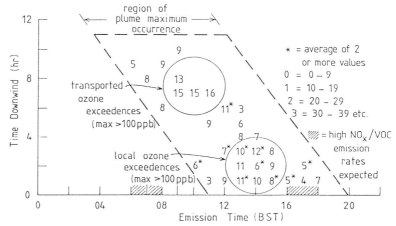

N.B. Actual time of plume maximum = emission time + travel time

ozone generation.

An expanding box model of the urban plume has been used to give more quantitative data on the factors influencing plume concentrations at the end of one day's travel[78]. A 2^6 factorial experiment was carried out using urban and rural emission rates of NO, CO and hydrocarbons as variables. It shows the dominance of urban emissions in the determination of plume maxima 100 km downwind of the upwind urban boundary. A sample of the model output is shown in Figure 6.11. The effect of urban hydrocarbon emissions is seen by comparing the two graphs. The effect of urban NO emissions is seen in the two broad bands of model results within each graph. The breadth of each band is due to other, less important factors. Changing urban hydrocarbon emissions from low to high levels results in an increase in plume ozone of 10-25 ppbv, depending on the level of NO emissions. That is, the NO and VOC emissions interact strongly and non-linearly[80]. This has been discussed in more detail in chapter 4.

The Harwell photochemical trajectory model, which employs a detailed description of the atmospheric physics and chemistry of photochemical ozone formation in the UK, but a highly simplified representation of the relevant transport processes, has been used to

252

Figure 6.10

Ozone and sulphur dioxide concentrations obtained by instrumented aircraft on 12 July 1983, during a period of northeasterly winds (After Ball[34]).

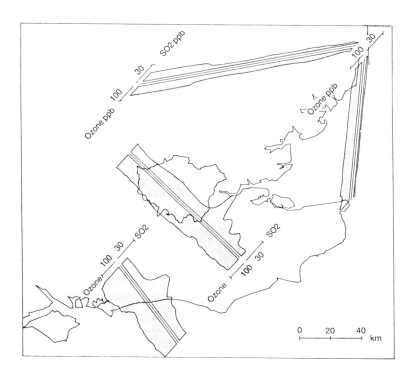

study, amongst other things, the role of different hydrocarbons on ozone formation, mechanisms for photochemical ozone production, and sensitivity and uncertainty analysis[81-89]. The model gives good agreement between predicted and measured levels of ozone, PAN and aerosol downwind of London. It shows that it is possible to form significant ozone concentrations, up to 100 ppbv, downwind of London, but the formation of higher levels requires persistent and slow-moving anticyclonic systems. Aromatic hydrocarbons were found to make the largest contribution to ozone and PAN formation during the first day's photochemistry and, under most circumstances, during the first day's photochemistry, control of hydrocarbon emissions would decrease ozone formation whereas control of NO emissions may increase or decrease ozone concentrations. Both models conclude that total hydrocarbon emissions, NO_x emissions and photolysis rates are the most important factors

Figure 6.11

Effect of emissions on plume concentrations downwind of London.

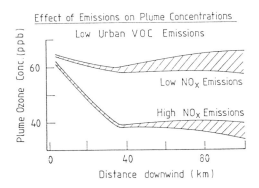

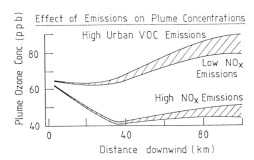

in determining ozone concentrations downwind of London[78,87]. Tracing ozone-rich air masses back to specific urban plumes is difficult. Backward air mass trajectories provide one method[78], aircraft measurements another. Models may be used in such a manner that the modelled air mass follows the atmospheric mean motion, as discussed above, but vertical wind shear may present a significant problem to such an approach[90].

6.4   AMBIENT OZONE CONCENTRATIONS IN THE UNITED KINGDOM AND IRELAND

In Britain, elevated ozone concentrations were first observed at Harwell, Oxfordshire in 1971[91] and subsequently at a number of sites, the majority of which are situated in south-east England. The Harwell measurements showed that ozone concentrations exceeded 80

254

ppbv on 6 days and 100 ppbv on 2 days in June/July. Measurements of ozone and halocarbons at three sites (Sibton, Suffolk; Harwell; Adrigole, County Cork) in August 1973 provided evidence that ozone and/or its precursors, under suitable meteorological conditions, can be transported over distances of the order 100-1000 km, and that continental emissions provide a major contribution to photochemical pollution in Britain[92]. The site in Adrigole (SW coast of Ireland) is often taken as a remote. In air advected from the Atlantic, daily ozone maxima range between 40-60 ppbv, however 80 ppbv ozone are quite often exceeded during the summer months. The occurrence of these higher mixing ratios coincides with maxima of CCl_3F concentrations. This halocarbon may be used as a tracer of anthropogenic pollution.

The summer of 1976 was an extreme example of the conditions required for photochemical ozone formation. The meteorological conditions were particularly favourable and persisted for a long period. Apling et al.[38] measured ozone at sites in southeastern England. Peak hourly concentrations reached 212 ppbv in central London and 258 ppbv in Harwell, the highest level ever reported in Britain. They pointed out that the primary pollutant plume requires several hours of solar radiation before photochemical ozone can be detected. In this time the plume may travel 50 km or more downwind of the primary source, so identification of the source is difficult in densely populated and highly industrialized regions.

Elevated ozone levels have also been observed in north-west England, around Lancaster[69,93-95]. This part of England is most geographically remote from the European continent. It has been shown that in this area there are three possible sources of photochemical ozone: advection from distant sources (continental Europe); advection within the pollution plumes from British urban areas; and local formation from local emitted and advected precursors. Harrison and Holman[69] attributed 50-80 ppbv ozone to long-range transport with an additional 20-50 ppbv from middle distance sources within Britain. More recently, Colbeck and Harrison[93] concluded that UK sources could contribute up to 60 ppbv to ozone levels at this same site. These values are similar to those reported by Simpson and Davies[96] who found that the passage of clean air over a number of urban-industrial regions of England could enhance ozone levels at Sibton, a rural site in East Anglia, by 50-130 ppbv. Colbeck and Harrison[94] also found that ozone concentrations exceeded 80 ppbv on 4.2% of the days monitored and nearly half of these days fell in May and July. They concluded that photochemical pollution was the most frequent cause of these elevated concentrations.

Evidence for the variability of ozone from year to year was shown in data from Bottesford, a rural site in central England[97]. In 1978 the hourly ozone concentration exceeded 80 ppbv for 10 days while in 1979 this concentration was exceeded for 25 days. These summers have been noted for the poor weather thus, even in bad summers, levels in excess of accepted air quality standards have been recorded.

Until 1986 only a limited number of ambient air quality monitoring stations were operated in the UK. Measurements were sporadic, uncoordinated and covered only a small area of the country. These measurements, from 47 sites, were summarised in a comprehensive report prepared at the request of the Department of the Environment by the UK Photochemical Oxidants Review Group, PORG[98]. Figure 6.12 illustrates the geographical coverage of the sites. Most of these sites operated for limited periods only.

The Royal Commission on Environmental Pollution in their 10th report, recommended that since available data on ozone and NO_x were limited and incomplete; the UK's baseline rural monitoring networks should be extended with a substantially increased number of sites measuring photochemical pollutants. The Warren Spring Laboratory was made responsible for developing the network, establishing the bulk of its monitoring stations, and coordinating the smooth running of the network as a whole. A network of 16 stations, together with an additional station at Mace Head in the Republic of Ireland is now established[99,100]. Figure 6.13 illustrates that wide geographical distribution has been attained, with a good mix of coastal and inland locations. Details of the site environment are listed in Table 6.7, while details of site selection criteria, quality assurance and organisational features are given by Bower[101].

The availability of data from the new network, and other sites, has presented new opportunities for studies of ozone episodes over the UK. During late April 1987 an ozone episode occurred during which mid-afternoon hourly ozone concentrations reached between 80 and 100 ppbv at many sites during a 5 day period. Table 6.8 shows maximum hourly concentrations for various sites. Maximum concentrations were recorded at Bush on 26th April (101 ppbv) and 28th April (98 ppbv), the smallest maxima were monitored at central London. Ozone budgets, derived by following air columns between well-mixed phases on successive days along trajectories indicate a net ozone generation of about 4 ppbv her hour of bright sunshine[102]. This implies maximum daily generation of 45 ppbv ozone in April and up to 60 ppbv in mid-summer. A photochemical model was employed to estimate the upper limit or potential for ozone generation along the trajectories[103]. There was good correlation between the observed and model ozone formation potential although the latter

Figure 6.12

The locations of the ozone monitoring sites in the United Kingdom and Republic of Ireland.

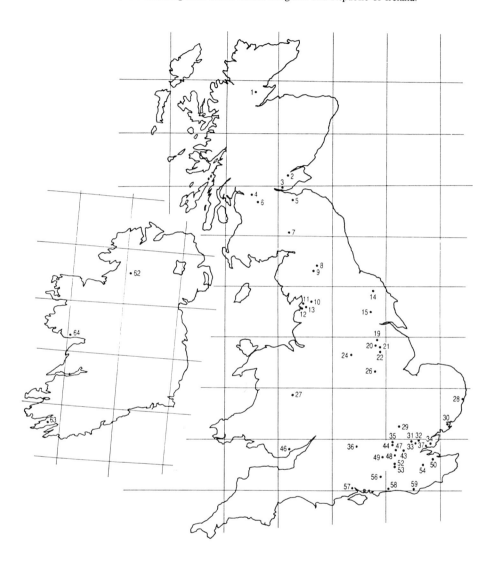

overestimated the former by a factor of 2.8 (a sharp reminder, perhaps, that a good general understanding of a non-linear process, such as ozone formation, is not equivalent to its accurate quantification at a specific time and place).

Data collated in the PORG report[98] together with that of the national network[104] will

Figure 6.13

The UK ozone monitoring network.

now be used to illustrate seasonal variations, temporal variations and magnitude and duration of ozone episodes in the UK, as a paradigm for other countries and regions.

Ozone measurements at rural sites typically follow a broad envelope which shows a maximum in April and May. Peak hourly mean concentrations of over 100 ppbv are recorded as episodes or isolated events and these are superimposed on a background distribution. Figures 6.14-6.16 illustrate methods in which the seasonal variation may be presented. In Figure 6.14 curves B and C are representative of background levels and indicate a value between 20 and 50 ppbv. Figures 6.15 and 6.16 indicate that remote sites are less susceptible to photochemical pollution, have less annual variability, but still exhibit a spring maximum in ozone concentrations. The source of ozone from the stratosphere might be expected to give rise to a late winter or spring maximum in tropospheric ozone, since stratosphere-troposphere exchange is most effective during this period[105-107]. Peak spring ozone levels have generally been considered to be due to stratosphere ozone

Table 6.7: United Kingdom - site information

Site	Grid Reference	Altitude	Commencement Date	Site Environment	Sampling Height, m
Stevenage	TL237225	90	1976	A suburban site on the edge of a residential new town near a light industrial estate. 100m east of A1(M) motorway. Topography flat.	8
Central London	TQ291790	20	1976	An urban background station and also part of the NO_2 Directive Network. Situated facing a quiet back street within a busy city centre location in Victoria, Central London.	10
Sibton	TM364719	46	1976	Open flat cereal farmland. Woodland to the north west.	4
Aston Hill	SO298901	370	1986	On the summit of a hill with clear views of the surrounding arable farmlands.	3
Lullington Heath	TQ538016	120	1986	On a high plateau 5 km from the south coast. Immediate area is a NCC heathland.	4
Strath Vaich	NH347750	270	1987	Remote hilly moorland, used for sheep-grazing.	3
High Muffles	SE776939	267	1987	Hilly moorland and forestry plantation.	3
Yarner Wood	SX86789	119	1987	Undulating farmland with forestry plantations.	4
Ladybower	SK164892	420	1988	Adjacent to forested area overlooking reservoir in Peak National Park.	4
Mace Head*	IL740320	15	1987	Open flat land on the Atlantic Coast.	3
Lough Navar	HO65545	130	1987	Clearing within a forestry plantation.	2
Harwell	SU474863	137	1972	Site is adjacent to Harwell research laboratories which is surrounded by flat cereal fields. The busy A34 is nearby.	2
Bush	NT245635	180	1986	On a flat plain between hills. Site surrounded by open and forested land. ITE buildings nearby.	4
Eskdalemuir	NT235028	269	1986	Situated on open moorland adjacent (500 m) to Met Office Laboratory.	3
Gt Dun Fell	NY11322	847	1984	Near the mountain summit and often above the cloud base.	3
Wharley Croft	NY698247	206	1985	Within a valley at the base of Great Dun Fell.	3
Glazebury	SJ690959	21	1988	On open, flat, ground in horticultural area between the conurbations.	2
Bottesford**	SK797376	32	1977	1 km from Bottesford village on open farmland.	2

* Established with the co-operation of the Department of the Environment of the Republic of Ireland.
** Not a DoE-funded site.

259

Table 6.8: Maximum hourly ozone concentrations (in ppbv) for each day during the episode (After Weston et al.[102])

Site	24 April	25 April	26 April	27 April	28 April	29 April	30 April
Strath Vaich	41(21)	65(18)	89(13)	92(14)	65(01)	60(16)	37(18)
Bush Estate	60(16)	74(14)	101(15)	84(16)	98(15)	66(11)	45(12)
Eskdalemuir	65(16)	80(11)	73(18)	84(16)	70(19)	54(10)	37(02)
Great Dun Fell	85(14)	82(12)	81(17)	90(10)	88(18)	61(20)	52(01)
Jenny Hurn	89(14)	83(16)	76(15)	91(16)	95(17)	64(16)	50(16)
Thorney	83(13)	78(16)	68(15)	80(14)	79(16)	50(19)	33(15)
Lincoln	80(13)	73(16)	52(14)	71(14)	70(16)	45(18)	38(12)
Syda	83(13)	70(15)	65(17)	78(17)	80(17)	53(16)	36(16)
Bottesford	77(14)	75(17)	67(16)	81(15)	93(17)	64(17)	42(18)
Aston Hill	65(20)	70(16)	56(17)	71(17)	75(19)	41(19)	37(21)
Stevenage	76(15)	53(14)	49(16)	69(13)	75(18)	37(15)	23(23)
Harwell	72(14)	45(13)	53(16)	87(16)	72(14)	55(14)	44(18)
Central London	33(16)	30(13)	35(15)	21(18)	21(18)	29(14)	28(15)
Lough Navar	72(19)	78(17)	88(17)	68(15)	79(17)	50(09)	46(19)
Macehead	53(19)	62(15)	55(24)	56(01)	57(17)	48(02)	42(12)

The hour of occurrence in GMT is shown in brackets

intrusions[7,108,109].

However, recent measurements of PAN and ozone[99] and estimates of the photochemical lifetime of ozone in winter[110], up to 200 days at 40°N, suggest that this spring increase could be due, in part, to tropospheric ozone production. As discussed in § 6.9 polluted air PAN concentrations peak in summer, while a spring maximum has been observed in clean air at Harwell[111] as shown in Figure 6.17. Since one molecule of PAN may produce approximately 50 molecules of ozone[112] the peak PAN concentration in Figure 6.17 could form up to 15 ppbv of ozone. Uncertainties exist in the ratio of formation rates of the two molecules at the end of winter and their differences in lifetimes in the atmosphere. Being thermally unstable, PAN has a much longer lifetime in winter than in the summer. The photochemical lifetime of ozone is also seasonally dependent in the same sense.

Concentration of pollutants, do of course, vary from year to year and reason for this will be discussed later. Table 6.9 lists the April to September average concentration for five sites in the UK.

During the summer months, daily ozone maxima do not always occur in the afternoon[113]. Figure 6.18 shows that maxima in concentration occurred during the night approximately 11% of the time at Stodday between May and September[114]. Just over 50%

Figure 6.14

Long-term ozone variations at Stodday, Lancs.

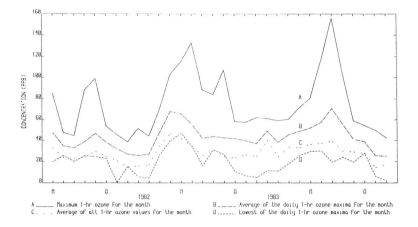

A ———— Maximum 1-hr ozone for the month   B ——— Average of the daily 1-hr ozone maxima for the month
C . . . Average of all 1-hr ozone values for the month D .... Lowest of the daily 1-hr ozone maxima for the month

Figure 6.15

Maximum hourly mean ozone concentrations for 1991 at (a) Lullington Heath and (b) High Muffles.

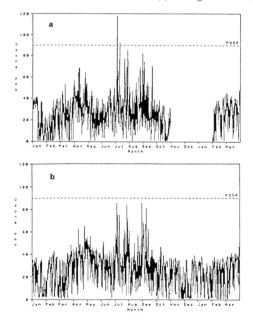

261

Figure 6.16

Ozone 1987-1988 seasonal variations of monthly averages and 99th percentile of hourly values in each month (From Bower et al.[104]).

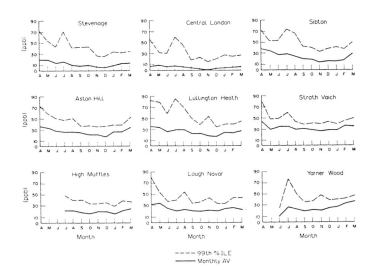

Figure 6.17

PAN concentrations between January 1980 and April 1981 at Harwell (From Penkett and Brice[111]).

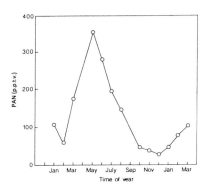

Figure 6.18

Frequency of occurrence of the daily maximum hourly ozone concentrations at each hour of the day.

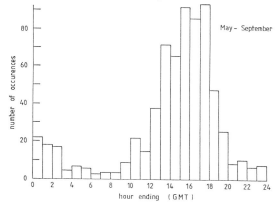

Figure 6.19

Ozone concentration and windspeed measured on 20-21 August 1982.

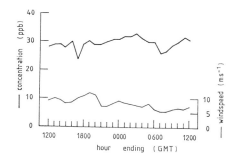

Figure 6.20

Ozone concentration , windspeed, relative humidity and rainfall measured on 20-21 September 1982.

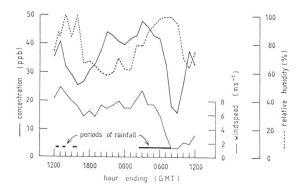

Table 6.9: April to September average ozone concentration (ppbv)

	Stevenage	Central London	Sibton	Harwell	Bottesford
1972		19			
1973		22	31		
1974		23	31		
1975		15	38		
1976	66	25	37	93	
1977	26	20	36		
1978	20	20	34		
1979	18	14	26		
1980	11	12	19		
1981	8	12	30		22
1982	24	14	30		28
1983	24	13	34	34	26
1985	19	24	24	32	25
1986	16	14		26	26
1987	14	6	25	25	21
1988	15	10	31	24	19
1989	16	14	34	25	20
1990	19	13	31	29	20
1991	17	13*	32	29	
1992	18	13*	31	29	19

* Site location changed, see Table 6.4

of these night-time maxima were on occasions when the ozone concentration stayed approximately constant throughout the day (Figure 6.19). On these days both ambient temperature and hours of sunshine, variables which determine the photochemical reaction rates, were below average and windspeeds were high, averaging in excess of 4.8 ms^{-1}. Such conditions inhibit the formation of the nocturnal stable layer and lead to a "well mixed" atmosphere. Throughout the night there may be isolated bursts of turbulence, when the boundary layer overturns; resulting in vertical downward mixing of air from the remnant daytime boundary layer. This mixing is forced by an increase in windspeed. Ground level ozone concentration is hence an excellent indicator of night-time atmospheric stability. Just over 30% of the nocturnal ozone maxima at this site could be correlated with an increase in windspeed and/or the onset of rain (Figure 6.20). Similar increases in ozone at night during heavy rain have been reported[115].

A night-time maximum is characteristic of mountain stations located above the nocturnal inversion, which isolates the site from the effects of surface deposition[116-119]. Other possible explanations include convective mixing of ozone from below and entrainment from above as the boundary layer grows, advection of higher concentrations of ozone in the free atmosphere when the mountain top is above the nocturnal boundary layer and a downslope wind regime with less chemical destruction during the night[120,121]. In the UK there are no 'mountain stations', the highest monitoring site is at Great Dun Fell (847 m above sea level), whilst close by is Wharleycroft (206 m above sea level), a valley site. Comparison between these two sites has shown that whilst daily maximum ozone concentrations, midday concentrations and peak valves are similar, the ozone dosage is significantly greater at the higher elevation site[352].

Table 6.10: Annual average ozone concentrations

Station	Latitude	Longitude	Altitude (meters)	Period	Annual Average Conc. (ppbv)
Point Barrow	71°N	157°W	11	1973-84	25.8
Mauna Loa	20°N	155°W	3397	1973-84	25.6
Arkona	55°N	13°E	42	1972-84	20.8
Hohenpeissenberg	48°N	11°E	975	1974-83	31.1
Fichtelberg	50°N	13°E	1213	1972-84	28.6
Bottesford	51°N	2°W	32	1978-85	22.7
Sibton	51°N	0°	50	1976-85	23.9
Samoa	14°S	171°W	82	1973-84	13.7

At remote sites in the UK (*e.g.* Strath Vaich), diurnal variations throughout the year are small. It therefore appears that chemical and physical sinks do not influence significantly the observed ozone levels. Such sites are probably providing a good indication of background levels in the boundary layer. The average concentration at rural sites in the UK is remarkably similar to many others in the North Hemisphere, indicating that pollution has very little influence on ozone averages (Table 6.10).

Ozone concentrations are generally made at ground-level *e.g.* about a height of 2 to 3 m. Vertical profiles of ozone, during the summer, have shown that the nocturnal ozone depletion is limited to the lowest tens of metres of the atmosphere. In the UK, only a few measurements of the variation of ozone with height have been made. The principal studies have used tethered balloon ascents[122,123], lidar[77] and fixed sites at two different heights in central London[98].

Concentrations show little vertical variation from about noon through to the late afternoon, consistent with a well-mixed atmosphere (Figure 5.4). During the night the ozone concentration is only depleted in the lowest layer of the atmosphere, the height of which corresponds to the surface-based radiation inversion. Variations with height are dependent upon meteorological conditions and the reactivity of ozone, both with surfaces and with other gases in the atmosphere. In rural areas, the inversion is generally less than 100m in height, whereas the heat island effect in urban areas raises the height of the inversions to between 100 and 200m. Diurnal variations of ground-level ozone are particularly pronounced during ozone episodes, since conditions favouring low-level temperature inversions are similar to those favouring the formation and build-up of ozone during summer months. It should be mentioned that two sites experiencing the same regional ozone distribution can exhibit markedly different patterns of ozone exposure depending on site factors and local scale meteorological influences.

In Figure 6.14 we saw that peaks in mean hourly ozone concentrations were recorded as isolated events or episodes superimposed on a background distribution. It is this peak concentration, and the length of time that concentrations persist above a given level, that is important in determining air quality standards.

Tables 6.4, 6.9 and 8.4 summarise some of measurements of ozone available for the UK, further details may be found in the PORG report[98]. The US primary national ambient air quality standard, designed to protect the health of the population at large, is an hourly average of 120 ppbv which should not be exceeded more than once a year (chapter 8). Since 1982 the site with the greatest number of hours with levels over 80 and 120 ppbv was Stevenage in southeast England (Figure 6.13). The summer of 1976 stands out for its record ozone concentrations. At Teddington, also in the southeast, ozone levels exceeded 80 ppbv for 363 hours, spread over 51 days. Overall, hours above the thresholds are significantly smaller at urban sites compared with suburban and rural sites; ozone levels exceeded 120 ppbv in at least one site in the UK every year bar one; and there is considerable year-by-year variability reflecting the different meteorological circumstances.

Using data prior to the establishment of a national monitoring network, it has been shown that on many occasions a large number of widely separated sites experience episodes on the same day, while at other times, sites experience episodes individually[98]. Generally sites in the north experience episodes later and for shorter durations than southerly areas. With the national network a clearer picture of the temporal and spatial variability of ozone concentrations is emerging (Figure 6.21). For example, during the period 19-26 May 1989 elevated ozone levels were observed across the UK (Figure 6.22, Table 6.11). The highest and most sustained concentrations were recorded in the south and west of England. At Lullington Heath there were 26 exceedences of 80 ppbv hourly ozone and 3 of 100 ppbv during this period. At Yarner Wood there were 59 and 19 exceedences of these thresholds.

Table 6.11: Maximum hourly average ozone concentrations (ppbv)

May 19 - 26, 1989

Site	May 19	May 20	May 21	May 22	May 23	May 24	May 25	May 26
Stevenage	48	71	70	54	80	68	50	42
Central London	56	69	77	40	47	64	44	41
Sibton	78	66	62	54	88	107	61	45
Ashton Hill	67	82	81	75	75	79	59	44
Lullington Heath	82	106	86	72	87	89	98	61
Strath Vaich	43	69	80	55	76	46	45	47
High Muffles	45	62	59	53	89	80	52	42
Lough Navar	30	83	79	73	76	44	44	46
Yarner Wood	99	130	112	99	74	96	90	51
Lady Bower	41	66	61	52	64	78	41	40
Harwell	61	97	88	66	31	53	34	33

Geostrophic trajectories for the 20th and 21st show southeasterly flow and hence a continental air mass over the UK. The synoptic situation during this episode is shown in Figure 6.23. Note that on the 21st, a classical situation for photochemical ozone formation had appeared, with a blocking anticyclone centred over Scandinavia, accompanying slack pressure gradients and generally hot, sunny conditions. This anticyclone gradually moved south and a cold front brought clearer cooler weather to Scotland. Ozone levels at northerly sites fell accordingly. As the front crossed the remainder of the UK ozone levels fell to background concentrations.

6.5   AMBIENT OZONE CONCENTRATIONS IN EUROPE

As discussed in chapter 1, ozone has been measured with varying degrees of success since the mid-nineteenth century. Probably the longest recent record, which is still actively being collected, is that from the network of what was the German Democratic Republic Meteorological Service. This commenced in 1952 and has been maintained at 5 stations until the present[124].

It was initially assumed that photochemical air pollution was unlikely to occur in

267

Figure 6.21

Ozone hours above (a) 60ppbv, (b) 80ppbv and (c) 100ppbv for 1987 - 1990.

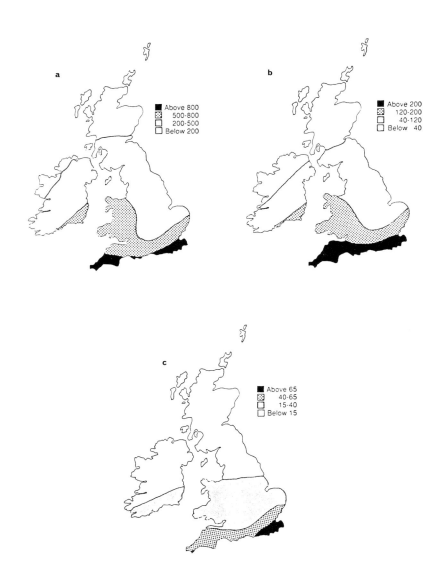

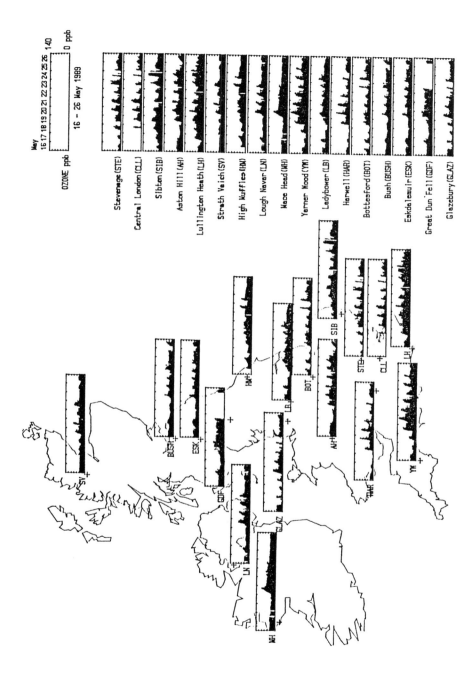

Figure 6.22 Ozone concentrations during the episode of 16 - 26 May 1989.

Figure 6.23

Synoptic charts and backward airmass trajectories for 20 - 25 May 1989.

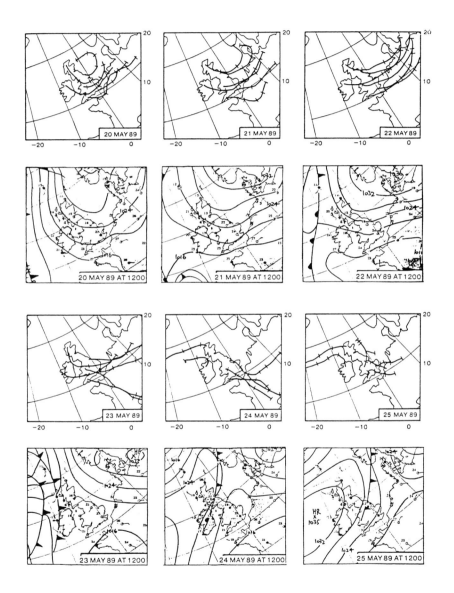

northwestern Europe as the essential prerequisites, such as high temperatures, adequate sunlight and ample emission of precursors were largely absent. However, it is now well established that the insolation is quite strong enough for production of ozone. In chapter 4 we saw that photochemistry starts with the absorption of light and that the actinic light intensity (290-400 nm) is one of the major parameters governing the formation of photochemical smog. From the zenith angle of the sun as a function of latitude and time of year it can be concluded that a sufficient amount of radiation is available for photochemical processes to occur, at least during a certain period of the year, at latitudes up to 70°N. At moderate and higher latitudes, the periods become progressively shorter of course. This means that, under certain meteorological conditions and precursor concentrations, virtually any country on earth might now or in the future be faced with *in situ* photochemical air pollution.

Evidence of photochemical ozone in Europe was first reported in the early 1970's. Measurements of oxidant in Frankfurt/Mainz during the summer of 1967 showed levels in excess of 80 ppbv[125]. Exploratory measurements in a highly industrialized region of the Netherlands showed oxidant levels in excess of 100 ppbv on a number of occasions during 1968[126]. Since these early measurements, photochemical air pollution has been observed throughout Europe[127]. While episodes are often associated with long-range transport of ozone and its precursors in northern and western Europe, they can be of a more local character in southern and central Europe.

The United Kingdom is lagging behind much of Europe in ozone monitoring. The measurement of ozone in meteorological stations and remote areas has been a long tradition in Germany[128-130]. During 1974/75 a large programme was started to measure ozone and its precursors, additionally numerous measuring sites for atmospheric pollutants have been built up by meteorological, hygienic and control institutions in Germany[131,132]. Data from these sites are described and summarised by Becker *et al.*[132]. Several stations are situated on mountains, and are often taken to be representative of the free troposphere. Table 6.12 provides an overview of surface ozone concentrations in Germany at different attitudes between 1975 and 1983. From this table we may conclude

(i)   average concentrations in urban areas are low compared with rural sites. This is similar to that found in the UK.

(ii)   At high altitudes, maximum concentrations and average summer ozone concentrations show little variation from year to year.

(iii)   highest ozone concentrations occur in the Rhine Valley.

(iv)   the summer of 1976 again stands out for its extreme ozone concentration.

Following sporadic measurements in Vlaardigen in 1968[126], continuous ozone measurements in the Netherlands have been made in Delft since 1969. Concentrations often exceed 120 ppbv[127,133] with maximum hourly value of 270 ppbv in 1976 at Vlaardigen. Extensive data on hourly mean ozone and precursor concentrations is available from the Netherlands National Network operated by RIVM-Bilthoven. Monitoring sites are based on a grid system, and a comprehensive and consistent set of observations provide excellent

Table 6.12: Annual maximum ozone concentrations (ppbv) in the Federal Republic of Germany and, in parentheses, average concentration between April and September.

	1975	1976	1977	1978	1979	1980	1981	1982	1983
Cologne[a]	140 (18)	195 (19)	150 (14)	125 (33)	95 (16)	120 (18)	273 (25)	163 (26)	152 (25)
Bonn[a]	160 (23)	185 (21)	198 (15)	174 (14)	110	137	-	-	-
Ölberg (460 m)[a]	130 (31)	170 (39)	130 (28)	170 (34)	175 (33)	125 (30)	159 (32)	-	-
Mannheim[b]	149 (28)	271 (26)	127 (21)	118 (17)	122 (23)	96 (19)	128 (22)	204 (28)	93 (26)
Karlsruhe[b]	-	179 (23)	204	149 (22)	100 (19)	143 (16)	122 (17)	130 (29)	111 (19)
Frankfurt[b]	-	204 (34)	105 (16)	115 (17)	139 (17)	157 (24)	173 (21)	183 (27)	135 (20)
Hohenpeeißenberg[c] (1000 m)	101 (41)	92 (43)	85 (37)	99 (40)	94 (37)	83 (36)	84 (39)	120 (43)	91 (41)
Zugspitze[c] (2964 m)	-	-	56 (30)	43 (27)	53 (30)	55 (30)	60 (32)	68 (38)	90 (53)

[a]	Half hour average
[b]	Three hour average
[c]	One hour average

data to study spatial variations on the scale of several hundreds of kilometres. Table 6.13 indicates that there are significant variations in peak concentrations from 92 to 144 ppbv. Elevated ozone concentrations appear to be widely distributed, but this does not imply that photochemical production is evenly distributed. On a spatial scale of 10 to 100 km there is significant variability in peak ozone levels.

In France, ozone measurements were carried out at Nice in 1973, and at Martigues. The maximum hourly average at Nice was 170 ppbv ozone, while 200 ppbv at Martigues in 1975 has been reported. In Vert-le-Petit, 35 km south of Paris, ozone has been monitored since 1976. Concentrations up to 215 ppbv have been observed and local precursor regions

Table 6.13: Percentiles of the distribution of hourly mean ozone concentrations at various locations in the Netherlands for the summer of 1983

1-hour values, ppbv

Location	50%	95%	98%	Maximum
Afferden	28	67	83	134
Posterholt	28	65	79	118
Oost-Maarland	24	57	71	102
Helvoirt	27	67	85	121
Heijningen	29	56	67	110
Heeze	27	68	85	126
Waarde	31	65	80	124
Niellar-Namen	30	68	83	126
Axel	27	68	82	122
Voorschoten	33	74	92	144
Brandwijk	25	64	82	138
Der Helder	34	57	64	94
Hem	34	65	75	132
Zaandam West	30	59	67	92
Biddinghuizen	26	62	74	125
Cabauw	30	68	83	176
Loenen	26	61	73	118
Puiflijk	32	73	89	126
Winterswijk	30	70	83	116
Vollenhove	30	58	68	98
Hellendoorn	28	60	73	114
Buurse	33	73	88	129
Kloosterbuten	36	63	74	114
Nijega	30	55	62	112
Sappeneer	30	56	65	115
Balk	34	63	75	115
Hoogersmilde	31	63	74	121
Sellingen	32	65	73	103
Weijerswold	29	59	69	116

(*e.g.* Paris) could not account for these high levels[134]. Ozone measurements in southwest France during 1978/79 are reported by Lopez *et al.*[135]. Marerico[136] concludes from measurements made in SW France at an altitude of 3000 m that the photochemical ozone formation rate is up to 2 ppbv h^{-1} in polluted air. Several air pollution monitoring stations have been established in eastern France since 1984. The majority of high ozone concentration episodes take place simultaneously across this region and are associated with easterly winds[137].

The Nordic countries have monitored ozone since the early 1970s. Norway commenced in 1975 with several stations in the southeast of the country. Due to the variable climate the total number of ozone episodes can vary considerably from one year to another. The episodes may be caused by long-range transport from the continent or Britain, by local and mesoscale production, or by a combination of both[27,138,139]. Concentrations of up to 200 ppbv have been recorded. Ozone has been measured at four sites in Sweden since 1973. During the summer ozone levels often exceed 100 ppbv, and 210 ppbv has been observed at a rural coastal site[140]. Measurements in Finland have been undertaken since 1985. High ozone concentrations were favoured when the wind direction was in the range 120°-210°; reflecting air masses originating from European industrialized areas[141,142,143].

Table 6.14: 95 percentiles points in the annual distributions of the half hourly mean concentrations ($\mu g\ m^{-3}$)

	Dubendorf suburban	Zurich urban	Basel suburban	Sion rural	Payerne rural	Lugano urban	Tanikon rural	Jungfrauhoch alpine
1981	79		68	118				
1982	117		98	131				
1983	83		145	139				
1984	84	70	124	156			107	
1985	110	93	137	135	139		90	
1986	114	93	147	165	157		98	
1987	112	86	107	113	111	127	103	
1988	101	94	98	105*	137	109	99	102

* Incomplete data set

Ozone in Austria, around Vienna, has been monitored since 1976. At Illmitz, 65 km south of Vienna maximum hourly ozone levels exceeded 100 ppbv for 164 days and 150 ppbv for 90 days during 1979. Switzerland now has a national network of 8 monitoring stations including one at Jungfraujoch, 3580 m above sea level[144,145]. Two monitoring stations, one in a rural area, the other in an urban area, have been operating in Geneva since 1974. In addition to ozone NO_x, CO, SO_2 and total nonmethane hydrocarbons are measured[146]. These two sites are not part of the national network. Table 6.14 summarises the results of the national network. Annual ozone concentrations are clearly dependent on elevation. The seasonal behaviour is governed by the strength and closeness of sources, while diurnal variations reflect the characteristics of topography and local sources of primary pollutants[147].

Data from Italy and Spain are not so extensive. Measurements of ozone began in 1974 in a suburban district of Rome. In the summer of 1977, maximum hourly ozone levels exceeded 80 ppbv on 32% of the days at a suburban site and 29% of the days within the centre of Rome 120 ppbv was exceeded 1% and 4% of the days respectively[148]. The maximum hourly mean in 1974 was 140 ppbv. Data has been reported for Ispra in northern Italy[149] and near Ravenna, an industrial area on the Adriatic coast[150]. At the latter site, the mesoscale circulation of pollutants was shown to be important. Ozone levels in Madrid throughout 1976 were lower than elsewhere in northwest Europe[113]. The Spanish Department of Environmental Health provide data on ozone levels in Madrid, made since 1982. The diurnal profiles show a peak between 1600 and 1800 h.

Greece has had the cooperation of Germany in establishing a measurement campaign in and around Athens during the summer of 1982[48,50,47]. Symptoms of photochemical ozone, such as reduction of visibility and the appearance of a brown cloud are frequently observed over Athens during sunny, anticyclonic weather. A concentration of nearly 300 ppbv was recorded at Lycabetos, a hill (~ 250 m) in the centre of Athens. The Athens region is particularly prone to sea breeze circulations and nocturnal ozone maxima are often observed.

Of the eastern European countries, the former state of Yugoslavia had a good track record for ozone measurements. The first measurements were made in Zagreb in the summer of 1975. The maximum hourly average concentration was 140 ppbv[151-153] and the hourly average concentration exceeded 80 ppbv for 41 days during this period. Measurements have also been conducted on the isle of Krk in the northern Adriatic Sea (October 1976-September 1977), and in the city of Split on the Adriatic coast (April-October 1979)[154-156]. Between May and September 1981 and 1982 oxidant was monitored at several stations in and around Belgrade. In June 1981, the daily (0800-1600h) maximum oxidant concentration exceeded 500 μg m^{-3}. The complex topographic conditions together with the distribution of precursor emissions results in high oxidant concentrations during warm anticyclonic periods[32]. However, at Bratislava, Czechoslovakia the annual mean surface concentration between May 1980 and April 1982 was only 9 ppbv and the maximum hourly mean ozone concentration in this period was 40 ppbv[157]. Ozone has been measured at Preila, Lithuania since 1980[354]. Mean hourly concentrations typically range form 5 to 50 ppbv and can rise to between 60 and 85 ppbv during photochemical episodes.

It is evident that photochemical pollution is a regional scale phenomenon, and that successful control of the oxidant problem will only be reached by the joint effects in several

Table 6.15: Number of hours (h) and days (d) with hourly ozone concentrations exceeding 120, 160, 200, 240 and 280 µg/m³, and maximum hourly and daily ozone concentration (µg/m³), April - September 1986

STATION	Total h	Total d	>120 h	>120 d	>160 h	>160 d	>200 h	>200 d	>240 h	>240 d	>280 h	>280 d	Maximum Ozone Concentrations (µg/m³) h	d
Austria														
Illmitz	3449	146	1918	138	892	108	384	56	135	31	28	8	348	198
Belgium														
Gent														
St Kruiswinkel	3614	159	54	13	5	2							194	89
Denmark														
Ulborg	3309	139	49	21									140	104
Germany														
Brotjacklriegel	4253	179	175	32									154	137
Deuselbach	4384	183	170	30	16	7	2	1					222	135
Langenbrügge-Waldhof	4365	183	170	30	73	14	4	3					208	129
Schauinsland	4315	182	1036	81	160	25	37	7	2	1			260	204
Westerland	3594	149	181	28	14	6							198	130
Finland														
Utö	3373	150	128	22									154	133
France														
Andrezel	4070	171	53	13	8	2	6	2					230	97
Montagny	4111	172	25	9	1	1							167	83
Pinceloup	3927	163	248	41	54	13	9	6	2	2	1	1	285	139
Netherlands														
Eibergen	4306	180	218	39	49	11	19	5	6	2			280	130
Witteveen	4038	168	120	21	23	5	3	1					213	106
Norway														
Birkenes	4090	172	14	4									144	97
Jeloya	4231	177	75	19	13	3	3	1	1	1			268	124
Langesund	2553	107	14	4									149	99

Table 6.15(contd.): Number of hours (h) and days (d) with hourly ozone concentrations exceeding 120, 160, 200, 240 and 280 µg/m³, and maximum hourly and daily ozone concentration (µg/m³), April - September 1986

STATION	Total		>120		>160		>200		>240		>280		Maximum Ozone Concentrations (µg/m³)	
	h	d	h	d	h	d	h	d	h	d	h	d	h	d
Sweden														
Aspvreten	2359	99	1	1									124	89
Norra Kvill	2636	111	66	8	9	2							176	141
Ringamåla -Sännen	1572	65	21	5									142	99
Rörvik	3769	167	186	32	15	4							191	133
Vavihill	4321	182	144	28	13	3							202	148
Vindeln	3211	136	21	4			1	1					145	114
Ammarnäs	2414	101											114	88
Stormyrberget	2097	88											104	81
Switzerland														
Payerne	4180	179	859	101	277	46	91	23	34	10	17	6	325	191
Sion	4223	179	832	97	377	55	183	30	79	17	32	10	361	193
Tänikon	4359	183	132	29	6	3							174	93
UK														
Bottesford	4364	182	93	18	14	4	1	1					206	120
Harwell	4192	175	77	10	27	5	9	2	3	1			272	133
Wray (Lancaster)	4074	173	14	3									150	97
Sibton	3424	146	4	1									136	85

Number of hours and days

277

countries. The first assessment of ozone data from several stations in Europe was made by Guicherit and Van Dop[127]. Many new stations for ozone and precursors have been established in Europe following an Environmental Directorate of the OECD in 1973. A pilot project was undertaken to compile ozone measurements from 24 monitoring stations in eight countries (Austria, Belgium, Federal Republic of Germany, Finland, Netherlands, Norway, Sweden and the UK) for the years 1976 to 1979[158]. The relations between photochemical episodes, large scale weather and air trajectories were analyzed. As part of the OECD Control of Major Air Pollutants (MAP) programme measurements of ozone, NO_2 and PAN from European OECD countries, since 1985, are being collated by institutions in Sweden and Norway under the OXIDATE project[159,160]. In addition to the eight countries in the earlier study, data from Denmark, France and Switzerland, 36 stations in all, were now included. The ozone data for some of these sites classified by the number of days and hours when hourly ozone concentrations exceeded certain levels is given in Table 6.15. In summer 1986, seven stations had hourly concentrations above 120 ppbv (240 μg m^{-3}). The highest hourly concentrations occurred at Sion, Switzerland and the highest daily concentration at Illmitz, Austria. This region of Europe is particularly prone to photochemical ozone. Backward air mass trajectories for Illmitz, with days in excess of 120 ppbv (240 μg m^{-3}) indicate that transport from the southeast, west and north is important, for days in excess of 80 ppbv (160 μg m^{-3}) transport from the north dominates. Plots of 90, 95 and 98 percentiles for summer 1986 indicate a northwest to southwest gradient in ozone concentration in Europe[160].

A new European network has recently been established as part of Tropospheric Ozone Research (TOR) which is designed to investigate the potential for year round tropospheric ozone formation on a sub-hemispheric scale. In addition to ozone, the stations (Table 6.16) will monitor its precursor molecules and other peroxy species characteristic of tropospheric chemical activity[161]. Vertical profiles for ozone and a limited number of other components will be made and occasional aircraft experiments will study the chemical change within well-defined synoptic situations. Observations have already shown that ozone and its precursors could be transported over long distances[92] and this was confirmed by theoretical studies[162]. Several research institutions are involved in mathematical modelling work to study the formation, transport and removal of ozone in the atmospheric boundary layer over north west Europe. In a model of atmospheric transport and chemistry the following aspects are involved: source characteristics; advection; dispersion; composition changes; loss processes out of the mixing layer; dry deposition processes; wet deposition

Table 6.16: Status of TOR, 1991

No.	Station	Location
1.	Ny Alesund	Sptizbergen
2.	Birkenes	near Kristiansand
3.	Utä	Finnish Archipelago
4.	Areskutan	Central Sweden
5.	Rorvik	Swedish West Coast
6.	Great Dun Fell	North West England
7.	Norwich	East Anglia
8.	Mace Head	Ireland
9.	Kollumerwaard	Holland
10.	Brennilis	France
11.	Schauinsland	Black Forest
12a.	Garmisch - P	German/Austian boarder
12b.	Wank	German/Austian boarder
12c.	Zugspitze	German/Austrian boarder
13.	Sonnblick	Central Austr. Alps
14.	Pic du Midi	French Pyrenees
15.	K-Puszta	Hungary
16.	RBI	Zagreb
17.	Izana	Tenerife
18.	Jungfraujoch	Swiss/French boarder
19.	Aspvreten	Sweden
20.	Donon	France
21.	Riso	Denmark
22.	Penkas Dourades	Portugal
23.	Functial	Portugal
24.	Lisbon	Portugal
25.	UK Network	UK

NB Station 12a no longer provides relevant TOR data
Station 6 to cease in favour of full TOR station in North Norfolk

processes; and the fate of the pollutant once it has reached the surface. A rigorous combined treatment of all these factors is beyond reach and simplifications are made in the formulation of models. This may involve simplifying the transport (*e.g.* neglecting variations with height, averaging over significant spatial scales) or the photochemistry (*e.g.* neglecting particular reactions or species, representing a class of hydrocarbons by one member of it). Models may be subdivided as Eulerian or Lagrangian depending on the frame of reference used. In the former case, the concentrations and deposition are estimated on a grid basis, while in the latter a succession of trajectories is followed and the chemical development of individual air parcels is followed in time.

The Dutch-German PHOXA project aims to establish a relationship between emissions with an area of 2000 km x 2000 km (grid cells approximately 30 km x 30 km) and the concentrations of photochemical oxidants and acidifying species of NW Europe. An emission data base has been established for SO_2, CO, NO_x, NH_3 and hydrocarbons split into reactivity classes, and with information about point sources and area sources. Further details of the PHOXA model are given in references[163-165].

Figure 6.24

Calculated mean of daily maximum ozone concentrations (ppb) July 1985 (After Simpson[171]).

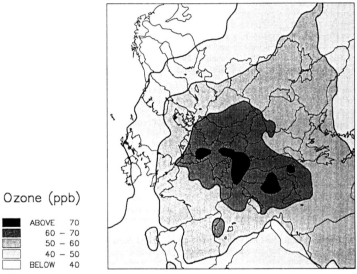

Ozone (ppb)

ABOVE	70
60 –	70
50 –	60
40 –	50
BELOW	40

Regional scale modelling of photochemical oxidants in the atmospheric boundary EMEP (European Monitoring and Evaluation Programme) and adding a comprehensive description of the photochemistry of photooxidants[166-169]. The grid element size is 150 km x 150 km at 60°N latitude. A number of episodes where high ozone levels have been recorded in southern Scandinavia have been interpreted using this model[165,169]. Recently a new model has been developed which has combined the previous EMEP activities on photochemical oxidants with the long period models for sulphur and nitrogen[170]. This model has been designed with the purpose of simulating ozone formation over long periods over the whole of Europe. The modelled ozone concentrations agree favourably with measured concentrations at most sites[170,171]. Figure 6.24 illustrates the mean daily maximum ozone concentration calculated across the EMEP grid for July 1985. The steady increase in ozone concentration from northwest Europe to central Europe agrees well with that suggested by the OXIDATE data.

The Harwell Photochemical Trajectory Model describes the chemical development in a parcel of boundary layer air using a Lagrangian approach[85]. It has been utilised to elucidate some of the basic features of photochemical ozone formation using large photochemical schemes but with relatively simple treatment of the meteorological processes[172]. More recently the model has been used to study the photochemical ozone creation potential (POCP) and the photochemical PAN creation (PPCP) of various hydrocarbons[173]. Trajectories across southern England, from Germany to the Republic of Ireland and from France to Sweden indicate that aromatic and olefinic hydrocarbons yield the highest POCP and PPCP values. Aliphatic hydrocarbons show POCP and PPCP values which increase in multi-day photochemical situations and hence make a significant contribution to ozone and PAN formation on the long range transport scale.

The output from models is of particular importance in assessing the impact of control policies on ozone and other secondary pollutants. It is clear that in the context of the European continent where long range transboundary transport is an important aspect of the photochemical chemical air pollution phenomenon that control or abatement strategies for ozone must be enforced internationally.

6.6   **OZONE CONCENTRATIONS IN THE REST OF THE WORLD**

The largest data set of ambient ozone concentrations in any one country exists for the USA. Much of the research has concentrated on the issue of natural background ozone levels and oxidant and precursor transport. As in Europe, ozone in many nonurban sites may be the result of either naturally generated ozone or transported manmade pollutants, or combinations of both. It is therefore not possible to categorize nonurban sites arbitrarily as being indicative of the natural atmospheric background. Many rural or even remote sites have been shown to be strongly affected by upwind urban pollution sources (see above).

Most analyses of data from rural sites in the US have focused either on individual case studies of ozone episodes[51,174-177] or on seasonal and diurnal behaviour and frequency distributions[67,178-187]. Among the more commonly recognized patterns of ozone concentrations in ambient air are (i) maximum concentrations between March and August, (ii) average maximum concentrations in the east in summer (60-85 ppbv) which are higher

281

Figure 6.25

Seasonal and diurnal distribution of ozone at rural sites in the United States. The upper panels show the seasonal distribution of daily average values; the middle panels show monthly averages of the daily maximum values and the lower panels show the diurnal behaviour of ozone in July. The left panels show results for sites in the eastern US : Montague, MA (solid): Scranton, PA (dashed); Duncan Falls, OH (dot-dash); and Rockport, IN (dotted). The right panels show results for sites in the western US : Cluster, MT (short dashes); Ochoco, OR (long dashes) and Apache, AZ (solid). After Logan[184].

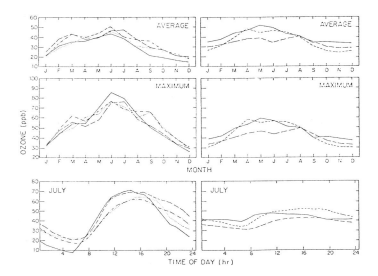

than those in the west (45-60 ppbv), while mean concentrations are quite similar in the east and west, (iii) a more pronounced diurnal variation in the east and (iv) a peak in concentration from about noon to 1800 h except for mountain sites where little diurnal variation is exhibited. Additionally, NO_x concentrations are frequently high enough (> 1 ppbv) in rural areas of the east to allow photochemical ozone formation during favourable meteorological conditions (the ubiquity of biogenic hydrocarbon emissions in the continental USA means that there will generally be a plentiful supply of fuel for ozone production). In remote regions of the west NO_x is much lower. Examples of seasonal and diurnal variations of ozone in the US are given in Figure 6.25. These are typical for stations between 35°N to 45°N. At coastal stations near 30°N, a summer minimum with maxima in spring and late autumn is observed[188].

Ozone episodes occur over a large spatial scale, often > 600,000 km^2 [183]. Considering data collected between May and August (1978 and 1979) in eastern US, Logan[183] concludes that ozone episodes occurred on 23% of the days. From this limited data set it appears that ozone episodes are more common in the eastern US than in western Europe. This may be the result of differences in emission of precursors or meteorology. Hydrocarbon to NO_x ratios are similar in both Europe and the US but the region with the densest anthropogenic emissions of precursors is approximately 10°-15° further south in the US. Hence meteorological conditions are more favourable for photochemical ozone formation there. Additionally, it has been shown that isoprene, in combination with anthropogenic sources of NO_x, could provide an important source of ozone in rural areas of the eastern United States[189]. As illustrated in Figure 6.26 high ozone days tend to occur over a longer season for southern cities than for the northeastern cities. In fact, approximately 40% of the US nonattainment areas for ozone are found in the southeast[351].

Ozone concentrations in Canada are well documented. Ozone was identified as an air pollutant of concern as early as 1960 when a link was discovered between damage to tobacco crops in SW Ontario and elevated ozone levels. A national ozone monitoring network commenced in 1979. By 1986 there were 57 sites in the network. Data from these sites is discussed by Dann[190]. In Alberta, mean monthly concentrations range from 11 to 44 ppbv with maxima in Spring and minima around September[191-195]. Canada's maximum acceptable level for one year (15 ppbv) is exceeded all of the time at rural stations while its maximum desirable level for 1 hour (50 ppbv) is exceeded 11 times more often at rural than urban sites. In a mountainous region, hourly average concentrations exceeded 80 ppbv with a frequency of 0.3%. Average concentrations of ozone in late spring and summer in Europe and rural United States exceed ozone concentrations in remote regions of Canada by between 6 and 22 ppbv or between 20% and 100% for late spring and summer, respectively[177].

Ontario has the most comprehensive monitoring network of any of the Canadian provinces. Southern Ontario has recorded Canada's highest ozone levels and has the greatest frequency of days with ozone concentrations greater than 80 ppbv than any other region. Long range transport is implicated as the significant contributor to ozone levels, although local emissions, especially downwind of Toronto cannot be ruled out[196-198].

Many other countries throughout the world have observed elevated ozone levels at rural sites, *e.g.* Bahrain[199], Israel[200], Japan[57] and Australia[201]. The measurements at rural sites which have been discussed are representative of many on this topic. They serve to give

Figure 6.26

The average number of reports of ozone concentrations ≥120 ppb at the combined cities of (a) New York and Boston and (b) Dallas and Houston, from 1983 to 1985 (1 April= week 14, 1 October = week 40). A representation of the annual variation in solar radiation reaching the earth's surface at (a) 40°N and (b) 30°N is shown (After NRC[211]).

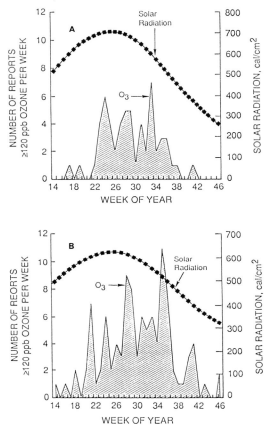

some indication as to the extent and level of photochemical air pollution and transport.

6.7   METEOROLOGICAL CONDITIONS AND OZONE EPISODES

In chapter 4, we saw how ozone is formed by the action of sunlight on the precursors, NO_2 and hydrocarbons. The accumulation of appreciable ozone concentrations is also dependent upon the prevailing meteorology, which will influence the local dispersion of precursors, the photochemical activity of the atmosphere, and the trajectory of the air mass. Below we list the typical meteorological conditions under which ozone episodes are observed to occur, and then proceed to explain why these conditions are so favourable.

Figure 6.27

Position of air pressure configurations associated with ozone formation (high concentration '+', low concentration '-')

In Europe, elevated ozone levels generally occur in association with stagnating anticyclones over central Europe or Scandinavia. Anticyclones promote photochemical pollution by increasing insolation, reducing ventilation of the source region, increasing temperature dependent emissions such as those from solvent evaporation and trees, and increasing most photochemical rate coefficients. The location of the high pressure areas determine which geographical regions will experience the high concentrations. If a high pressure centre is located over Great Britain and the North Sea, ozone levels will be high in Great Britain and the European continent and low over Scandinavia. If the anticyclone is located over north-central or north eastern Europe, high concentrations can simultaneously be observed in Scandinavia, Great Britain and the European continent. The influence of the location of high pressure areas on ozone levels in Europe is illustrated in Figure 6.27. Typical wind directions will obviously depend on the country concerned: for England and the Netherlands the airflow is often Easterly, for Germany north-easterly and for Sweden westerly and south-westerly.

In Britain an ozone episode may develop as follows: an anticyclone situated over

285

the North Sea, North Germany or Scandinavia results in light easterly winds. Cloud cover breaks up and warm sunny spells persist for several days. Only on the first day of such a situation is there any clear evidence that precursor emissions from the UK affects ozone in the UK. On subsequent days, long-range transport provides an additional source of ozone. Elevated levels of ozone will persist as long as the anticyclone prevails. By analysing the synoptic situation at 1200 h for each day from May to September Colbeck and Harrison[93] found that, in 1978, there were 24 days when an anticyclone was situated over the British Isles; 16 days when an anticyclone/ridge was situated over NW Europe; and 8 days with an anticyclone/ridge over the North Sea and Scandinavia. In 1981 the corresponding number of days for each situation was 7, 6 and 6 days respectively. This variation in anticyclonic activity was reflected in the air temperature, hours of sunshine and ozone levels for each year. 1978 was a "good" year for ozone production (16 days in excess of 80 ppbv near Lancaster, 10 days at Bottesford) and 1981 a "poor" year (6 days in excess of 80 ppbv near Lancaster, 0 days at Bottesford). Over the period July 1972 to September 1975, there were at least twenty-six episodes of successive days with hourly mean concentrations in excess of 80 ppbv, in southern England. In nearly every case anticyclonic weather conditions prevailed over the British Isles, with high pressure to the north-east[202]. During June and August 1976 there were 40 days with an anticyclone over NW Europe, the North Sea and Scandinavia and just under 30 days with hourly ozone concentrations in excess of 80 ppbv in London[41].

With the appearance of a slow moving anticyclone somewhere over Europe elevated ozone levels may occur over extensive areas of NW Europe. The location of these regions depends on the trajectories of the incoming air masses and hence on the position of the dominant anticyclone[203]. A good example of this type of episode is that of 25-28 June 1986. The maximum ozone levels recorded at a number of sites in Europe during this period are shown in Figure 6.28 and illustrate the widespread nature of ozone formation. High pressure was stable over north Europe, wind speeds were low and temperature over much of Europe were over 30°C. Most photochemical episodes in NW Europe are broken up abruptly by the passage of a frontal system or depression. Any rain accompanying these events will remove much of the water-soluble oxidation products that have accumulated in the boundary layer. Wind directions may well shift so, for example, polluted continental air is replaced by clean maritime air. Any precipitation associated by boundary layer breakdown following a photochemical episodes could cause an episode of acidic deposition since much of the emitted nitrogen and sulphur compounds will have been oxidised to nitric and sulphuric

Figure 6.28

Maximum hourly ozone concentrations in μg m⁻³ measured during 25 -28 June 1986.

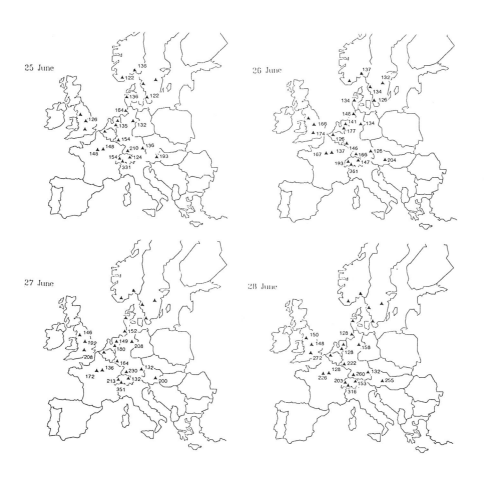

acid. A climatological study has been carried out of the occurrence and properties of a large number of dry and sunny spells in the summers of 1976-1983 over Europe[204]. The frequency of the spells, which are favourable for photochemical activity, were found to be significantly higher in the investigated period than in previous years. Information on temperature, inversion height and cloud cover during episodes were provided as input to a numerical model[205]. Both measurements and predictions indicated elevated ozone

287

concentrations. One conclusion from this study was that 50% of high ozone concentrations (> 80 ppbv) were associated with a non stagnant weather situation. The climate of some areas of Europe is known to experience systematic changes on time-scales of years to decades. Over the UK anticyclonic weather conditions have been more frequent since the 1930s[206] and this has been associated with reductions in mean annual wind speeds[207]. Given this increase in summer atmospheric pressure over a large area of Europe, an assessment of the role of climate change in altering levels of surface ozone has been made[208]. A wind speed index, developed from surface isobaric gradients, can explain 60% of the variance over an eleven year period for data from Hohenpeissenberg (FRG) while at Sibton a close association between ozone values and a measure of the circulation character over the UK, based on existing weather type classification was found.

In the United States, ozone episodes occur preferentially in the presence of weak, slow moving and persistent high pressure systems, as they migrate from west to east, or from northwest to southeast, across eastern states[171,183,209]. Essentially a clean air mass moves out of Canada into midwestern and eastern United States where large numbers of anthropogenic sources exist. Pollutants are mixed into the air mass and transported eastwards. Theoretical investigations of the residence time of air parcels in a high pressure system show that they spend up to a week in the system[210]. Episodes usually end with the passage of a front that brings cooler, cleaner air to the region. Ozone episodes last from 3-4 days on average, occur as many as 7-10 times a year, and are of a large spatial scale[211]. Maximum values of rural ozone commonly exceed 90 ppbv during these episodes, compared with average daily maximum values of 60 ppbv in summer.

Early studies have found that high ozone occurred most often on the trailing side of high pressure systems[209,212]. More recent research concluded that this was the case for only 50% of the time over eastern states[183]. However, in southern Ontario Canada it has been shown that over the period 1976-1981 highest average concentrations occur with 'back of the high' situations. Additionally, 78% of the events with concentrations in excess of 80 ppbv occur under synoptic conditions indicative of back or centre of the high situations[213]. Vukovich and Fishman[214] found that the spatial pattern of ozone in the eastern United States was strongly related to the path of the anticyclones. In Pittsburgh, Pennsylvania, high ozone events are associated with either the western side of a slowly migrating anticyclone or a stagnating extended high pressure ridge[215]. Logan[183] suggests that the pattern of precursor emissions, in addition to the meteorological situation, influences the concentrations of ozone during an episode. It must be noted that at all locations, ozone episodes are associated with anticyclones, but the converse is not true. Additionally, yearly fluctuations in ozone are determined largely by the prevailing meteorology although non climatic factors, such as emissions, may be important[215]. 'Chronically' high ozone years are those which have a persistence of conducive meteorological conditions[216]. The build up of ozone in the Los Angeles basin illustrates the importance of meteorology. The Pacific high pressure system dominates the area and results in air subsidence and the formation of an inversion, trapping the pollutants. The San Gabriel Mountains exaggerate the problem since the prevailing air flow in the upper atmosphere is from the northeast, which enhances the sinking motion of air on the leeward side of the mountains into the basin. Low level air flow is controlled by a daytime sea breeze circulation which is channelled by the terrain into certain locations. Because of its low latitude and subsiding air flow in the upper atmosphere, the basin experiences long hours of small zenith angle sunlight and relatively few clouds.

The meteorological conditions associated with anticyclones, during the summer months, are a well defined boundary layer, subsidence inversion, light wind speeds, high temperatures and a high intensity of solar radiation. The first three parameters influence the degree of mixing of a pollutant in the atmosphere. Mixing and dilution processes proceed at a relatively slow rate in surface and elevated inversion layers. Pollution in stable air tends to spread slowly in the vertical while that emitted into a turbulent boundary layer mixes rather quickly to produce an almost uniform concentration. The vertical mixing profile through the lower layers of the atmosphere follows a typical and predictable cycle on a clear day[217]. There are several methods available for measuring the mixing height. Over the UK Simpson[218] report sodar measurements in London which confirm that photochemically active days coincide with those showing strong convective mixing, so that mixing heights tend to be above 600 m. Similar results in rural/suburban and urban sites, strongly support a summer average midday mixing depth of 1000 m[219].

Strong winds will dilute pollution concentrations even if there is a low level inversion while light or calm wind conditions may lead to an accumulation of pollution even if the afternoon mixing depth is large. Boundary layer depth is determined by a combination of windspeed, cloud cover and solar radiation. Over the UK, during July-September the mean surface windspeed increases from the SW to the NE. Surface wind speeds during ozone episodes are generally in the range 1 to 5 ms^{-1}. Light windspeeds mean that the construction of back-trajectories are liable to large uncertainties but also, of course, that long-range transport is less important. Guicherit and van Dop[127] limited the construction of trajectories to 27 hours because of the light wind speeds.

Clouds play an important role in the vertical redistribution of ozone and its precursors (see Figure 5.5 above). The effect of clouds on vertical transport depends on their size and type. The net ozone flux in cumulus clouds is a linear function of the difference in ozone concentrations between the boundary layer and the cloud layer. The role of cumulonimbus clouds in transporting polluted boundary layer air to the upper troposphere has been demonstrated[220,221]. Pickering *et al.*[222] argued that convective redistribution of ozone precursors may lead to an increase in the production rate of ozone averaged over the troposphere. Venting NO_x from the boundary layer leads to lower concentrations, and the efficiency of ozone production per molecule of NO_x is higher for lower NO_x[110].

The intensity and spectral distribution of solar radiation have direct effects on photochemistry. The intensity varies with season and geographical latitude. In the UK, the daily total of global radiation differ by a factor of three between January and June. At higher latitudes, periods with a sufficient amount of radiation for photochemical processes, become progressively smaller. Models have shown that the potential for photochemical pollution is similar for all latitudes from Los Angeles to Alaska[223]. The UV-B radiation, important for the formation of ozone, increases considerably with height[224]. There is a high correlation between ozone concentration and solar radiation at mountain stations[225,226]. In Germany solar radiation could explain between 6 and 21% of the climatological ozone variance at five stations ranging in altitude from 42 to 1213 m above sea level[353].

Photochemical activity generally occurs simultaneously with strong surface heating. This heating will produce a substantial temperature gradient through the mixed layer. The meteorological conditions giving rise to high surface temperatures are just those which would also encourage the generation of ozone. Temperature not only influences the rate of many chemical reactions, but also the emission rate of precursors, especially that of evaporative emissions of hydrocarbons. In weather situation involving extensive cloud and rainfall the formation of photochemical ozone in the boundary layer is generally not very significant. However the occurrence of thunderstorms is frequent in weather situations which favour photochemical ozone formation. Thunderstorms can transfer large quantities of polluted boundary layer air into the free troposphere with a transport time of 1 hour or less. There it is acted upon by transport and chemistry for a much longer time than is usually the case in the boundary layer. Research has shown that pollutants may be transported and deposited into the free troposphere from the boundary layer by non precipitating cumulus convective clouds. These clouds are produced by vigorous rising thermals during periods of strong solar insolation. As well as removing ozone from the boundary layer, anomalously large short term ozone concentrations have been observed in the vicinity of thunderstorms[227,228]. It has been postulated that ozone may result from chemical reactions activated by lighting associated with thunderstorms. Lightning provides an effective source of NO_x to catalyse ozone production[229,230]. Lightning activity is strongly concentrated over the continents, and also maximises in the summer hemisphere of the tropics.

Many workers have investigated statistical relationships between daily hourly maximum ozone concentrations and meteorological parameters[40,201,231-235]. Ignoring ozone levels the previous day, the most effective variables are generally air temperature and air parcel trajectory. Correlations are often improved by using temperatures on the 850 mbar level rather than those at the ground[216,229]. Statistical models are, however, of limited value, because they allow no statements on cause and effect to be made. They are only valid for the location for which they were developed, and the conditions at that time.

6.8   **TRENDS IN OZONE CONCENTRATIONS**

The determination of a secular trend of ozone requires continuous or quasi-continuous measurements over a sufficiently long time period. Additionally, there is often uncertainty in the data due to the measurement technique used, artifacts, calibration and siting problems.

When considering trends one should make a distinction between free tropospheric levels, levels in the less polluted background or rural areas and levels in urban areas which are to a great extent influenced by anthropogenic emissions. Tropospheric ozone originates from both downward transport of ozone from the stratosphere and in situ photochemical production. Ozone concentrations in the boundary layer on the other hand are governed by the exchange of boundary layer air with tropospheric air, photochemical formation and destruction and dry deposition. As we have seen, polluted air masses from urban and industrial areas can affect rural areas at considerable distances, and even affect ozone concentrations in parts of the free troposphere. It must be clear that the free troposphere and different regimes of the boundary layer may influence one another.

The longest sequence of modern measurements are those obtained by the German Democratic Republic (GDR) since 1956. In addition to the recent tropospheric ozone measurements, there are a few historic records of surface ozone measurements. Unfortunately, most of the historic data is unreliable because of the uncertainties in the measurement method used. It has been established, however, that towards the end of the nineteenth century surface ozone measurements were approximately half of the mean of daily maximum of observations taken during the last 15 years at similar locations in Europe and North America[236]. The most important set of historical data is a series of daily means collected at the Montsouris Observatory in Paris between 1876 and 1905[237,238]. The Montsouris results, corrected for SO_2 interference, together with data from Arkona, GDR (situated on the Baltic coast) for a similar time period starting in 1956 are plotted in Figure 1.1 Disregarding the apparent trend in the Arkona data, the annual average ozone concentration during the 30 year period following 1956 was about 21 ppbv. This shows that today's background ozone level in central Europe are approximately 100% higher than those 100 years ago. The Montsouris data indicates considerable year to year fluctuations but no discernable trend. The Arkona data show a marked increase in ozone towards the end of the 1970s of 2-3% per year and a tendency for ozone to decrease after 1979[124]. This decrease is not easily explained from current knowledge of ozone loss and formation processes[239]. There is evidence that ozone levels have been decreasing in part of the Netherlands[240] in recent years. The data from Montsouris and Arkona has been thoroughly investigated although questions concerning the absolute accuracy need to be answered. Current ozone concentrations at rural sites are not known to be as low as the Montsouris values. It is reasonable to assume background ozone levels in the late nineteenth century were about a factor of two lower than present levels but further confirmatory evidence is required. Concern about the positive trend of tropospheric ozone at mid-latitudes of the Northern Hemisphere has focused attention on historical ozone observations made by the Schonbein technique one century ago (see Chapter 1). Reevaluated historical data from South America show that one century ago the surface ozone levels at mid-latitudes of the Southern Hemisphere were comparable to those observed in the Northern Nemisphere. At tropical latitudes, historical data are lower than those observed at mid-latitudes in both hemispheres. Currently, mid-latitude ozone levels of the Southern Hemisphere are lower than those of the Northern Hemisphere, indicating a change with respect to the past century. Levels at tropical latitudes are still lower that those observed at mid-latitudes, showing a situation qualitatively similar to that of the preindustrial period.

In addition to Arkona, measurements at Hohenpeissenberg (elevation 975 m) in southern Germany cover a long time period. Continuous ozone measurements commenced in 1971. Over the full period of the records (up to 1988) both stations, which are about 800 km apart, show a positive trend of about 1% per year. There are marked differences in the fluctuations over various subperiods at the two locations[241]. A larger increase in mean surface ozone (~ 2.1% per year) in 1970s compared with that in the 1980s (~0.5% per year) is evident at Hohenpeissenberg. The Arkona data show a linear increase from 1956 to 1979 (~2.4% per year) but a linear decrease (~-2.4% per year) in the 1980s. These differences may result from a combination of many factors which include photochemistry, changes in surface deposition, local changes in atmospheric circulation and data quality. Low et al.[242] have shown that the uncertainty due to SO_2 interference, at Hohenpeissenberg during pre-1976 period, could have a significant effect on the apparent upward trend, and possibly lead to its over estimation by a factor of 3. The decline in ozone observed during the early 1980s at Arkona has been reported for other maritime sites in Europe[243], whilst sites further into the continental interior exhibit positive trends. It should be noted that very few trends reported for data from sites in Europe over the period 1978-1988 are statistically significant.

In the US two kinds of measures are commonly used to monitor year-to-year trends: concentration indicators (e.g. second highest 1h maximum ozone concentration) and threshold indicators (e.g. number of days on which the 1h daily maximum of 120 ppbv is exceeded). To complement its annual report on the number of areas out of attainment (i.e. above its guidelines, see Chapter 8), the US Environmental Protection Agency (EPA) issues an annual analysis of trends for ozone and other criteria pollutants over the preceding decade. Figure 6.29 shows the trend in the annual second highest daily maximum concentrations between 1981 and 1990[244]. The 1990 composite average for the 471 sites was 10% lower than the 1981 average. The relatively high concentrations in both 1983 and 1988 may be attributed to meteorological conditions being more favourable to ozone formation than in other years. In 1990 ozone levels were low as a result of the meteorological conditions being less favourable and to a reduction of precursor emissions. The meteorological contribution to the trend could, in theory at least, be removed using a correlation of ozone concentrations and some meteorological indicator of hot, sunny weather. The actual trend due to emission reductions would then become apparent but to date this has not been attempted

The strong seasonality of ozone levels makes it possible for areas to limit their ozone monitoring to a certain portion of the year, termed the ozone season. The length of the ozone season varies from one area to another and is typically from May to October in many regions of the US, although certain cities may have to monitor throughout the entire year. Figures 6.30 and 6.31 show the average daily hourly ozone maximum over the ozone season as well as the average of the ten highest daily hourly ozone concentrations for six

Figure 6.29

Boxplot comparisons of trends in annual second highest daily maximum 1 hour ozone concentration at 471 monitoring sites, 1981-1990. The trends in ambient air quality, presented as boxplots, display the 5th, 10th, 25th, 50th, 75th, 90th and 95th percentile of the data as well as the composite average. The 5th, 10th and 25th percentiles depict the cleaner sites; the 75th, 90th and 95th percentiles the more polluted sites.

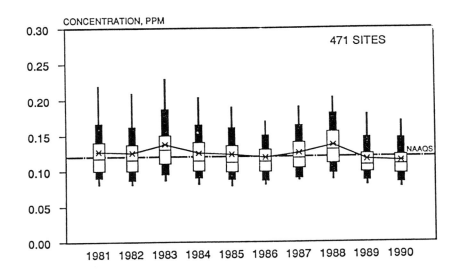

cities. The ozone season for each city is San Francisco (January to December), Denver (May to September), Boston (April to October), Chicago (April to October), New York (April to October) and Los Angeles (January to December). Average ozone levels show a significant downward trend over the period for the ten highest ozone days at New York, Denver and Los Angeles.

Some observations on current trends in ozone at or near the Earth's surface are summarised in Table 6.17. Industrial regions of Europe and the United States show an increase of approximately 20-100% in ozone over the last 40 years[185]. From Table 6.17 we see that all European sites show statistically significant positive trends while the Canadian sites exhibit negative trends, of which only one is significant. Surface ozone behaviour is different throughout the Northern Hemisphere, depending on the site's location compared to the main source regions. Southern Hemisphere sites do not show statistically significantly

Figure 6.30

Average daily hourly ozone maximum over the ozone season.

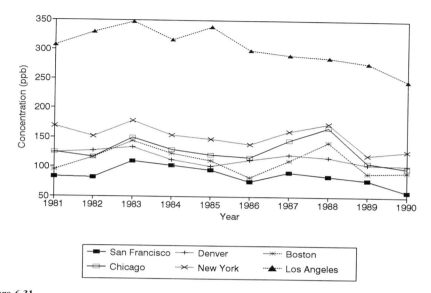

Figure 6.31

Average of the ten highest daily hourly maximum ozone concentrations over the ozone season.

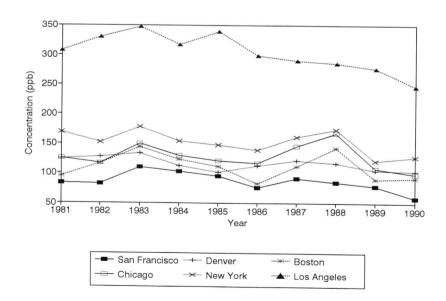

Table 6.17: Surface ozone concentrations and trends deduced from ground based and balloon borne instruments (After WMO[248])

Site	Latitude	Period	Average Mixing Ratio (ppbv)	Trend %/yr (± 2σ)
Northern Hemisphere				
Resolute [a]	75°N	1966-85		-0.1(±0.1)
Barrow	71°N	1973-87	26[b]	0.8(±0.4)
Churchill [a]	59°N	1974-85		-0.7(±1.3)
Arkona	55°N	1956-84	19	1.4(±0.8)
Edmonton [a]	54°N	1973-85		-0.1(±1.4)
Goose [a]	53°N			-1.1(±1.0)
Dresden	51°N	1952-84	10	2.6(±1.6)
Kaltennordheim	51°N	1955-83	16	3.1(±2.1)
Gr. Inselberg	51°N	1972-83	21	3.1(±2.4)
Fichtelberg	50°N	1954-84	23	1.1(±2.0)
Hohenpeissenberg	48°N	1974-86	31[b]	1.1(±0.3)
Hohenpeissenberg [a]	48°N	1969-86		2.3(±0.5)
Payerne [a]	47°N	1968-86		2.4(±0.8)
Sapporo [a]	43°N	1969-86		0.9(±1.1)
Tateno [a]	36°N	1969-86		1.3(±0.6)
Kagoshima [a]	32°N	1969-86		2.5(±1.0)
Mauna Loa	20°N	1973-87	27[b]	0.8(±0.4)
Southern Hemisphere				
Samoa	14°S	1976-87	14[b]	-0.3(±0.7)
Cape Point	34°S	1982-88	20	0.3(±2.0)
Cape Grim	41°S	1982-86	24	0.6(±0.7)
South Pole	90°S	1975-87	20[b]	-0.5(±0.4)

[a] sonde data

trends except for the South Pole, which exhibits a significant negative trend. Interestingly measurements in the austral winter show an increase and those in the austral summer a decrease[245]. Overall this decrease, due to reduced advection of ozone from lower latitudes, outweighs the increase.

Ozone data has been collected at remote, clean air sites by the NOAA GMCC program since 1973[5]. Barrow and Mauna Loa in the Northern Hemisphere and Samoa, Cape Point, Cape Grim and the South Pole in the Southern Hemisphere are sited such that ozone concentrations can be considered to be representative of the region at large. Positive trends in the Northern Hemisphere are observed at sites remote from industrial emissions[5,246].

Balloon-borne observations indicate that tropospheric ozone has increased over large parts of the Northern Hemisphere (Table 6.17). The negative trends for Canadian stations may result from an instrument change that occurred in the early 1980s[247]. A significant trend of 0.8($\pm$0.5)% per year in the first three kilometres of the atmosphere in the Northern Hemisphere is found by averaging over all nine stations shown in Table 6.17[248]. Average trends in the troposphere above this altitude were not significant. Above the tropopause ozone trends become significant again but are negative. This is probably due to the action of chlorine compounds derived from chlorofluorocarbons.

NO_x, CO, CH_4 and non-methane hydrocarbons are known to have increased, mainly due to anthropogenic emissions, in the Northern Hemisphere[249,250]. An increase in tropospheric ozone is not unexpected. Model studies have indicated that a significant portion of the increase in ozone may be attributed to the increase in global anthropogenic NO_x emissions while the contribution of CH_4 to the increase is small[239,251]. Further increases in NO_x will contribute to a continued rise in free tropospheric ozone, while the occurrence of ozone episodes in the boundary layer may not change much[252]. The height of the NO_x injection is also important. NO_x emissions from aircraft may well have a disproportionately large impact since the emissions are directly into the region where ozone production is limited by lack of the NO_x catalyst. Pre-industrial emissions of NO_x are approximately 35% lower than those of present day levels[253]. Results of models of the pre-industrial atmosphere agree well with the Montsouris ozone data. Estimates of the future growth in NO_x emissions indicated that tropospheric ozone will continue to increase at a rate faster than during the past 100 years.

Although increases in surface and tropospheric ozone over the past century seems to be well established, its magnitude and possible causes are still a controversial issue due mainly to a lack of enough reliable data. Experimental techniques and observational frequencies used have differed in different places and even at different times in the same place. Some stations have been operative for over a decade, whereas others have had a transitory existence. Typical tropospheric ozone variability and instrumental precision means that the current ozonesonde network, and presumably surface ozone sites, are insufficient to detect a trend in tropospheric ozone of about 1% per year at the 2σ level, even at stations with records a decade in length[254]. At least a doubling of the current network is necessary to detect an ozone trend of 1% per year on a decadal time frame[254].

6.9 URBAN, RURAL AND REMOTE PAN COMPOUND CONCENTRATIONS

Unlike ozone, PAN does not have large natural sources in the stratosphere and offers advantages over ozone as an indicator of photochemical air pollution. PAN is phytotoxic[255], a lachrymator[256], and has been suggested as a possible etiological agent in the growing incidences of skin cancer[257]. It is of relevance to precipitation chemistry since it dissociates in rainwater to nitrate and organic species[258]. Recently, Altshuller[259] reviewed the sources of PANs, the atmospheric distribution of PANs and their fate in the atmosphere.

PAN was first observed in the 1950s during photochemical smog episodes in Los Angeles[260,261]. It has now been observed throughout the world at urban, rural and 'unpolluted sites'. Observations at the rural sites indicate that PAN is formed and transported in urban plumes on a regional scale[262,263], while those at remote sites provide justification for its consideration as a ubiquitous global trace species.

Published data on the concentrations in ambient air of photochemical oxidants other than ozone are not comprehensive or abundant. Measurement techniques for non-ozone oxidants are not yet routine (chapter 3). However, the common features of PAN behaviour in ambient air have been identified. The atmospheric distributions of PAN in ambient air qualitatively resembles those of ozone. The extant literature, up to 1983, on PAN concentrations in ambient air has been reviewed by Temple and Taylor[264] and Altshuller[265]. Altshuller analyzed the relationships to ozone of PAN and other photochemical reaction products while Temple and Taylor reviewed PAN concentrations in the context of phytotoxicity.

While a large number of PAN measurements have been made in polluted urban areas, fewer have been reported over rural areas and only an extremely limited data set is available for remote regions. In the Northern Hemisphere (Pacific Ocean, south of 48°N) PAN levels are highly variable ranging from <10 pptv to >70 pptv with an average level of 38 pptv. In the Southern Hemisphere (Pacific Ocean, north of 48°S) PAN levels are approximately 5 pptv[266]. However, most of these measurements were made not far from the west coast of America and, thus, continental influence on the sampled air masses cannot be excluded. Recently, measurements at tropical and subtropical latitudes over the open Atlantic have indicated that no significant background levels of PAN (i.e. < 0.4 pptv) exist at sea level and lower latitudes[267]. This is compatible with the short atmospheric PAN lifetime and low concentrations of PAN precursors at these latitudes (35°N-30°S). In contrast to polluted atmospheres, PAN concentrations at remote sites in winter greatly exceed those in summer.

In recent years, the role of PAN as a carrier and reservoir of NO_x has received considerable attention[268-271]. PAN can be separated effectively from other compounds which contribute to NO_y. The thermal decomposition of PAN is strongly temperature dependent. It is highly stable in the colder regions of the middle and upper troposphere and provides a mechanism for NO_x storage and transport. In lower altitudes and lower latitudes where warmer temperatures prevail, PAN and its homologues can easily release NO_x [272]. Model calculations have indicated that PAN is important in the chemistry of the remote troposphere[273,274]. Measurements show that in clean conditions the contribution of PAN to total odd-nitrogen concentration can be as much or more abundant than the inorganic forms[275]. At Alert, in the Canadian Arctic in 1985, PAN was observed at concentrations of over 200 pptv, about an order of magnitude higher than concurrently observed total inorganic nitrate[276]. During severe Arctic haze episodes PAN was even more abundant, with concentrations in excess of 500 pptv[277,278]. It has now been demonstrated that PAN dominates the NO_y budget in the Arctic [268,279,280]. This is due largely to the advection of mid-latitude polluted air into the Arctic during late winter/spring. Hence, ample opportunity exists for the chemical conversion of NO_x into inorganic and organic nitrates. The preferential wet and dry removal of inorganic nitrates over organic nitrates then leads to an air mass enriched in organic nitrates.

The number of published reports on PAN is growing at a steady rate both at urban and rural sites. Although there are many short series of PAN measurements very few observations of periods longer than one year are known. Measurements of PAN in southern California have been carried out for over 25 years. Ambient levels of PAN in excess of 50 ppbv have been recorded during severe smog episodes[281,255]. Recently details of the spatial and seasonal variations of PAN in this region have been reported[332]. Results from up to nine locations indicated that summertime levels of PAN were consistent with photochemical formation during transport, and increased substantially from coastal to inland locations. Daily maxima (up to 30 ppbv) coincided with those of ozone at all locations and shifted from midday at coastal sites to late afternoon inland. Figure 6.32 shows the diurnal profiles for 3 sites. This indicates that PAN levels in the coastal and central areas were consistently as high or higher during autumn than during summer. The lower temperatures, increased stagnation and higher NO_2/NO ratios may all contribute to the elevated PAN concentration during autumn.

PAN at numerous other urban sites, predominantly in the US and Canada has been published[269,283-287]. The Canadian cities, with PAN up to 7 ppbv in summer and 1-2 ppbv

Figure 6.32

Comparison of composite summer and autumn profiles for ambient PAN at coastal (Long Beach), and central locations (Los Angeles and Burbank) (After Williams and Grosjean[332]).

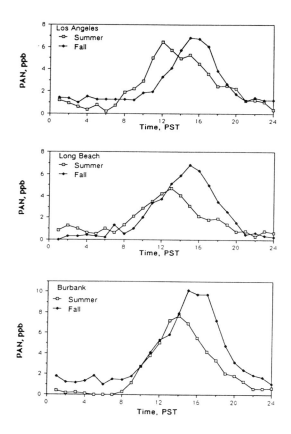

in winter, indicate the potential for photochemical production of PAN on a local scale even at relatively northern latitudes (51°N).

Fewer measurements have been reported in Europe. The earliest date back to 1972 and were made in Delft[288]. A linear correlation between ozone and PAN indicated that both secondary pollutants were generated by photochemical smog reactions. PAN has since been measured in several European cities including London, Oslo, Gothenburg, Rome and Bonn[289]. In all these locations the maximum PAN levels were in the low ppbv range. Recent measurements give similar values. In 1985 the highest recorded PAN value in Athens was 3.7 ppbv with an overall mean of 0.65 ppbv[290,291]. On the outskirts of Zurich the maximum PAN value reported for 1987-88 was 4.4 ppbv[292]. However at Creteil, a suburb of Paris, 33.6 ppbv of PAN was observed in the summer of 1986[293,294]. Both these sets measurements revealed a complex relation between PAN and temperature, with a minimum in the range 4-12°C and an increase in the lower and higher temperatures arising from the competition between photochemical production and thermal decomposition.

Data on PAN in rural areas have been limited, although recently several papers on PAN and peroxypropionyl nitrate (PPN) have been published[269,295,283]. Generally at rural sites in the US, the average PAN concentration is of the order 0.1 to 0.6 ppbv[269,265]. Canadian data indicate a late winter/early spring maxima (0.3 ppbv) and a summer minimum (0.1 ppbv)[296]. An early spring maxima is also observed in Switzerland[292] while a late spring maxima is seen at Harwell, England[297]. Earlier results from Harwell showed a pronounced seasonal variation with maximum levels during the summer months. The site is susceptible to air pollution from London and hence concentrations will be strongly influenced by the meteorological conditions, which as we have shown for ozone, are extremely variable from one year to another. For most of Europe, a PAN concentration of 0.2-0.5 ppbv can be considered to be the background level. Substantial PAN levels (1-4 ppbv) may be recorded during photochemical pollution episodes[297,292,298]. PAN is now measured routinely at many sites in the TOR network from Spitzbergen at 80°N to Izania at 30°N. Provisional results show that it is a very common component of the atmosphere throughout the year and comprises a large fraction of the total NO_y measured at remote sites.

Table 6.18 summarizes PAN concentrations at urban and non-urban sites. Non-urban PAN levels tend to be lower than urban levels but some overlap occurs in the ranges in PAN concentrations at urban and non-urban locations.

Since ozone and PAN have a common photochemical origin and both are phytotoxic, it is of interest to compare their concentrations. The relationship between PAN and ozone, mainly for urban atmospheres has been reviewed by Altshuller[265]. The average O_3/PAN ratio for a number of US cities was 14, whilst for remote sites in the continental US, this ratio was 120. Similar results have been found in the Netherlands[288] where O_3/PAN ratios of approximately 14 were found for air masses impacted by anthropogenic pollution, and much higher ratios ($\approx$ 30-50) for clean air. Similar ratios have been reported for rural and urban

Table 6.18: Range of average concentrations and maximum concentrations for PAN (After Altshuller[259]).

	Average concentrations (ppbv)		Maximum concentrations (ppbv)	
	Low values	High values	Low values	High values
Urban	0.1-0.5	20-30	0.5-1	50-70
Non Urban	0.1-0.2	0.5-1	0.2-0.5	4-6

sites in Ontario[284] with the ratio increasing, from 25-75 in Toronto to 100-150 in Dorset, a rural site approximately 200 km north of Toronto. Recent measurements at southern Californian mountain forest locations produced O_3/PAN ratios, in agreement with the earlier studies in the US, i.e. 14-22. Diurnal variations of ozone/PAN exhibit a nighttime minima and morning maxima[28]. This variation reflect several factors that influence the relative abundance of ozone and PAN: temperature (thermal decomposition of PAN is slower at lower temperatures); deposition velocities (ozone is lost more rapidly by dry deposition), relative rates of formation by chemical reactions and relative rates of removal by chemical reactions.

The two most abundant peroxycarboxylic nitric acid anhydrides, PAN and PPN have been determined at selected urban, rural and remote sites. At the remote background site of Mauna Loa, Hawaii, the average concentrations of PAN, PPN and ozone were 17 pptv, 0.3 pptv and 43 ppbv respectively, measurements were made in May and June 1988 and can be taken to be representative of the free troposphere. PPN is generally a small fraction of PAN (i.e. 0-10%)[299]. Reported PPN concentrations in urban sites have an average value in the range 20-700 pptv although maximum values can exceed 1 ppbv[269,284].

PPN levels at rural sites are substantially below the urban values. Southern Californian levels of PPN are higher than those measured elsewhere. 5.1 ppbv PPN has been reported[282] and 24 hour average values between August and October range from 0.24 to 1.86[282,283]. PPN/PAN ratios of 3-14% have been reported at five urban and two remote sites in the US[269], about 20% in Rio de Janeiro, Brazil[300], 9-14% downwind of Rome[301], 20% in St. Louis[302], 5-15% downwind of Toronto[284] 4-15% at Niwot Ridge, Colorado[295] and up to 28% in southern California[282,286]. At the remote site of Mauna Loa, Hawaii, the

PPN/PAN ratio in the free troposphere averages 1.5%[303]. We may conclude that there is a general tendency for PPN to PAN ratios to decrease from urban to remote sampling sites. The lower PPN/PAN ratios for clean air are a combination of several factors: variation in anthropogenic fluxes of hydrocarbons: the PPN precursors (*e.g.* 1-butene) have a shorter lifetime than the PAN precursors: PPN contains two secondary C-H bonds and hence has a shorter reactive lifetime with respect to OH; and PAN, but not PPN, can be produced from the oxidation of isoprene[295]. In both urban as well as rural locations PAN and PPN, can show a great deal of variability associated with meteorological conditions, however the mean diurnal behaviour of both is virtually identical. It is inferred from atmospheric measurements that PAN and PPN are produced and destroyed by very similar processes.

Information regarding ambient levels of higher molecular weight PAN compounds such as peroxy-n-butyryl nitrate (PnBN) is very limited. At Tanbark Flat, 35 km from Los Angeles an average concentration of 0.05 ppbv PnBN has been recorded for the period August-October 1989[282]. The highest level recorded was 2.3 ppbv and 24 hour averages were in the range 0.04-0.51 ppbv[304]. There was an overall trend towards higher concentrations during October. This is similar to that observed for both PAN and PPN at the same location. Diurnal variations of PnBN exhibited mid afternoon maxima and coincided with those of PAN and PPN. The average PnBN/PAN ratio was 0.077. Like PPN, PnBN is expected to react faster with OH than PAN does, thus resulting in a more rapid removal during long-range transport. In view of the short transport time from the urban area to this site, removal of PnBN by OH is probably not important in this case. The PnBN/PAN ratio actually exhibited an afternoon maxima which is consistent with a higher in-situ production rate for PnBN relative to that for PAN during episodes of higher photochemical activity. This may be explained if hydrocarbons that are precursors to PnBN are more reactive than those that are precursors to PAN.

6.10   URBAN, RURAL AND REMOTE PEROXIDE CONCENTRATIONS

Like PAN, H_2O_2 is an important trace species produced by photochemical processes in the atmosphere. The measurement of H_2O_2 in ambient air is fraught with difficulties (chapter 5). The early measurements suffered from interference by ozone, SO_2 and transition metals. [305-307]. Altshuller[265] reviewed data obtained in the South Coast Air Basin of California in the late 1970s for possible consistency in the interference of ozone in H_2O_2 measurements. While laboratory measurements had indicated that 1 ppbv H_2O_2 would be generated per 100 ppbv ozone[305]. Altshuller's analysis showed that this relationship did not hold in ambient air once H_2O_2 levels exceed 5 to 10 ppbv.

Numerous investigations regarding gaseous atmospheric H_2O_2 have been conducted during the 1980s and these have been reviewed by Sakugawa *et al.*[308]. Here we will summarize the findings of this review and investigate some of the recent H_2O_2 measurements. Field measurements of atmospheric gaseous H_2O_2 at various locations in North America, Europe, Brazil and Japan have shown that the concentrations range from the 1 pptv level to more than 1 ppbv, but generally remain between 5 pptv and 10 pptv.

Models of global tropospheric chemistry predict H_2O_2 levels between a few pptv to about 5 pptv as a function of latitude, altitude and season[309]. Measurements have confirmed these calculations. Published data show a positive gradient of H_2O_2 with altitude[310] as well as from higher latitudes to the equator[311], and a clear seasonal variation. The altitudinal variation is consistent with a switch from hydrocarbon-limited ozone production at the ground to NO_x-limited ozone destruction higher up (chapter 4). The shipboard measurements by Jacob and Klockow[311] indicate an increase of approximately 45 pptv per degree latitude between 50°N and 0°, with a maximum of 3.5 ppbv at the equator. This compares well with measurements in Brazil at 13°S[312] and in the marine troposphere in the eastern part of the Atlantic close to the equator[313] with maxima H_2O_2 levels between 3 and 4 ppbv. The latitude dependency of H_2O_2 has been detected by aircraft measurements over the US. During the winter of 1987, van Valin *et al.*[314] determined that the concentration of H_2O_2 increases by 40 to 50 pptv per degree latitude from north to south. In June 1987, they also observed that H_2O_2 concentrations decreased from lower (40°N) to higher (44°N) latitudes over the northeastern United States[315]. These results strongly suggest that photochemical processes, primarily the recombination of HO_2 radicals, are largely responsible for the observed levels of H_2O_2. This conclusion is supported by the fact that no significant emissions of H_2O_2 from natural and industrial sources have been detected.

Diurnal variations indicate peak H_2O_2 levels in the afternoon and lowest levels at night[308,316,317]. The diurnal variation is greatest in the spring and summer. Figure 6.33 shows the diurnal trends in both H_2O_2 and PAN and illustrates typical differences between summer and winter time periods in terms of absolute concentrations. It is clear that H_2O_2 exhibits trends similar to PAN, but with peak concentrations tending to occur later than those of PAN. Significant H_2O_2 concentrations can be sustained overnight and under specific meteorological conditions; at certain locations (*e.g.* San Bernadino Mountains, Southern

Figure 6.33

Diurnal trend in hydrogen peroxide (a) summer, (b) winter and PAN (c) summer, (d) winter for Harwell (After Dollard et al.[297]).

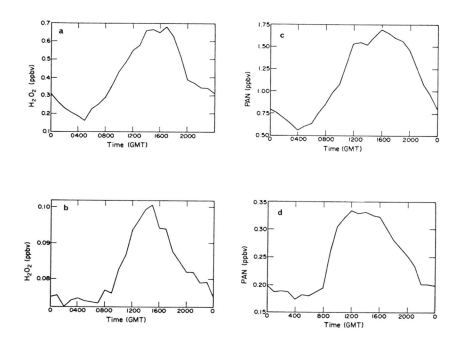

California) nighttime levels can exceed those during the day. Nocturnal chemistry initiated by the NO_3 radical, favours peroxide production, even in polluted air masses (chapter 4).

Most studies of H_2O_2 have focused on a single season, usually summer because of the role of H_2O_2 in the acidification of precipitation. The studies that span a yearly cycle clearly indicate highest concentrations in the summer[318,297,308]. Boatman et al.[319] measured H_2O_2 at an altitude of 1500 m in the central US and reported average concentrations of 0.3 ppbv in winter and 4.1 ppbv in summer. In Los Angeles a mean winter concentration of 0.2 ppbv and a mean summer concentration of 1 ppbv have been measured[320]. In Harwell, England, average concentrations for July and December 1987 were 0.4 ppbv and < 0.01 ppbv respectively. Calculations and measurements have shown how the H_2O_2 concentration is influenced by emission rates, temperature, solar intensity and humidity[308]. Recently it has been postulated that the seasonal variation is due to a transition between two qualitatively different photochemical states of the atmosphere: low NO_x and high NO_x[318]. The low NO_x regime occurs when radical production is greater than the NO_x emission rate, the high NO_x regime occurs when it is less.

The relationship between the concentration of H_2O_2, ozone and meteorological parameters has been the subject of several investigations[321,320,297]. Olszyna et al[321] indicated that H_2O_2 is strongly correlated with O_3, ambient temperature and dew point, and only weakly correlated with the intensity of solar radiation. These measurements were made on the summit of Whitetop mountain, UA, USA. Results from Los Angeles showed that H_2O_2 was correlated with solar intensity, whereas O_3 was correlated with temperature. In the UK, H_2O_2 has been shown to be correlated with O_3 and ambient temperature[297]. For August 1988 a ratio of 21 ppbv O_3 to 1 ppbv H_2O_2 was found. Earlier measurements gave a O_3/PAN ratio of 26[322], whilst modelling studies for photochemical episodes predict a ratio of approximately 9[323].

A significant effect of water vapour on the formation of gaseous H_2O_2 has been suggested[324,315]. In theory, water vapour is involved in the photochemical formation of gas phase H_2O_2, and high water vapour content in air accelerates the formations of gas phase H_2O_2[325]. In the free troposphere H_2O_2 has been shown to be strongly correlated with the atmospheric water vapour content[310], whereas no correlation between H_2O_2 and water vapour was found in the San Bernadino Mountains, California[326].

1. Fishman, J. (1985). Ozone in the troposphere. In : *Ozone in the Free Atmosphere*, R.C. Whitten and S.S. Prasad (Eds.), Van Nostrand Reinhold, New York, 161-194.

2. Levy, H., Mahlman, J.D., Moxim, W.J. and Liu, S.C. (1985) Tropospheric ozone : the role of transport, J. Geophys. Res., 90, 3753-3772.

3. Oltmans, S.J., Komhyr, W.D., Franchois, P.R. and Mathews, W.A. (1989) Tropospheric ozone : variations from surface and ECC ozonesonde observations. In : *Ozone in the Atmosphere*, R.D. Bojkov and P. Fabian (Eds.), A Deepak Pub., Hampton, VA, 539-543.

4. Schmitt, R., Schreiber, B. and Levin, I. (1988) Effects of long-range transport on atmospheric trace constituents at the baseline station Tenerife (Canary Islands), J. Atmos. Chem., 7, 335-351.

5. Oltmans, S.J and Komhyr, W.D. (1986) Surface ozone distributions and variations from 1973-1984 measurements at the NOAA geophysical monitoring for climatic change baseline observatories, J. Geophys. Res., 91, 5229-5236.

6. Ayers, G.P., Penkett, S.A., Gillett, R.W., Bandy, B., Galbally, I.E., Meyer, C.P., Elsworth, C.M. Bentley, S.T., and Forgan, B.W. (1992) Evidence for photochemical control of ozone concentrations in unpolluted marine air, Nature, 360, 446-449.

7. Fabian, P. and Pruchniewicz, P.G (1977) Meridional distribution of ozone in the troposphere and its seasonal variations, J. Geophys. Res., 82, 2063-2073.

8. Pratt, R. and Falconer, P. (1979). Circumpolar measurements of ozone and carbon monoxide from a commercial airliner, J. Geophys. Res., 84, 7876-7882.

9. Routhier, F., Dennett, R., Davies, D.D., Wartburg, A., Haagenson, P. and Delany, A.C. (1980) Free tropospheric and boundary layer airborne measurements of ozone over the latitude range for 58°S and 70°N, J. Geophys. Res., 85, 7307-7321.

10. Seiler, W. and Fishman, J. (1981) The distribution of carbon monoxide and ozone in the free troposphere, J. Geophys. Res., 86, 7255-7265.

11. Stallard, R.F., Edmond, J.M. and Newell, R.E. (1975) Surface ozone in the southeast Atlantic between Dakar and Walvis Bay, Geophys. Res. Lett., 2, 289-292.

12. Winkler, P. (1988) Surface ozone over the Atlantic Ocean, J. Atmos. Chem., 7, 73-91.

13. Kirchhoff, V.W.J.H. (1988) Surface ozone measurements in Amazonia, J. Geophys. Res., 93, 1469-1476.

14. Logan, J.A. and Kirchhoff. V.W.J.H. (1986) Seasonal variations of tropospheric ozone at Natal, Brazil, J. Geophys. Res., 91, 7857-7882.

15. Browell, E.V., Gregory, G.L., Harriss, R.C. and Kirchhoff, V.W.J.H. (1988) Tropospheric ozone and aerosol distributions across the Amazon basin, J. Geophys. Res., 93, 1431-1451.

16. Gregory, G.L., Browell, E. V. and Warren, L.S. (1988) Boundary layer ozone : an airborne survey above the Amazon basin, J. Geophys. Res., 93, 1452-1468.

17. Cros, B., Delmas, R., Nganga, D., Clairac, B. and Fontan, J. (1988) Seasonal trends of ozone in equatorial Africa : experimental evidence of photochemical formation, J. Geophys. Res., 93, 8355-8366.

18. Fishman, J., Minnis, P. and Reichle, H.G. (1986) Use of satellite data to study tropospheric ozone in the tropics, J. Geophys. Res., 91, 14,451-14,465.

19. Fishman, J. and Browell, E.V. (1988) Comparison of satellite total ozone measurements with the distribution of tropospheric ozone obtained by an airborne UV-DIAL system over the Amazon basin, Tellus, 40B, 393-407.

20. Fishman, J. and Larsen, J.C. (1987) Distribution of total ozone and stratospheric ozone in the tropics : implications for the distribution of tropospheric ozone, J. Geophys. Res., 92, 6627-6634.

21. Winkler, P., Brylka, S. and Wagenbach, D. (1992) Regular fluctuations of surface ozone at Georg-von-Neumayer station, Antarctica, Tellus, 44B, 33-40.

22. Lefohn, A.S., Krupa, S.V. and Winstanley, D. (1990) Surface ozone exposures measured at clean locations around the world, Environ. Pollut., 63, 189-224.

23. Evans, G.F., Finkelstein, P., Martin, B., Possiel, W. and Graves, M. (1983) Ozone measurements from a network of remote sites, J. Air Pollut. Control. Assoc., 33, 291-296.

24. Lefohn, A.S. and Jones C.K. (1986) The characterization of ozone and sulfur dioxide air quality data for assessing possible vegetation effects, J. Air Pollut. Control Assoc., 36, 1123-1129.

25. Haagen-Smit, A.J. (1952) Chemistry and physiology of Los Angeles Smog, Ind. Eng. Chem., 44, 1342-1346.

26. Haagen-Smit, A.J., Bradley, C.E and Fox, M.M. (1953) Ozone formation in photochemical oxidation of organic substances, Ind. Eng. Chem., 45, 2086-2089.

27. Schjoldager, J. (1979) Observations of high ozone concentrations in Oslo, Norway, during the summer of 1977, Atmos. Environ., 13, 1689-1696.

28. Ganor, E., Beck, Y.(R.E.) and Donagi, A. (1978) Ozone concentrations and meteorological conditions in Tel-Aviv, 1975, Atmos. Environ., 12, 1081-1085.

29. Nasralla, M.M. and Shakour, A.A. (1981) Nitrogen oxides and photochemical oxidants in Cairo city atmosphere, Environment International, 5, 55-60.

30. Murao, N., Ohta, S., Furuhashi, N. and Mizoguchi, I. (1990) The causes of elevated concentrations of ozone in Sapporo, Atmos. Environ., 24A, 1501-1507.

31. Kanbour, F.I., Faiq, S.Y., Al-Taie, F.A., Kitto, A.M.N. and Bader, N. (1987) Variation of ozone concentrations in the ambient air of Baghdad, Atmos. Environ., 21, 2673-2679.

32. Vukmirović, Z., Spasova, D., Marković, D., Veselinović, D., Vukmirović, D., Stanojević, Č., Popović, M. and Hadžipavlović, A. (1987) Some characteristics of oxidant occurrence in the atmosphere of Belgrade, Atmos. Environ., 21, 1637-1641.

33. Lalas, D.P., Tombrou-Tsella, M., Petrakis, M., Asimakopoulous, D. N. and Helmis, C. (1987) An experimental study of the horizontal and vertical distribution of ozone over Athens, Atmos. Environ., 21, 2681-2693.

34. Ball, D.J. (1986) A history of secondary pollutant measurements in the London region, 1971-1986, Sci. Total Environ., 59, 181-206.

35. Varshney, C.K. and Aggarwal, M. (1992) Ozone pollution in the urban atmosphere of Delhi, Atmos. Environ., 26, 291-294.

36. Liu, G.M., Liu, S.C. and Shen, S.H. (1990) A study of Taipei ozone problem, Atmos. Environ., 24, 1461-1472.

37. Inouye, R. and Abidin, A.Z. (1986) Diurnal variations and frequency distribution of air pollutants concentration in Kuala Lumpur and its outskirts - a preliminary analysis, Pertanika, 9, 201-208.

38. Apling, A.J., Sulivan, E.J., Williams, M.L., Ball, D.J., Bernard, R.E., Derwent, R.G., Eggleton, A.E.J., Hampton, L. and Waller, R.E. (1977) Ozone concentrations in south-east England during the summer of 1976, Nature, 269, 569-573.

39. Derwent, R.G. and Stewart, H.N.M. (1973) Elevated ozone levels in the air of central London, Nature, 241, 342-343.

40. Stewart, H.N.M., Sullivan, E.J. and Williams, M.L. (1976). Ozone levels in central London, Nature, 263, 582-584.

41. Ball, D.J. and Bernard, R.E. (1978) An analysis of photochemical pollution incidents in the Greater

London area with particular reference to the summer of 1976, Atmos. Environ., 12, 1391-1401.

42. Ball, D.J. (1976) Photochemical ozone in the atmosphere of Greater London, Nature, 263, 580-582.

43. Trainer, M., Buhr, M.P., Curran, C.M., Fehsenfeld, F.C., Hsie, E.Y., Liu, S.C., Norton, R.B., Parrish, D.D. and Williams, E.J. (1991) Observations and modelling of reactive nitrogen photochemistry at a rural site, J. Geophys. Res., 96, 3045-3063.

44. Stasiuk, W.N. Jr. and Coffey, P.E. (1974) Rural and urban ozone relationships in New York state, J. Air Pollut. Control Assoc., 24, 564-568.

45. Jacobson, J.S. and Salottolo, G.D. (1975) Photochemical oxidants in the New York - New Jersey metropolitan area, Atmos. Environ., 9, 321-332.

46. Chan, M.W., Allard, D.W. and Possiel, M.C. (1981) Relative contributions of transported and locally produced ozone downwind of Philadelphia, Pennsylvania. Paper 81-21.1, 74th Annual Meeting, Air Pollution Control Association, Philadelphia, PA.

47. Cvitaš, T., Güsten, H, Heinrich, G., Klasinc, L., Lalas, D.P. and Pentrakis, M. (1985) Characteristics of air pollution during the summer in Athens, Greece, Stuab Reihhaltung der Luft, 45 (6), 297-301.

48. Lalas, D.P., Asimakopoulous, D.N. and Deligiorgi, D.G. (1983) Sea-breeze circulation and photochemical pollution in Athens, Greece, Atmos. Environ., 17, 1621-1632.

49. Prezarakos, N.G. (1986) Characteristics of the sea breeze in Athens, Greece, Boundary Layer Met., 245-266.

50. Gunsten, H, Günther, H., Cvitaš, T., Klasnic, L., Ruŝćić, B., Lalas, D.P. and Pentrakis, M. (1988) Photochemical ozone formation and transport of ozone in Athens, Greece, Atmos. Environ., 22, 1855-1861.

51. Blumenthal, D. L., White, W. H. and Smith, T. B. (1978) Anatomy of a Los Angeles smog episode : pollutant transport in the daytime sea breeze regime, Atmos. Environ., 12, 893-207.

52. McElroy, J.L. and Smith, T.B. (1986) Vertical pollutant distributions and boundary layer structure observed by airborne lidar near the complex southern California coastline, Atmos. Environ., 20, 1555-1566.

53. Galbally, I.E. (1972) Ozone and oxidants in the surface air near Melbourne, Victoria. Proc. Int. Clean Air Conf., Melbourne, 192-201.

54. Evans, L.F., Weeks, I.A. and Eccleston, A.J. (1981) An aerial survey of Melbourne's photochemical smog, Proc. Int. Clean Air Conf., Sydney, 93-110.

55. Post, B.E. and Bilger, R.W. (1978) Trajectory measurements of photochemical smog formation, Proc. Int. Clean Air Conf., Brisbane, 135-144.

56. Uno, I., Wakamatsu, S., Suzuki, M. and Ogawa, Y. (1984) Three dimensional behaviour of photochemical pollutants covering Tokyo metropolitan area, Atmos. Environ., 18, 751-61.

57. Mizuno, T. and Yoshikado, H. (1983) On some characteristics of the diurnal variation of O_3 observed in island, urban and rural areas, Atmos. Environ., 17, 2575-2582.

58. Steinberger, E.H. and Ganor, E. (1980) High ozone concentrations at night in Jerusalem and Tel-Aviv, Atmos. Environ., 14, 221-225.

59. Dimitriades, B., Dodge, M.C., Bufalini, J.J., Dermerjian, K.L. and Altshuller, A.P. (1976) Comments on

ozone formation from NO$_x$ in clear air, Environ. Sci. Technol., 10, 934-935.

60. Isaksen, I.S.A., Hov, O and Hesstvedt, E. (1978) Ozone generation over rural areas, Environ. Sci. Technol., 12, 1279-1284.

61. Cox, R.A., Eggleton, A.E.J., Derwent, R.G., Lovelock, J.E. and Pack, D.H. (1975) Long range transport of photochemical ozone in north western Europe, Nature, 255, 118-121.

62. Auer, R. (1939) Uber den taglichen gang des Ozongehalts der bodennahenluft Gerlands, Beirtrage Z. Geophysik, 54, 137-145.

63. Derwent, R.G. and Hov, O. (1982) The potential for secondary pollutant formation in the atmospheric boundary layer in a high pressure situation over England, Atmos. Environ., 16, 655-665.

64. Garland, J.A. and Derwent, R.G. (1979) Destruction at the ground and the diurnal cycle of ozone and other gases, Quart. J. R. Met. Soc., 105, 169-183.

65. Kelly, N.A., Wolf, G.T. and Ferman, M.A. (1982) Background pollutant measurements in air masses affecting the eastern United States - I air masses arising from the northwest, Atmos. Environ., 16, 1077-1088.

66. Fehsenfeld, F.C., Bollinger, M.J., Liu, S.C., Parrish, D.D., McFarland, M., Trainer, M., Kley, D., Murphy, P.C. and Albritton, D.L. (1983) A study of ozone in the Colorado mountains, J. Atmos. Chem., 1, 87-105.

67. Pratt. G. C., Hendrickson, R.C., Chevone, B.I., Christopherson, D.A., O'Brien, M.V. and Krupa, S.V. (1983). Ozone and oxides of nitrogen in the rural upper-midwestern USA, Atmos. Environ., 17, 2013-2023.

68. Kelly, N.A., Wolff, G.T. and Ferman, M.A. (1984) Sources and sinks of ozone in rural areas, Atmos. Environ., 18, 1251-1266.

69. Harrison, R.M. and Holman, C.D. (1979) The contribution of middle and long range transport of tropospheric photochemical ozone to pollution at a rural site in north west England, Atmos. Environ. 13, 1535-1545.

70. Bohm, M., McCune, B., and Vandetta, T. (1991) Diurnal curves of tropospheric ozone in the western United States, Atmos. Environ., 25A, 1577-1590.

71. Altshuller, A.P. (1986) Review paper : the role of nitrogen oxides in non-urban ozone formation in the planetary boundary layer over N. America, W. Europe and adjacent areas of ocean, Atmos. Environ., 20, 245-268.

72. Altshuller, A.P. (1988) Some characteristics of ozone formation in the urban plume of St Louis, MO, Atmos. Environ., 22, 499-510.

73. Stevens, C.S. (1987) Ozone formation in the greater Johannesburg region, Atmos. Environ., 21, 523-550.

74. Cleveland, W.S., Kleiner, B., McRae, J.E. and Warner, J.L. (1976) Photochemical air pollution : transport from the New York City area into Connecticut and Massachusetts, Science, 191, 179-181.

75. Varey, R.H., Ball, D.J., Crane, A.J., Laxen, D.P.H., and Sandalls, F.J. (1988). Ozone formation in the London plume, Atmos. Environ., 22, 1335-1346

76. Sexton, K. and Westberg, H. (1980) Elevated ozone concentrations measured downwind of the Chicago-Gary urban complex, J. Air Pollut. Control. Assoc., 30, 911-914.

77. Varey, R.H., Clark, P.A., Crane, A.J., Ellis, R.M., Sutton, S., Sprague, P.J., Ball, D.J., Laxton, D.P.H.,

Woods, P.T., Jolliffe, B.W., Michelson, E., Swan, N.R.W., Cox, R.A. and Sandalls, F.J. (1987) The London ozone survey 1982-85. Central Electricity Generating Board, Report TPRDL/3142/R87.

78. MacKenzie, A.R. (1989). Ph.D Thesis, University of Essex

79. Glover, G.M. and Lightman, P. (1983) Airborne measurements of atmospheric pollutants in London and the south east, London Environ. Bull., 1, 11-13.

80. MacKenzie, A.R., Harrison, R.M. and Colbeck, I. (1990) The impact of local emissions on the formation of secondary pollutants in urban plumes, Sci. Tot. Environ., 93, 245-254.

81. Derwent, R.G. and Hov, O. (1979) Computer modelling studies of photochemical air pollution in north-west Europe, AERE-R-12615, Harwell Laboratory.

82. Derwent, R. G. and Hov, O. (1980) Computer modelling studies of the impact of vehicle exhaust emission controls on photochemical air pollution formation in the United Kingdom, Environ. Sci. Tech., 14, 1360-6.

83. Hough, A.M. and Reeves, C. (1988) Photochemical oxidant formation and the effect of vehicle exhaust emission controls in the UK. The result from 20 different chemical mechanisms, Atmos. Environ., 22, 1121-1135.

84. Hough, A.M. and Derwent, R.G. (1987) The impact of motor vehicle control technologies on future photochemical ozone formation in the United Kingdom, Environ. Poll., 44, 109-118.

85. Hough, A.M. and Derwent, R.G. (1987) Computer modelling studies of the distribution of photochemical ozone production between different hydrocarbons, Atmos. Environ., 21, 2015-2033.

86. Hough, A. (1988) An intercomparison of mechanisms for the production of photochemical oxidants, J. Geophys. Res., 93, 3789-3812.

87. Derwent, R.G. (1990) The long range transport of ozone within Europe and its control, Environ. Poll., 63, 299-318.

88. Hough, A.M. and Derwent, R.G. (1987) Computer modelling studies of the distribution of photochemical ozone production between different hydrocarbons, Atmos. Environ., 21, 2015-2033.

89. Derwent, R.G. (1990) Evaluation of a number of chemical mechanisms for their application in models describing the formation of photochemical ozone in Europe, Atmos. Environ., 24A, 2615-2624.

90. Lin, M.K. and Seinfeld, J.H. (1975) On the validity of grid and trajectory models of urban air pollution, Atmos. Environ., 9, 555-574.

91. Atkins, D.H.F., Cox R.A. and Eggleton, A.E.J. (1972) Photochemical ozone and H_2SO_4 aerosol formation in the atmosphere over southern England, Nature, 235, 372-376.

92. Cox, R.A., Eggleton, A.E.J., Derwent, R.G., Lovelock, J.E. and Pack, D.H. (1975) Long range transport of photochemical ozone in north western Europe, Nature, 255, 118-121.

93. Colbeck, I. and Harrison, R.M. (1985) The frequency and causes of elevated concentrations of ozone at ground level at rural sites in north west England, Atmos. Environ., 19, 1577-1587.

94. Colbeck, I. and Harrison, R.M. (1985) The photochemical pollution episode of 5-16 July 1983 in north-west England, Atmos. Environ., 19, 1921-1929.

95. Harrison, R.M and McCartney, H.A. (1980) Ambient air quality at a coastal site in rural north-east

England, Atmos. Environ., 14, 233-244.

96. Simpson, D. and Davies, S. (1985) Trajectory analysis of ozone episodes at a rural site in Sibton, Suffolk - 1980-1983. Warren Spring Laboratory Report LR536 - Stevenage.

97. Martin, A. and Barber, F.R. (1981) Sulphur dioxide, oxides of nitrogen and ozone measured continuously for 2 years at a rural site, Atmos. Environ., 15, 567-568.

98. Photochemical Oxidant Review Group (1987) *Ozone in the United Kingdom.*

99. Bower, J.S., Broughton, G.F.J. and Dando M.T. *et al.* (1988) Ozone monitoring in the UK : a review of 1987/8 data from monitoring sites operated by Warren Spring Laboratory. Report LR 678 (AP)M, Warren Spring Laboratory.

100. Bower, J.S., Stevenson, K.J., Broughton, G.F.J., Lampert, J.E., Sweeney, B.P., Dando, M.T., Lees, A.J., Clark, A.C., Driver, G.S., Parker, V.J. and Waddon, C. (1989) Ozone and oxides of nitrogen monitoring in the UK : elements of data quality control and assurance. Laboratory Report 718 (AP), Warren Spring Laboratory.

101. Bower, J.S. (1989) A pragmatic view on optimizing performance of air quality monitoring networks, Measurement and Control, 22, 142-145.

102. Weston, K.J., Kay, P.J.A., Fowler, D., Martin, A. and Bower, J.S. (1989) Mass budget studies of photochemical ozone production over the U.K, Atmos. Environ., 23, 1349-1360.

103. Derwent, R.G. (1989) A comparison of model photochemical ozone formation potential with observed regional scale ozone formation during a photochemical episode over the United Kingdom in April 1987. Atmos. Environ., 23, 1361-1371.

104. Bower, J.S., Broughton, G.F.J., Dando, M.T., Stevenson, K.J., Lampert, J.E., Sweeney, B.P., Parker, V.J., Driver, G.S., Clark, A.G., Waddon, C.J., Wood, A.J. and Williams, M.L. (1989) Surface ozone concentrations in the U.K. in 1987-1988, Atmos. Environ., 23, 2003-2016.

105. Singh, H.B., Viezee, W., Johnson, W.B. and Ludwig. F.L. (1980). The impact of stratospheric ozone on tropospheric air quality, J. Air Pollut. Control Assoc., 30, 1009-1017.

106. Johnson, W.B. and Viezee, W. (1981) Stratospheric ozone in the lower troposphere (I) : presentation and interpretation of aircraft measurements, Atmos. Environ., 15, 1309-1323.

107. Viezee, W., Johnson, W.B. and Singh, H.B. (1983) Stratospheric ozone in the lower troposphere (II) : assessment of downward flux and ground-level impact, Atmos. Environ., 17, 1979-1993.

108. Junge, C.E. (1962) Global ozone budget and exchange between stratosphere and troposphere, Tellus, 14, 363-377.

109. Liu, S.C., Kley, D., McFarland, M., Mahlman, J.D. and Levy H., II. (1980) On the origin of tropospheric ozone, J. Geophys. Res., 85, 7546-7552.

110. Liu, S.C., Trainer, M., Fehensfeld, F.C., Parrish, D.D., Williams, E.J., Fahey, D.W., Hübler, G. and Murphy, P.C. (1987) Ozone production in the rural troposphere and the implications for regional and global ozone distributions, J. Geophys. Res., 92, 4191-4207.

111. Penkett, S.A. and Brice, K.A. (1986) The spring maximum in photo-oxidants in the northern Hemisphere, Nature, 319, 655-657.

112. Brice, K.A., Penkett, S.A., Atkins, D.H.F., Sandalls, F.J., Bamber D.J., Tuck A.F. and Vaughan, G. (1984) Atmospheric measurements of PAN in rural south-east England : seasonal variations, winter photochemistry and long-range transport, Atmos. Environ., 18, 2691-2702.

113. Zurita, E. and Castro, M. (1983) A statistical analysis of mean hourly concentrations of surface ozone at Madrid (Spain) Atmos. Environ., 17, 2213-2220.

114. Colbeck, I. (1988) The occurrence of nocturnal ozone maxima at a rural site in north-west England, Environ. Technol. Lett., 9, 75-80.

115. Martin, A. (1984) Estimated washout coefficients for sulphur dioxide, nitric oxide, nitrogen dioxide and ozone, Atmos. Environ., 18, 1955-1961.

116. Samson, P.J. (1978) Nocturnal ozone maxima, Atmos. Environ., 12, 951-955.

117. Broder, B. (1981) Late evening ozone maxima, PAGEOPH, 119, 978-989.

118. Wolff, G.T., Lioy, P.J. and Taylor, R.S. (1987) The diurnal variations of ozone at different altitudes on a rural mountain in the eastern United States, J. Air Pollut. Control Assoc., 37 (1), 45-48.

119. Angle, R.P. and Sandhu, H.S. (1986) Rural ozone concentration in Alberta, Canada, Atmos. Environ., 20, 1221-1228.

120. Aneja, V.P., Businger, S., Li, Z., Claiborn, C.S. and Murthy, A. (1991) Ozone climatology at high elevations in the southern Appalachians, J. Geophys. Res., 96, 1007-1021.

121. Aneja, V.P. and Li, Z. (1992) Characterization of ozone at high elevation in the eastern United States : trends, seasonal variations and exposure, J. Geophys. Res., 97, 9873-9888.

122. Colbeck, I. and Harrison, R.M. (1985) Dry deposition of ozone : some measurements of deposition velocity and of vertical profiles to 100 metres, Atmos. Environ., 19, 1807-1818.

123. Harrison, R. M., Holman, C.D., McCartney, M.A. and McIlveen, J.F.R. (1978) Nocturnal depletion of photochemical ozone at a rural site, Atmos. Environ., 12, 2021-2026.

124. Feister, U. and Warmbt., W. (1987) Long-term measurements of surface ozone in the German Democratic Republic, J. Atmos. Chem., 5, 1-21.

125. Jost, D. (1970) Survey of the distribution of trace substances in pure and polluted atmospheres, Pure and Applied Chemistry, 24, 643-654.

126. Wisse, J.A. and Velds, C.A. (1970) Preliminary discussion of some oxidant measurements at Vlaardingen, The Netherlands, Atmos. Environ., 4, 79-85.

127. Guicherit, R. and van Dop, H. (1977) Photochemical production of ozone in western Europe (1971-1975) and its relation to meteorology, Atmos. Environ., 11, 145-155.

128. Regener, V.H. (1938) Neue messungen der vertikalen verteilung des ozons in der atmosphare, Z. Phys., 109, 462-670.

129. Regener, V.H. (1964) Measuring the atmospheric ozone with the chemiluminescent method, J. Geophys. Res., 69, 3795-3800.

130. Ehmert, A. (1949) Uber den oz ongehalt der unteren atmosphare bei winterlichem hochdruckwetter nach messungen, Ber. Dtsch. Wetterdeinst US-Zone, No. 11, 63-66.

131. Gunsten, H. (1986) Formation, Transport and Control of Photochemical Smog, In : *Environmental Chemistry, 4 (Part A)* Hutzinger, O. (Ed.), Springer-Verlag. Berlin, 53-105.

132. Becker, K.H., Ficke, W., Lobel, J. and Schurath, U. (1985) Formation Transport and Control of Photochemical Oxidants, In : *Air Pollution by Photochemical Oxidants*, Guderian, R. (Ed.) Springer-Verlag, Berlin, 3-125.

133. Guicherit, R., Jeltes, R. and Lindqvist, F. (1972) Determination of the ozone concentration in outdoor air near Delft, The Netherlands, Environ. Pollut., 3, 91-110.

134. Benarie, M., Benec'h, A. Chuong, B.T. and Menard, T. (1979) Study of photochemical pollution at a rural site - measurement of the ozone and nitrogen oxides at Vert-le-Petet, Pollut. Atmos., 81, 44-52.

135. Lopez, A., Prieur, S., Fontan, J. and Kim, P.S. (1982) Study of the ozone-source in the planetary boundary layer, In : *Physico-chemical Behaviour of Atmospheric Pollutants*, Versino, B. and Ott, H. (Eds.) D. Reidel, Dordrecht, 362-376.

136. Marenco, A. (1986) Variations of CO and O_3 in the troposphere : evidence of O_3 photochemistry. Atmos. Environ., 20, 911-918.

137. Proyou, A.G., Toupance, G. and Perros, P.E. (1991) A two year study of ozone behaviour at rural and forested sites in eastern France, Atmos. Environ., 25A, 2145-2153.

138. Schjoldager, J., Sivertesen, B. and Hanssen, J.E. (1978) On the occurrence of photochemical oxidants at high latitudes, Atmos. Environ., 12, 2461-2467.

139. Schjoldager, J. (1981) Ambient ozone measurements in Norway 1975-1979, J. Air Pollut. Control Assoc., 31, 1187-1191.

140. Grennfelt, P. (1977) Ozone episodes on the Swedish west coast, Prof. Int. Conf. Photochem. Oxidant Pollut. and Its Control. EPA-600/3-77-001 a, 329-377.

141. Hakola, H., Taalas, P., Joffre, S. and Lattila, H. (1989) Ozone concentration behaviour at marine and forested sites in Finland, In : *Ozone in the Atmosphere*. R.D. Bojkov and P. Fabian (Eds.), A Deepak Publishing, Hampton, 481-485.

142. Joffre, S.M., Lättilä, H., Hakola, H, Taalas, P. and Plathan, P. (1988) Measurements of ozone at two background stations in Finland, In : *Tropospheric Ozone*. Isaksen, I.S.A. (Ed.) D. Reidel Publishing Company, 137-146.

143. Hakola, H, Joffre, S., Lattila, H. and Taalas, P. (1991) Transport, formation and sink processes behind surface ozone variability in north European conditions, Atmos. Environ., 25A, 1437-1447.

144. Messresultate des National Beobachtungsnetzes fur Luftfreindstroffe Schritftenreihe Umweltshutz. (1989). Nr. 105.

145. Messresultate des National Beobachtungsnetzes fur Luftfreindstroffe Schritftenreihe Umweltshutz. (1988). Nr. 94.

146. Landry, J-Cl. and Cupelin, F. (1981) The monitoring of ozone immissions in rural and urban areas, Intern J. Environ. Anal. Chem., 9, 169-187.

147. Wunderli, S. and Gehrig, R. (1990) Surface ozone in rural, urban and alpine regions of Switzerland, Atmos. Environ., 24A, 2641-2646.

148. Cecinato, A., Possanzini, M., Liberti, A. and Brocco, D. (1979) Photochemical oxidants in the Rome metropolitan area, Sci. Tot. Environ., 13, 1-8.

149. Stangl, H., Lohse, C., Payrissat, M., Versino, B., Nicollin, B., Ottobrini, G. and Rau, H. (1980) Discussion of field measurements of pollutants made at Ispra and other sites of the Po valley, In : *Physico-chemical Behaviour of Atmospheric Pollutants*, Versino, B. and Ott, H. (Eds.) D. Reidel, Dordrecht, 472-478.

150. Giovanelli, G., Fortezza, F., Minguzzi, L., Strocchi, V. and Vandini, W. (1982) Photochemical ozone transport in an industrial coastal area, In : *Physico-chemical Behaviour of Atmospheric Pollutants*, Versino, B. and Ott, H. (Eds.) D. Reidel, Dordrecht, 492-501.

151. Božičević, Z., Klasinc, L., Cvitaś, T. and Güsten, H. (1976) Photochemische ozonbildung in der unteren atmosphäre über der Stadt Zagreb, Staub-Reinhalt Luft, 36, 363-366.

152. Božičević, Z., Butković, V., Cvitaś, T. and Klasinc, L. (1978) Statistička obrada podataka fotokemijskog zagadenja u Zagrebu, Kem. Ind., 27, 177-181.

153. Cvitaś, T., Güsten, H. and Klasinc, L. (1979) Statistical association of the photochemical ozone concentrations in the lower atmosphere of Zagreb with meteorological variables, Staub-Reinhalt Luft, 39, 92-95.

154. Cvitaś, T. and Klasinc, L. (1979). ,,Trenutni snimak" zagradenja zraka u Kvarnerskom zaljevu, Zast Atmos., 15, 13-16.

155. Novak, I and Sabljić, A. (1981). "Trenutni snimak" zagradenja zraka u Kvarnerskom zaijevu 1979, Kem. Ind., 30, 5-8.

156. Butković, V., Cvitaś, T., Gotovac, V. and Klasnic, L. (1983) Variation of tropospheric ozone concentrations in selected areas of Croatia, Proc. VIth World Congr. Air Qual., May 1983. Paris, 3, 175-180.

157. Závodská, E. and Babušík, I. (1985) Surface ozone in Bratislava, Contrib. Geophys. Inst. Slovak. Acad. Sci. Ser. Meteo., 6, 31-40.

158. Schjoldager, J., Dovland, H., Grennfelt, P. and Saltbones, J. (1981) Photochemical Oxidants in North-western Europe, 1976-79, A Pilot Project. NILU OR 19/81.

159. Grennfelt, P., Hoem, K., Saltbones, J. and Schjoldager, J. (1989) Oxidant Data Collection in OECD-Europe 1985-87. NILU OR 63/89.

160. Grennfelt, P., Saltbones, J. and Schjoldager, J. (1988) Oxidant Data Collection in OECD-Europe 1985-87. NILU OR 31/88.

161. Penkett, S.A., Isaksen, I. and Kley, D. (1988) A program of tropospheric ozone research (TOR) In : *Tropospheric Ozone*, Isaksen, I. (Ed.) D. Reidel, Dordrecht, 345-363.

162. Hov, O., Hesstvedt, E. and Isaksen, I.S.A. (1978a) Long-range transport of tropospheric ozone, Nature, 273, 341-344.

163. van Ham, J. and Builtjes, P.J.H. (1985) Study on photochemical oxidants and precursors. Phase IV. Long-range transport phenomena. TNO, Apeldorn, The Netherlands.

164. Builtjes, P.J.H. (1985) The PHOXA project, photochemical oxidants and acid deposition model application. Proceedings of COST 611-Workshop at RIVM, Bilthoven, The Netherlands.

165. Air Pollution Research Report 1 - Evaluation of the photooxidants precursor relationship in Europe.

(1986). Hov, O., Becker, K.H., Builtjes, P., Cox, R.A. and Kley, D. Commission of the European Communities.

166. Eliassen, A., Hov, O., Isaksen, I.S.A., Saltbones, J. and Storday, F. (1982) A lagrangian long-range transport model with atmospheric boundary layer chemistry, Stud. Environ. Sci., 21, 347-356.

167. Hov, O., Stordal, F. and Eliassen, A. (1985) Photochemical Oxidant Control Strategies in Europe: A 19 Days' Case Study using a Lagrangian Model with Chemistry, NILU TR 5/85.

168. Hov, O. (1988) Photochemical Oxidant Episodes, Acid Deposition and Global Atmospheric Change. NILU OR 12/88.

169. Hov, O., Eliassen, A. and Simpson, D. (1988) Calculations of the distribution of NO_x compounds in Europe, In : *Tropospheric Ozone*, Isaksen, I. (Ed.) D. Reidel, Dordrecht, 239-261.

170. Simpson, D. (1992) Long period modelling of photochemical oxidants in Europe. Model calculations for July 1985, Atmos. Environ., 26A, 1609-1634.

171. Simpson, D. (1993) Photochemical model calculations over Europe for two extended summer periods : 1985 and 1989. Model results and comparison with observations, Atmos. Environ., 27A, 921-943.

172. Derwent, R.G. (1985) Modelling the formation of secondary pollutants in the atmosphere over north-west Europe, In : *Pollutants and Their Ecotoxicological Significance*, Nürnberg, H.W. (Ed.) John Wiley and Sons Ltd., 85-95.

173. Derwent, R.G. and Jenkins, M.E. (1990) Hydrocarbon involvement in photochemical ozone formation in Europe. AERE Report R 13736. HMS0.

174. Wolff, G.T., Lioy, P.J., Wight, G.D., Meyers, R.E. and Cederwall, R.T. (1977) An investigation of long-range transport of ozone across the midwestern and eastern United States, Atmos. Environ., 11, 797-802.

175. Clarke, J.F. and Ching, J.K.S. (1983) Aircraft observations of regional transport of ozone in the northeastern United States, Atmos. Environ., 17, 1703-1712.

176. Wolff G.T. and Lioy, P.J. (1980) Development of an ozone river associated with synoptic scale episodes in the eastern U.S, Environ. Sci. Technol., 14, 1257-1260.

177. Vukovich, F.M., Fishman, J. and Browell, E.V. (1985) The reservoir of ozone in the boundary layer of the eastern United States and its potential impact on the global tropospheric ozone budget, J. Geophys. Res., 90, 5687-5698.

178. Wolff, G.T., Lioy, P.J. and Taylor, R.S., (1987) The diurnal variations of ozone at different altitudes, J. Air Pollut. Control Assoc., 37, 45-8.

179. Meagher, J.F., Lee, N.T., Valente, R.J. and Parkhurst, W.J. (1987) Rural ozone in the southeastern United States, Atmos. Environ., 21, 605-615.

180. Parrish, D.D., Fahey, D.W., Williams, E.J., Liu, S.C., Trainer, M., Murphy, P.C., Albritton, D.L. and Fehsenfeld, F.C. (1986) Background ozone and anthropogenic ozone enhancement at Niwot Ridge, Colorado, J. Atmos. Chem., 4, 63-80.

181. Fehsenfeld, F.C., Bollinger, M.J., Liu, S.C, Parrish, D.D., McFarland, M., Trainer, M., Kley, D., Murphy, P.C. and Albritton, D.L. (1983) A study of ozone in the Colorado mountains, J. Atmos. Chem., 1, 87-105.

182. Singh, H.B., Ludwig, F.L. and Johnson, W.B. (1978) Tropospheric ozone : concentrations and

variabilities in clean remote atmospheres, Atmos. Environ., 12, 2185-2196.

183. Logan, J.A. (1989) Ozone in rural areas of the United States, J. Geophys. Res., 94, 8511-8532.

184. Logan, J.A. (1988) The ozone problem in rural areas of the United States, In : *Tropospheric Ozone*, Isaksen, I. (Ed.) D. Reidel, Dordrecht, 327-344.

185. Logan, J.A. (1985) Tropospheric ozone : seasonal behaviour, trends, and anthropogenic influence, J. Geophys. Res., 90, 10,463-10,482.

186. Tilton, B.E. and Meeks, S.A. (1989) Concentrations and patterns of photochemical oxidants in the United States, In : *Atmospheric Ozone Research and its Policy Implications*, Schneider, T., Lee, S.D., Wolters, G.J.R. and Grant, L.D. (Eds.) Elsevier, Amsterdam, 177-194.

187. Environmental Protection Agency (1986) Air Quality Criteria for Ozone and Other Photochemical Oxidants, Rep. EPA 600/8-84-020-CF and Rep. EPA 600/8-84-020-EF, Research Triangle Park, NC.

188. Chatfield, R. and Harrison, H. (1978) Tropospheric ozone I. evidence for higher background values, J. Geophys. Res., 82, 5965-8.

189. Trainer, M., Williams, E.J., Parrish, D.D., Buhr, M.P., Allwine, E.J., Westberg, H.H., Fehsenfeld, F.C., Lui, S.C. (1987) Models and observations of the impact of natural hydrocarbons on rural ozone, Nature, 329, 705-707.

190. Dann, T.F. (1988) Trends in ambient ozone concentrations in Canada, 1979-86, Proc. APCA Ann. Meet., Paper 88/45.2.

191. Sandhu, H.S. (1975) A study of photochemical air pollutants in the urban airsheds of Edmonton, Alberta, Canada, Atmos. Environ., 22, 973-981.

192. Sandhu, H.S. (1977) Oxidant levels in Alberta airsheds, In Proc. Int. Conf. on Photochemical Oxidant Pollution and its Control, EPA 600/3-77-001 a, US Environmental Protection Agency, 299-305.

193. Peake, E., Maclean, M. A. and Sandhu, H.S. (1983) Surface ozone and peroxyacetyl nitrate (PAN) observations at rural locations in Alberta, Canada, J. Air Pollut. Control Assoc., 33, 881-883.

194. Angle, R.P. and Sandhu, H.S. (1986) Rural ozone concentrations in Alberta, Canada, Atmos. Environ., 20, 1221-1228.

195. Angle, R.P. and Sandhu, H.S. (1989) Urban and rural ozone concentrations in Alberta, Canada, Atmos. Environ., 23, 215-221.

196. Anlanf, K.G., Lusis, M.A., Wiebe, H.A. and Stevens, R.D.S. (1975) High ozone concentrations measured in the vicinity of Toronto, Canada, Atmos. Environ., 9, 1137-1139.

197. Heidorn, K.C. and Yap, D. (1986) A synoptic climatology for surface ozone concentrations in southern Ontario, 1976-1981, Atmos. Environ., 20, 695-703.

198. Yap, D., Ning, D.T. and Dong, W. (1988) An assessment of source contributions to the ozone concentrations in southern Ontario 1979-1985, Atmos. Environ., 22, 1161-1168.

199. Zainal, A.J.M. (1990) Relationship of non-methane hydrocarbons and NO_x to ozone formation in the atmosphere of Bahrain, Atmos. Environ., 24, 1915-1922.

200. Lifshitz, B., Peleg, M. and Luria, M. (1988) The influence of medium-range transport on O_3 at rural sites

in Israel, J. Atmos. Chem., 7, 19-33.

201. Galbally, I.E., Miller, A.J., Hov, R.D., Ahmet, S., Joynt, R.C. and Attwood, D. (1986) Surface ozone at rural sites in the Latrobe valley and Cape Grim, Australia, Atmos. Environ., 20, 2403-2422.

202. Derwent, R.G., McInnes, G., Stewart H.N.M. and Williams, M.L. (1976) The occurrence of air pollution by photochemically produced oxidant in the British Isles, 1972-1975, Warren Spring Lab. LR (227) AP.

203. Dechaux, J.C., Galloo, J.C., Apling A.J., Rogers, F., Peperstraete, H.J., Claes, W., Magdonelle, F. and Zierock, K.H. (1986) L'épisode Européen de "Smog" photochimique de septembre 1982, Pollution Atmosphérique, July-September, 184-189.

204. van Dop, H., den Tonkelaar, J.F. and Briffa, F.E.J. (1987) A modelling study of atmospheric transport and photochemistry in the mixed layer during anticyclonic episodes in Europe, Part I: meteorology and air trajectories, J. Climate Appl. Met., 26, 1305-1316.
205. Selby, K. (1987) A modelling study of atmospheric transport and photochemistry in the mixed layer during anticyclonic episodes in Europe, Part II: calculations of photo-oxidant levels along air trajectories, J. Climate Appl. Met., 26, 1317-1338.

206. Jones, P.D. and Kelly, P.M. (1982) Principal component analysis of the Lamb catalogue of daily weather types: Pt. 1, annual frequencies. J. Clim., 2, 147-157.

207. Palutikof, J.P., Davies, T.D., Kelly, P.M. and Halliday, J.A. (1986) Meteorological and statistical characteristics of high and low wind over the British Isles : a preliminary analysis, In : *Wind Energy Conversion 1987*, Anderson, M.B. and Powles, S.R.J. (Eds.) Mechanical Engineering Publications, London, 71-78.

208. Davies, T.D., Kelly, P.M., Low, P.S. and Pierce, C.E. (1992) Surface ozone concentrations in Europe : links with the regional scale atmospheric circulation, J. Geophys. Res., 97, 9819-9832.

209. Vukovich, F.M., Bach, W.D., Crissman, B.W. and King, W.J. (1977) On the relationship between high ozone in the rural boundary layer and high pressure systems, Atmos. Environ., 11, 967-983.

210. Vukovich, F.M. (1979) A note on air quality in high pressure systems, Atmos. Environ., 13, 255-265.

211. National Research Council. (1991) Rethinking the Ozone Problem in Urban and Regional Air Pollution, National Academy Press, Washington, DC.

212. Ripperton, L.A, Worth, J.J.B., Vukovich, F.M. and Decker, C.E. (1977) Research Triangle Institute studies of high ozone concentration in non-urban areas, Proc. Int. Conf. on Photochemical Oxidant Pollution and Its Control, EPA 600/3-77-001A, Research Triangle Park, NC, 413-425.

213. Heidorn, K.C. and Yap, D. (1986) A synoptic climatology for surface ozone concentrations in southern Ontario, 1976-1981, Atmos. Environ., 20, 695-703.

214. Vukovich, F.M. and Fishman, J. (1986) The climatology of summertime O_3 and SO_2, Atmos. Environ., 20, 2423-2433.

215. Comrie, A.C. and Yarnal, B. (1992) Relationships between synoptic scale atmospheric circulation and ozone concentrations in metropolitan Pittsburgh, Pennsylvania, Atmos. Environ., 26B, 301-321.

216. Pagnotti, V. (1990) Seasonal ozone levels and control by seasonal meteorology. J. Air Waste Manage. Assoc., 40, 206-210.

217. Simpson, D. and Timmis, R.J. (1985) Meteorological and related land use data for use in photochemical

air pollution models, Warren Spring Laboratory, LR 546(AP)M.

218. Simpson, D. (1987) An analysis of NO_2 episodes in London, Warren Spring Laboratory, LR 575(AP)M.

219. Spanton, A.M. and Williams, M.L. (1988) A comparison of the structure of the atmospheric boundary layers in central London and a rural/suburban site using acoustic sounding, Atmos. Environ., 22, 211-223.

220. Dickerson, R.R., Huffman, G.J., Luke, W.T., Nunnermacker, L.J., Pickering, K.E., Leslie, A.C.D., Lindsey, C.G., Slinn, W.G.N., Kelly, T.J., Daum, P.H., Delany, A.C., Greenberg, J.P., Zimmerman, P.R., Boatman, J.F., Ray, J.D. and Stedman, D.H. (1987) Thunderstorms : an important mechanism in the transport of air pollutants, Science, 235, 460-465.

221. Pickering, K.E., Dickerson, R.R., Luke, W.T. and Nunnermacker, L.J. (1989) Clear-sky vertical profiles of trace gases as influenced by upstream convective activity, J. Geophys. Res., 94, 14879-14892.

222. Pickering, K.E., Thompson, A.M., Dickerson, R.R., Luke, W.T., McNamara, D.P., Greenberg, J.P. and Zimmerman, P.R. (1990) Model calculations of tropospheric ozone production following observed convective events, J. Geophys. Res., 95, 14049-14062.

223. Nieber, H., Carter, W.P.L., Lloyd, A.C. and Pitts, J.N. (1976) The effect of latitude on the potential formation of photochemical smog, Atmos. Environ., 10, 731-734.

224. Reiter, R., Carnuth, W. and Sladkovic, R. (1972). Utraviolettstrahlung in alpinen Hohenlagen, Wetter and Leben, 24, 231-247.

225. Schmidt, M. (1989) Relationships between tropospheric ozone and global solar radiation in Germany, In : *Ozone in the Atmosphere*. Bojkov, R.D. and Fabian, P. (Eds.) A Deepak Publishing, Hampton, 451-454.

226. Feister, U. Grasnick, K.H. and Jakobi, G. (1989) Surface ozone and solar radiation, In : *Ozone in the Atmosphere*. Bojkov, R.D. and Fabian, P. (Eds.) A Deepak Publishing, Hampton, 441-444.

227. Shlanta, A. and Moore, C.B. (1972) Ozone and point discharge measurements under thunderstorms, J. Geophys. Res., 77, 4500-4510.

228. Clarke, J.F. and Griffing, G.W. (1985) Aircraft observations of extreme ozone concentrations near thunderstorms, Atmos. Environ., 19, 1175-1179

229. Lui, S.C., McFarland, M., Kley, D., Zafiriou, O. and Huebert, B, Tropospheric NO_x and O_3 budgets in the equatorial Pacific (1983) J. Geophys. Res., 88, 1360-8.

230. Chatfield, R.B. (1989) How lightning and human activity determine tropical ozone levels, In : *Ozone in the Atmosphere*. Bojkov, R.D. and Fabian, P. (Eds.) A Deepak Publishing, Hampton, 572-575.

231. Karl, T.R. (1979) Potential application of model output statistics (MOS) to forecasts of surface ozone concentration, J. Applied Met., 18, 256-265.

232. Clark, T.L. and Karl, T.R. (1982) Application of prognostic meteorological variables to forecasts of daily maximum one-hour ozone concentrations in the northeastern United States, J. Applied Met., 21, 1662-1671.

233. Chock, D.P., Kumar, S. and Herrmannn, R.W.(1982) An analysis of trends in oxidant air quality in the south coast air basin of California, Atmos. Environ., 16, 2615-2624.

234. Thornes, J.E. (1977) Applied climatology : ozone comes to London, Progress in Physical Geography, Volume I, 506-517.

235. Mukammal, E.I., Neumann, H.H. and Gillespie, T.J. (1982) Meteorological conditions associated with ozone in southwestern Ontario, Canada, Atmos. Environ., 16, 2095-2106.

236. Bojkov, R.D. (1986) Surface ozone during the second half of the nineteenth century, J. Climate Applied Met., 25, 343-352.

237. Kley, D., Volz, A. and Mulheins, F. (1988) Ozone measurements in historic perspective, In : *Tropospheric Ozone*, Isaksen I.S.A. (Ed.) D. Reidel, Dordrecht, 63-72.

238. Volz, A. and Kley, D. (1988). Evaluation of the Montsouris series of ozone measurements made in the nineteenth century, Nature, 332, 240-242.

239. Isaksen, I.S.A. (1988) Is the oxidising capacity of the atmosphere changing? In : *The Changing Atmosphere*, Rowland, F.S. and Isaksen, I.S.A. (Eds.) John Wiley, Chichester, 141-157.

240. Guicherit, R. (1988) Ozone on an urban and regional scale, In : *Tropospheric Ozone*, Isaksen I.S.A. (Ed.) D. Reidel, Dordrecht, 49-62.

241. Low, P.S., Davies, T.D., Kelly, P.M. and Farmer, G. (1990) Trends in surface ozone at Hohenpeissenberg and Arkona, J. Geophys. Res., 95, 22441-22453.

242. Low, P.S., Davies, T.D., Kelly, P.M. and Reiter, R. (1991) Uncertainties in surface ozone trend at Hohenpeissenberg, Atmos. Environ., 25A, 511-515.

243. Low, P.S., Kelly, P.M. and Davies, T.D. (1992) Variations in surface ozone trends over Europe, Geophys. Res. Lett., 19, 1117-1120.

244. Environmental Protection Agency. (1991) National Air Quality and Emissions Trends Report 1990, US EPA, Research Triangle Park, EPA/450/4-91/023.

245. Komhyr, W.D., Oltmans, S.J. and Grass, R.D. (1988) Atmospheric ozone at South Pole Antarctica in 1986, J. Geophys. Res., 93, 5167-84.

246. Bodhaine, B.A. and Rosson, R.M. (Eds). (1988) Geophysical monitoring for climate change, Summary Report 1987, NOAA Air Resources Lab., Boulder 1988.

247. Bojkov, R.D. (1988) Ozone changes at the surface and in the free troposphere, In : *Tropospheric Ozone*, Isaksen I.S.A. (Ed.) D. Reidel, Dordrecht, 83-96.

248. Scientific Assessment of Stratospheric Ozone: 1989, Volume 1, WMO, (1989) Global Ozone Research and Monitoring Project, Report No. 20.

249. Gammon, R., Wofsy, S.C., Cicerone, R.J., Delany, A.C., Harriss, R.T., Khalil, M.A.K., Logan, J.A., Midgley, P. and Prather, M. (1986) Tropospheric trace species, In : *Atmospheric Ozone 1985*, 1, 57-116, WMO Global Ozone Research and Monitoring Project, Report Number 16.

250. Ehhalt, D.H., Fraser, P.J., Albritton, D., Makide, Y., Cicerone, R.J., Rowland, F.S., Khalil, M.A.K., Steele, L.P., Legrand, M. and Zander, R. (1989) Trends in source gases, In : *Report of the International Ozone Trends Panel - 1988*, WMO Report Number 18, WMO, Geneva.

251. Dignon, J. and Hameed, S. (1985) A model investigation of the impact of increases in anthropogenic NO_x emissions between 1967 and 1980 on tropospheric ozone, J. Atmos. Chem., 3, 491-506.

252. Hov, O. (1990) The role of nitrogen oxides in the long-range transport of photochemical oxidants, Sci. Tot. Environ., 96, 101.

253. Hough, A.M. and Derwent, R.G. (1990) Changes in the global concentration of tropospheric ozone due to human activities, Nature, 344, 645-648.

254. Prinn, R.G. (1988) Toward an improved global network for determination of tropospheric ozone climatology and trends, J. Atmos. Chem., 6, 281-298.

255. Taylor, O.C. (1969) Importance of peroxyacetyl nitrate (PAN) as a phytotoxic air pollutant, J. Air Pollut. Control Ass. 19, 347-351.

256. Mudd, J.B. (1975) Peroxyacyl Nitrates, In : *Responses of Plants to Air Pollution*, Mudd, J.B. and Kozlowski, T. (Eds.) Academic Press, New York.

257. Lovelock, J. E. (1977) PAN in the natural environment : its possible significance in the epidemiology of skin cancer, Ambio 6, 131-133.

258. Spicer, C.W., Holdren, M.W. and Kiegley, G.W. (1983) The ubiquity of peroxyacetyl nitrate in the continental boundary layer, Atmos. Environ., 17, 1055-1058.

259. Altshuller, A.P. (1993) PANs in the atmosphere, J. Air Waste Manage. Assoc., 43, 1221-1230.

260. Stephens, E.R. (1987) Smog studies of the 1950s, EOS, 61, 89-93.

261. Stephens, E.R., Hanst, P.L., Doerr, R.C. and Scott, W.E. (1956) Reactions of nitrogen dioxide and organic compounds in air, Ind. Engng. Chem., 48, 1498-1504.

262. Brice, K.A., Penkett, S.A., Atkins, D.H.F., Sandalls, F.J., Bamber, D.J., Tuck, A.F. and Vaughan, G. (1984) Atmospheric measurements of peroxyacetylnitrate (PAN) in rural, south-east England : seasonal variations winter photochemistry and long-range transport, Atmos. Environ., 18, 2691-2702.

263. Nielsen, T., Sammnelsson, U., Grennfelt, P. and Thompsen, E.L. (1981) Peroxyacetylnitrate in long-range transported polluted air, Nature, 293, 553.

264. Temple, P.J. and Taylor, O.C. (1983) World-wide ambient measurements of peroxyacetyl nitrate (PAN) and implications for plant injury, Atmos. Environ., 17, 1583-1587.

265. Altshuller, A.P. (1983) Measurements of products of atmospheric photochemical reactions in laboratory studies and in ambient air : relationships between ozone and other products, Atmos. Environ., 17, 2382-2427.

266. Singh, H.B., Salas, L.J. and Viezee, W. (1986) Global distribution of peroxyacetyl nitrate, Nature, 321, 588-591.

267. Müller, K.P. and Rudolph, J. (1992) Measurements of peroxyacetylnitrate in the marine boundary layer over the Atlantic, J. Atmos. Chem., 15, 361-367.

268. Bottenheim, J.W., Barrie, L.A. and Atlas, E. (1993) The partitioning of nitrogen oxides in the lower Arctic troposphere during spring 1988, J. Atmos. Chem., 17, 15-27.

269. Singh, H.B. and Salas, L.J. (1989) Measurements of peroxyacetyl nitrate (PAN) and peroxypropionyl nitrate (PPN) at selected urban, rural and remote sites, Atmos. Environ. 23, 231-238.

270. Singh, H.B. (1987) Reactive nitrogen in the troposphere, Environ. Sci. Technol., 21, 320-327.

271. Singh, H.B., Herlth, D., O'Hara, D., Zahnle, K., Bradshaw, J.D., Sandholm, S.T., Talbot, R., Crutzen, P.J. and Kanakidou, M. (1992) Relationship of peroxyacetyl nitrate to active and total odd nitrogen at northern high latitudes : influence of reservoir species on NO_x and O_3, J. Geophys. Res., 97, 16523-16530.

272. Rudolph, J., Vierkorn-Rudolph, B. and Meixer, F.X. (1987) Large-scale distribution of peroxyacetylnitrate results from the Stratoz III flights, J. Geophys. Res., 92, 6653-6661.

273. Hov, O. (1984) Modelling of the long-range transport of peroxyacetylnitrate to Scandinavia, J. Atmos. Chem., 1, 187-202.

274. Pressman, A.Y., Galperin, M.V., Popov, V.A., Afinogenova, O.G., Subbotin, S.R., Grigoryan, S.A. and Dedkova, I.S. (1991) A routine model of chemical transformations and transport of nitrogen compounds, ozone and PAN within a regional scale, Atmos. Environ., 25A, 1851-1863.

275. Singh, H.B., Salas, L.J., Ridley, B.A., Shetter, J.D., Donahue, N.M., Fehsenfeld, F.C., Fahey, D.W., Parrish, D.D., Williams, E.J., Liu, S.C., Hübler, G. and Murphy, P.C. (1985) Relationship between peroxyacetyl nitrate and nitrogen oxides in the clean troposphere, Nature, 318, 347-349.

276. Bottenheim, J.W., Gallant, A.J. and Brice, K.A. (1986) Measurements of NO_y species and O_3 at 82°N latitude. Geophys. Res. Lett., 13, 113-116.

277. Bottenheim, J.W. and Gallant, A.J. (1989) PAN over the Arctic : observations during AGASP-2 in April 1986, J. Atmos. Chem., 9, 301-316.

278. Barrie, L.A., den Hartog, G., Bottenheim, J.W. and Landsberger, S.J. (1989) Anthropogenic aerosols and gases in the lower troposphere at Alert, Canada in April 1986, J. Atmos. Chem., 9, 101-127.

279. Barrie, L. A. and Bottenheim, J.W. (1991) Sulphur and nitrogen pollution in the Arctic atmosphere, In : *Pollution of the Arctic Atmosphere*, Sturges, W. (Ed.) Elsevier Press.

280. Jaffe, D.A., Honrath, R.E. and Herring, J.A. (1991) Measurements of nitrogen oxides at Barrow, Alaska : evidence for regional and north-hemispheric sources of pollution, J. Geophys. Res., 96, 7395-7406.

281. Grosjean, D. (1993) Distribution of atmospheric nitrogenous pollutants at a Los Angeles area smog receptor site, Environ. Sci. Technol., 17, 13-19.

282. Williams, E. L. and Grosjean, D. (1991) Peroxypropionyl nitrate at a southern California mountain forest site, Environ. Sci. Technol., 25, 653-659.

283. Grosjean, D., Williams, E.L. and Grosjean, E. (1993) Peroxyacyl nitrates at southern California mountain forest locations, Environ. Sci. Technol., 27, 110-121.

284. Shepson, P.B., Hastie, D.R., So, K.W. and Schiff, H.I. (1992) Relationships between PAN, PPN and O_3 at urban and rural sites in Ontario, Atmos. Environ., 26A, 1259-1270.

285. Peake E., MacLean, M.A., Lester, P.F. and Sandhu, H.S. (1988) Peroxyacetylnitrate (PAN) in the atmosphere of Edmonton, Alberta, Canada, Atmos. Environ., 22, 973-981.

286. Peake, E., MacLean, M.A. and Sandhu, H.S. (1985) Total inorganic nitrate (particulate nitrate and nitric acid) observations in Calgary, Alberta, J. Air Pollut. Control Assoc., 35, 250.

287. Peake, E. and Sandhu, H.S. (1983) The formation of ozone and peroxyacetyl nitrate (PAN) in the urban atmospheres of Alberta, Can. J. Chem., 61, 927-935.

288. Nieboer, H. and van Ham, J. (1976) Peroxyacetyl nitrate (PAN) in relation to ozone and some meteorological parameters at Delft in the Netherlands, Atmos. Environ., 10, 115-120.

289. Gusten, H. (1986). Formation, Transport and Control of Photochemical Smog, In : *The Handbook of Environmental Chemistry, 4, Part A, Air Pollution.* Hutzinger, O. (Ed.) Springer-Verlag, Berlin, 53-105.

290. Tsani-Bazaca, E., Glavas, S. and Güsten, H. (1988) Peroxyacetyl nitrate (PAN) concentrations in Athens, Greece, Atmos. Environ., 22, 2283-2286.

291. Tsalkani, N., Perros, P. and Toupance, G. (1988) Continuous atmospheric measurements of peroxyacetyl nitrate (PAN) in a mediterranean site (Athens, Greece) Environ. Tech. Letters, 9, 143-152.

292. Wunderli, S. and Gehrig, R. (1991) Influence of temperature on formation and stability of surface PAN and ozone, A two year field study in Switzerland, Atmos. Environ., 25A, 1599-1608,

293. Tsalkani, N., Perros, P. and Toupance, G. (1987) High PAN concentrations during nonsummer periods : a study of two episodes in Creteil (Paris), France, J. Atmos. Chem., 5, 291-299.

294. Tsalkani, N., Perros, P., Dutot, A.L. and Toupance, G. (1991) One year measurements of PAN in the Paris basin : effect of meteorological parameters, Atmos. Environ., 25A, 1941-1949.

295. Ridley, B.A. Shetter, J.D., Walega, J.D., Madronich, S., Elsworth, C., Parish, D.D., Hübler, G., Bruhr, M., Williams, E.J., Allwine, E.J. and Westberg, R.H. (1990) The Behaviour of some organic nitrates at Boulder and Niwot Ridge, Colorado, J. Geophys. Res., 95, 13949-13961.

296. Brice, K.A., Bottenheim, J.W., Anlauf, K.G. and Wiebe, H.A. (1988) Long-term measurements of atmospheric peroxyacetylnitrate (PAN) at rural sites in Ontario and Nova Scotia; Seasonal Variations and Long-range Transport, Tellus, 40B, 408-425.

297. Dollard, G.J., Jones, B.M.R. and Davies, T.J. (1991) Measurements of gaseous hydrogen peroxide and PAN in rural southern England, Atmos. Environ., 25A, 2039-2053.

298. Perros, P., Tsalkani, N. and Toupance, G. (1988) PAN measurements in a forested area (Donon, France), Environ. Tech. Letters, 9, 351-358.

299. Singh, H.B., O'Hara, D., Herlth, D., Bradshaw, J.D., Sandholm, S.T., Gregory, G.L., Sachse, G.W., Blake, D.R., Crutzen, P.J. and Kanakidou, M.A. (1992b) Atmospheric measurements of peroxyacetyl nitrate and other organic nitrates at high latitudes : possible sources and sinks, J. Geophys. Res., 97, 16511-16522.

300. Tanner, R.L., Miguel, A.H., de Andrade, J.B., Gaffney, J.S. and Streit, G.E. (1988) Atmospheric chemistry of aldehydes : enhanced peroxyacetyl nitrate formation from Ethanol fuelled vehicular emissions. Environ. Sci. Technol., 22, 1026-1034.

301. Ciccioli, P., Brancaleoni, E., DiPalo, V., Liberti, V., and DiPalo, C. (1986) Misura delle alterazoni della qualita dell'aria da fenomeni di smog fotochimico, Acqua Aria, 7, 675-683.

302. Lonneman, W.A., Bufalini, J.J. and Seila, R.L. (1976) PAN and oxidant measurement in ambient air, Environ. Sci. Technol., 10, 374-380.

303. Walega, J.G., Ridley, B.A., Madronick, S., Grahek, F.E., Shetter, J.D., Sauvain, T.D., Hahn, C.J., Merill, J.T., Bodhaine, B.A. and Robinson, E. (1992) Observations of peroxyacetyl nitrate, peroxypropionyl nitrate, methyl nitrate and ozone during the Mauna Loa Observatory photochemical experiment, J. Geophys. Res., 10311-87.

304. Grosjean, D., Williams, E.L. and Grosjean, E. (1993b) Ambient levels of peroxy-n-butyryl nitrate at a southern California mountain forest smog receptor location, Environ. Sci. Technol., 27, 326-331.

305. Heikes, B.G.A., Lazrus, G.L., Kok, S.M., Kunen, B.W., Gandrud, S.N., Gitlin, S.N. and Sperry, P.D. (1982) Evidence for aqueous phase hydrogen peroxide synthesis in the troposphere, J. Geophys. Res., 87, 3045-3051

306. Zika, R.G. and Saltzman, E.S. (1982) Interaction of ozone and hydrogen peroxide in water : implications for analysis of H_2O_2 in air, Geophys. Res. Lett., 9, 231-23

307. Ibusuki, T. (1983) Influence of trace metals on the determination of hydrogen peroxide in rainwater by using a chemiluminescent technique, Atmos. Environ., 17, 393-396.

308. Sakugawa, H., Kaplan, I.R., Tsai, W. and Cohen, Y. (1990) Atmospheric hydrogen peroxide, Environ. Sci. Technol., 24, 1452-1462.

309. Kleinman, L.I. (1986) Photochemical formation of peroxides in the boundary layer, J. Geophys. Res., 91, 10899-10904.

310. Tremmel, H.G., Junkerman, W., and Slemr, F. (1993) On the distribution of hydrogen peroxide in the lower troposphere over the northeastern United States during late summer 1988, J. Geophys. Res., 98, 1083-1099.

311. Jacob, P. and Klockow, D. (1992) Hydrogen peroxide measurements in the marine atmosphere, J. Atmos. Chem., 15, 353-360.

312. Jacob, P., Tavares, T.M., Rocha, V.C. and Klockow, D. (1990) Atmospheric H_2O_2 field measurements in a tropical environment : Bahia, Brazil, Atmos. Environ., 24A, 377-382

313. Harris, G.W., Burrows, J.P., Klemp, D. and Zenker, T. (1989) Measurement of trace species in the remote maritime boundary layer using TDLAS, In : *Monitoring of Gaseous Pollutants by Tunable Diode Lasers*, Grisar, R., Schmidtke, G., Tacke, M. and Restell, G. (Eds) Kluwer Academic Publishers, Dordrecht, 68-75.

314. Van Valin, C.C., Luria, M., Ray, J.D. and Boatman, J.F. and Gunter, R.L. (1987) Hydrogen peroxide in air during winter over the south-central United States. Geophys. Res. Lett., 14, 1146-1149.

315. Van Valin, C.C., Luria, M., Ray, J.D. and Boatman, J.F. (1990) Hydrogen peroxide and ozone over the northeastern United States in June 1987, Geophys. Res., 95, 5689-5695.

316. Ross, H.B., Johansson, C., de Serves, C. and Lind, J. (1992) Summertime diurnal variations of atmospheric peroxides and formaldehyde in Sweden, J. Atmos. Chem., 14, 411-423.

317. Dollard, G.J. and Davies, T.J. (1992) Observations of H_2O_2 and PAN in a rural atmosphere, Environ. Pollut., 75, 45-52.

318. Kleinman, L.I. (1991) Seasonal dependence of boundary layer peroxide concentration : the low and high NO_x regimes, J. Geophys. Res., 96, 20721-20733.

319. Boatman, J.F., Wellman, D.L., Van Valin, C.C., Gunter, R.L., Ray, J.D., Sievering, H., Kim, Y., Wilkison, S.W. and Luria, M. (1989) Airborne sampling of selected trace chemicals above the central United States, J. Geophys. Res., 94, 5081-5093.

320. Sakugawa, H. and Kaplan, I.R. (1989) H_2O_2 and O_3 in the atmosphere of Los Angeles and its vicinity : factors controlling their formation and their role as oxidants of SO_2, J. Geophys. Res., 94, 12957-12973.

321. Olszyna, K.J., Meagher, J.F. and Bailey, E.M. (1988) Gas-phase, cloud and rain-water measurements of hydrogen peroxide at a high-elevation site, Atmos. Environ., 22, 1699.

322. Dollard, G.J., Derwent, R.G. and Sandalls, F.J. (1989) Measurements of gaseous hydrogen peroxide in southern England during a photochemical pollution episode, Environ. Pollut., 58, 115-1124.

323. Derwent, R.G. and Hov, O. (1988) Applications of sensitivity and uncertainty analysis technique to a photochemical model, J. Geophys. Res., 93, 5158-5199.

324. Daum, P.H., Kleinman, L.I., Hills, A.J., Lazrus, A.L., Leslie, A.C.D., Busness, K. and Boatman, J. (1990). Measurement and interpretation of concentrations of H_2O_2 and related species in the upper midwest during summer. J. Geophys. Res., 95, 9857-9871.

325. Calvert, J. G. and Stockwell, W.R. (1983) Acid generation in the troposphere by gas-phase chemistry, Environ. Sci. Technol. 17, 428A-443A.

326. Sakugawa, H. and Kaplan, I.R. (1993) Comparison of H_2O_2 content in the atmospheric samples in the San Bernadino Mountains, Southern California. Atmos. Environ., 27A, 1509-1515.

327. Ghauri, B.M.K., Salam, M. and Mirza, M.I. (1991). Surface ozone in Karachi, In : *ozone Depletion, Implications for the Tropics*. M. Ilyas (ed.), University of Science Malaysia, 169-177.

328. Tsutsumi, Y., Makino, Y. and Hirota, M. (1991). Tropospheric ozone studies in Japan, In : *ozone Depletion, Implications for the Tropics*. M. Ilyas (ed.), University of Science Malaysia, 196-201.

329. Hov, O. (1983) One dimensional vertical model for ozone and other gases in the atmospheric boundary layer, Atmos. Environ., 17, 535-549.

330. Gabally, I. (1968) Some measurements of ozone variation and destruction in the atmospheric surface layer, Nature 218, 456-457.

331. Van Dop, H. Guicherit, R. and Lanting R.W. (1977) Some measurements of the vertical distribution of ozone in the atmospheric boundary layer, Atmos. Environ., 11, 65-71.

332. Williams, E.L. and Grosjean, D. (1990) Southern California air quality study : peroxyacetyl nitrate, Atmos. Environ., 24A, 2369-2377.

333. Sexton, K. (1983). Evidence of an additive effect for ozone plumes from small cities, Environ. Sci. Technol., 17, 402-407

334. White, W.H., Blumenthal, D.L., Anderson, J.A., Husar, R.B., and Wilson, W.E. Jnr. (1977). Ozone formation in the St.Louis urban plume, Proc. Int. Conf. on Photochemical Oxidant Pollution and its Control, 1, B. Dimitriades (Ed.), pp 237-247, EPA-600/3-77-00la.

335. Karl, T.R. (1978) Ozone transport in the St Louis area. Atmos. Environ., 12, 1421-1431

336. Spicer, C.W., Joseph, D.W. and Stickel, P.R. (1982) An investigation of the ozone plume of a small city, J. Air Pollut. Control Assoc., 32, 278-281.

337. Lindsay, R.W. and Chameides, W.L. (1988) High-ozone events in Atlanta, Georgia in 1983 and 1984, Environ. Sci. Technol., 22, 426-431.

338. Chung, Y-S. (1977) Ground-level ozone and regional transport of air pollutants, J. Appl. Meteor., 16, 1127-1136.

339. Van Valin, C.C. and Luria, M. (1988) O_3, CO, hydrocarbons and dimethyl sulphide over the western Atlantic ocean, Atmos. Environ., 22, 2401-2409.

340. Spicer, C.W., Koetz, J.R., Keigley, C.W., Sverdrup G.M. and Ward, G.F. (1982) Nitrogen oxides reactions within urban plumes transported over the ocean, US EPA contract No. 68-02-2957.

341. Zeller, K.F., Evans, R.B., Fitzsimmons, C.K., and Siple, G.W. (1977) Mesoscale analysis of ozone measurements in the Boston environs, J. Geophys. Res., 82, 5879-5887.

342. Siple, G.W., Fitzsimmons, C.K. Zeller, K.F. and Evans, R.B. (1977) Long-range airborne measurements of ozone off the coast of the northeastern United States, Proc. Int. Conf. on Photochemical Oxidants Pollution and its Control, 1, Dimitriades, B. (Ed.) 249-258, EPA-600/3-77-001a.

343. Spicer, C.W., Gemma, J.L. and Stickel, P.R. (1977) The transport of oxidant beyond urban areas, Data analysis and predictive models for the Southern New England study, EPA-600/3-77-041.

344. Spicer, C.W., Joseph, D.W. and Sticker, P.R. (1979) Ozone sources and transport in the north-eastern United States, Environ. Sci. Technol., 13, 975-984.

345. Cleveland, W.S., Kleiner, B., McRae, J.E. and Warner J.L. (1975) The analysis of ground-level ozone data from New Jersey, New York, Connecticut and Massachusetts: Transport from the New York City Metropolitan area, 4th Symp. Stat. Environ., 70-90.

346. Cleveland, W.S., Kleiner, B., McRae, J.E., Warner, J.L., and Pasceri, R.E. (1977) Geographical properties of ozone concentrations in the north eastern United States, J. Air Pollut. Control Assoc., 27, 325-328.

347. Wolff, G.T., Lioy, P.J. Wight, G.D. and Pasceri R.E. (1977) Aerial investigation of the ozone plume phenomenon, J. Air Pollut. Control Assoc., 27, 460-463.

348. Clark, T.L. and Clarke, J.F. (1984) A Lagrangian study of the boundary layer transport of pollutants in the northeastern United States, Atmos. Environ., 18, 287-297.

349. Possiel, N.C., Eaton, W.C., Saeger, M.L., Sickles, J.E., Bach, W.D. and Decker, C.E. (1979) Ozone-precursor concentrations in the vicinity of a medium sized city. 72nd Annual Meeting of the Air Pollution Control Association, paper 79-58.3.

350. Blake, N.J., Penkett, S.A., Clemitshaw, K.C., Anwyl, P., Lightman, P., Marsh, A.R.W. and Butcher, G. (1993) Estimates of atmospheric hydroxyl radical concentrations from the observed decay of many reactive hydrocarbons in well-defined urban plumes, J. Geophys. Res., 98, 2851-2864.

351. Aneja, V.P., Yoder, G.T. and Arya, S.P. (1992) Ozone in the urban southeastern United States, Environ. Pollut., 75, 39-44.

352. Gay, M.J. (1991) Meteorological and altitudinal influences on the concentration of ozone at Great Dun Fell, Atmos. Environ., 25A, 1767-1781.

353. Feister, U. and Balzer, K. (1991) Surface ozone and meteorological predictors on a subregional scale, Atmos. Environ., 25A, 1781-1790.

354. Girgždiene, R. (1991) Surface ozone measurements in Lithuania, Atmos. Environ., 25A, 1791-1795.

355. Parrish, D.D., Holloway, J.S., Trainer, M., Murphy, P.C., Forbes, G.L. and Fehsenfeld, F.C. (1993) Export of North American ozone pollution to the North Atlantic Ocean, Science, 259, 1436-1439.

CHAPTER 7
ENVIRONMENTAL EFFECTS

Whereas stratospheric ozone is generally seen as beneficial to the environment, tropospheric ozone and especially photochemically produced ozone poses several adverse effects. It not only plays a role in damaging plant species, but also leads to irritation of mucus membrane and hence affects human health. Furthermore, it has been shown to damage materials. Since ozone has an absorption band at 9.6 μm it absorbs upward directed terrestrial radiation and therefore contributes to the absorption in the atmosphere of longwave radiation emitted from the Earth, the so-called greenhouse effect.

It is the potentially serious human health effects that have caused numerous countries to adopt air quality standards and to introduce emission control strategies. International cooperation is clearly needed to study and control ozone damage. Polluted air masses do not respect international boundaries and hence have the potential to cause economic damage in neighbouring countries and in, regions of international importance, such as the North Pole.

7.1 IMPACT ON CLIMATE

Ozone indirectly exercises a controlling influence on the composition of the troposphere via its role as a source of hydroxyl radicals (see chapter 4) and also acts as a greenhouse gas. Tropospheric ozone concentrations are influenced by the distribution of CH_4, CO, NO_x and NMHCs, leading to indirect greenhouse contributions for these gases. The uncertainties connected with estimates of these indirect effects are larger than the uncertainties of those connected to estimates of the direct effects. Early estimates predicted that a doubling of ozone in the troposphere would produce a surface air temperature increase of approximately 1°C. However, due to the absence of data from which meaningful global trends can be derived, it is not possible at this stage to quantify the current contribution of tropospheric ozone to the global greenhouse radiative forcing [1].

7.2 EFFECTS ON NON BIOLOGICAL MATERIALS

Damage to materials by photochemical oxidants has been studied for many years. However, virtually all research has focused on ozone and so a number of the less abundant oxidants may equal or surpass ozone in reactivity with certain materials. Similarly most research has focused on economically important or abundant materials that are susceptible to oxidant damage. These include elastomers, textile fibres, dyes and paints. Various antioxidants and other protective measures have been formulated to reduce the rates of attack but at an economic cost.

Most elastomeric materials are composed of unsaturated, long-chain organic molecules which are particularly susceptible to ozone attack. The damage mechanisms involve the scission of the carbon-carbon bond (Figure 7.1). Synthetic elastomers with more saturated chemical structures have an inherent resistance to ozone damage[2]. Ozone damage usually takes the form of cracking leading to brittleness.

Figure 7.1

Postulated mechanism for damage to elastomers by ozone.

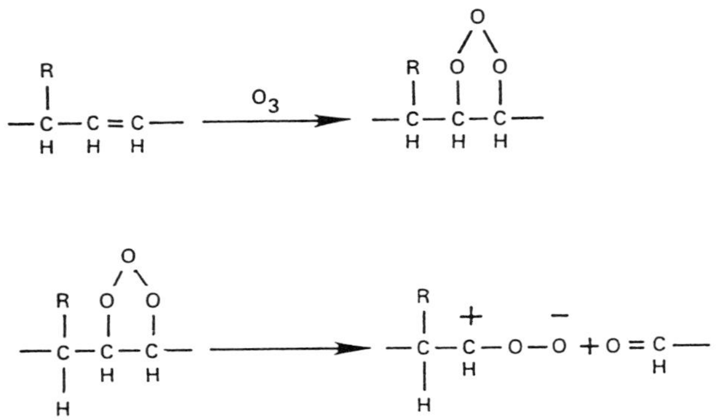

$$
\begin{array}{c}
\text{R} \\
| \\
-\text{C}-\text{C}=\text{C}- \\
\text{H}\ \ \text{H}\ \ \text{H}
\end{array}
\quad \xrightarrow{\text{O}_3} \quad
\begin{array}{c}
\qquad \overset{\text{O}}{\underset{\diagup\ \diagdown}{}} \\
\text{R}\quad \text{O}\quad \text{O} \\
|\qquad |\qquad | \\
-\text{C}-\text{C}-\text{C}- \\
\text{H}\ \ \text{H}\ \ \text{H}
\end{array}
\qquad (a)
$$

$$
\begin{array}{c}
\qquad \overset{\text{O}}{\underset{\diagup\ \diagdown}{}} \\
\text{R}\quad \text{O}\quad \text{O} \\
|\qquad |\qquad | \\
-\text{C}-\text{C}-\text{C}- \\
|\quad \text{H}\ \ \text{H} \\
\text{H}
\end{array}
\quad \longrightarrow \quad
\begin{array}{c}
\text{R} \\
|\quad + \qquad\quad - \\
-\text{C}-\text{C}-\text{O}-\text{O} +\text{O}=\text{C}- \\
|\quad \text{H} \qquad\qquad \text{H} \\
\text{H}
\end{array}
\qquad (b)
$$

For textiles, ozone results in a reduced breaking strength and increased rate of wear. Of greater importance is the role played by ozone in causing colour changes in dyed fibres. The surface area of the fibre has been shown to be related to the degree of fading[3,4]. This may be explained by ozone penetrating the fibre itself and subsequent diffusion to the surface[5]. Overall dye fading is a complex function of ozone levels, relative humidity and synergistic effects. Grosjean and coworkers[6,7,8] have studies the mechanism whereby certain dyes fade when exposed to ozone. For both curcumin and indigo electrophilic addition of ozone onto the unsaturated carbon-carbon bond of the dye is responsible. In such experiments ozone levels up to 10 ppm were used but the exposure time was short.

Damage to paint is similar to that of elastomers, that is embrittlement and cracking due to chain scission and cross-linking. The effects are small in comparison with those of other factors. Some of the colour pigments used in commercial paints and dyes are also used in artists paints. It has been shown that several pigments faded severely if exposed to ozone levels typical of a Los Angeles photochemical smog[9,10,11]. Additionally it has been shown that ozone may also attack photographic materials[12] and damage books[13].

Table 7.1: Ozone concentration limits for art galleries, archives and libraries (after Cass et al.[20])

	Ozone Concentration (ppbv)
British Museum Library	0
Canadian Conservation Institute	10
National Bureau of Standards	13
National Research Council	1
Newberry Library	1
G. Thomson	0.1

It is well documented that indoor ozone concentrations can be a significant fraction of outdoor levels[14,15] and hence damage to works of art in museums has recently received increasing attention[16]. In museums with high air exchange rates indoor/outdoor values approaching 0.7 have been reported[17,18], with low exchange rates indoor/outdoor rates may be small[19]. Ozone concentrations as high as 143 ppbv were measured in one museum. To protect works of art the ozone should be virtually zero. If a collection experiences 1 ppbv for 100 years the exposure is 8.8×10^5 ppbv-h which is approximately equal to 8.6×10^5 ppbv-h accumulated during exposure to 400 ppbv for 90 days, a figure that is known to produce fading. In fact several museums have recommended ozone concentration limits and these are summarised in Table 7.1. Several approaches have been used to protect collections[15,20]. These included activated carbon filters in the buildings ventilation system, ventilation system redesign to reduce outdoor air make-up, construction of display cases, framing of paints and prints behind glass and the application of binders and coatings.

It should, of course, be remembered, that other pollutants will also be found in museums and these may affect the contents. Levels of NO_2, PAN, SO_2, chlorinated hydrocarbons and HNO_3 have been measured in several museums in southern California[21,22] and indoor concentrations were similar to those outdoors. It has recently been shown that a mixture of photochemical oxidants including ozone (200 ppbv), NO_2 (75 ppbv) and PAN (15 ppbv) cause fading in artists' colorants[23].

7.3   EFFECTS ON HUMAN HEALTH

It is hardly surprising that pollutants in the atmosphere can be damaging to health. A fairly active person typically inhales between 10000 and 20000 litres of air over 24 hours, and this rises sharply in people taking more vigorous exercise; for instance a jogger may inhale up to 3000 litres of air over the course of an hour. Once in the lungs the air is exchanged via the alveoli which are in close contact with many tiny blood vessels. At any time the total surface area of blood vessel within the alveoli exposed to the atmosphere is as large as half a tennis court. Given the powerful oxidising capacity of ozone and the exothermicity of tissue oxidation, the exposure of large areas of tissue to ozone is clearly of some concern.

Ozone was recognized as a powerful lung irritant soon after its initial synthesis in 1851[24] and health effects among the general community were first reported in 1967 among high school athletes in California[25]. Numerous reviews on this topic have been published[26-29] so only a brief summary will be given here.

Ozone exposure affects the structure and function of the respiratory tract in several ways. Research on humans has mainly focused on its effects on respiratory function, especially on transient responses to acute exposures. Lung functional responses to acute and subacute exposures have been studied, but largely in animals. These responses include mucociliary and early alveolar zone particle clearance, functional responses in macrophages and epithelial cells and changes in lung cell secretions.

Animal studies indicate that when inhaled, ozone is capable of causing inflammation and impairment of local defences against other irritants, infections and allergens and at high doses irreversible changes in lung structure and function. In human subjects exposed to 400 ppbv ozone for 2 hours, while performing intermittent exercise, showed increased levels of inflammatory cells in the airways[30]. An 8.2-fold increase in polymorphonuclear leucocytes (a sign of lung inflammation) and a significant decrease in macrophages were found. Recent research has shown that effects can be produced with exposures as short as five minutes[31], and that various effects become progressively larger as exposures at a given concentration are extended[32]. Lippmann summarizes the nature of the various human responses to single exposures to ozone. The results clearly demonstrate functional and biochemical responses, which accumulate over multiple hours and persist for many hours after exposures at levels typically encountered in ambient air. The main respiratory symptoms were cough, breathlessness and pain on inspiration. These findings have been obtained from controlled chamber studies. The main advantage of this method is it allows the effects of ozone to be looked at in isolation or in combination with known proportions of other gases.

Epidemiological studies, on the other hand, involve investigating a sample of a population that has already been exposed to whatever factor is being studied. Most of these studies to date have been from North America and involve children or adolescents[33-40]. The results of these studies are given in Table 7.2. Negative associations are evident between moderately elevated levels of ozone (70-120 ppbv) and measures of lung function: usually FEV (volume of air expired during the first second of forced expiration) and FVC (forced vital capacity). For the children's studies shown in Table 7.2, the mean decrement in FEV, is about 0.77 ml/ppbv ozone which, at 100 ppbv would be 77 ml: about 3% of the predicted FEV, of a child aged 12. For adults a change in FVC of - 2.1 ml/ppbv was found[40], an equivalent reduction considering the larger lung size of adults. A closer examination of the studies show that a key factor in ozone responsiveness is the level of activity. The more active the subjects the more likely it is that he or she will experience decreased lung function[41].

Exposure to ozone alone in the atmosphere is rare, usually exposure occurs to a mixture of both primary and secondary pollutants. There may be confounding effects of other pollutants which means that it is difficult to infer human health impacts in one area from epidemiological studies performed elsewhere in the world. There is some evidence that photochemical pollution may be associated with small increases in asthma morbidity[42] and even mortality[43]. There is concern that asthmatics might be especially at risk of an exacerbation when exposed to ozone since airway hyperactivity is a feature of this condition. Most studies suggest that this is not true[26]. There is also no evidence that smokers, subjects with chronic obstructive pulmonary disease or the elderly are more sensitive to ozone than other subjects.

In summary, reproducible and statistically significant changes in indices of lung function will occur in many people taking exercise out of doors during photochemical pollution episodes. These changes are unlikely to produce irreversible damage although individuals who are sensitive to ozone may experience respiratory symptoms. However, it

Table 7.2: Summary of field studies carried out among healthy children to examine the effects of daily variations in levels of ambient ozone on lung function

Location (ref)	Number age range	Physical activity level	Max 1-hr Ozone conc(ppbv)	Change in FVC ml/ppbv	FEV$_1$ ml/ppbv	Comments
Indiana PA [23]	58 8-13 yrs	moderate	110	-1.06	-0.78	YMCA summer day camp. 2wk.
Mendham NJ [37,38]	39 7-13 yrs	low	185	-0.12	-0.28	Private summer day camp. 5wk. Evidence of persistence effects.
Fairview Lake, NJ [34]	91	moderate	113	-1.03	-1.42	YMCA residential summer camp. 2 or 4 wk.
Kingston and Harriman Tenessee [39]	154 7-13 yrs	low	78	-0.92	-0.99 (FEV$_{.75}$)	Grades 5 and 6 from one school. Also participants in larger epidemiological study. Weekly measurements for wk. Ozone measured 6 miles away.
San Bernadino, CA [39]	43 7-13 yrs	moderate	245	-0.44	-0.39	Residential summer camp. Continuous on site measurement of pollutants. Larger slopes if 1-2hr ozone used.
Lake Couchiching Ont (cited in [36])	52	moderate?	?	-0.14		Residential summer camp. 10 consecutive days July 1983. FVC males -0.65; females +0.34. 50% of children were asthmatic. FP and ozone highly correlated. No differences between asthmatics and non-asthmatics.

is clear that even the levels of ozone found in ambient air in rural areas can affect the lungs of healthy active children and adults.

It should be remembered that air quality standards are primarily based on human clinical studies in which young, healthy, exercising volunteers are exposed to ozone for single, short periods. High ozone levels in the atmosphere may last up to 12 hours a day for several consecutive days. Research has highlighted the importance of exposure time per day, physical activity and the potential exposure for multiple hours in outdoor air with respect to lung function decline and inflammation, as well as the progression of structural and biochemical effects in lung tissue. There is now a growing demand for a move from 1 hour to a multihour averaging time for an ambient ozone standard which adequately protects public health against acute pulmonary effect. This issue is addressed in chapter 8.

7.4   EFFECTS ON VEGETATION

Injury to vegetation is one of the earliest, in terms of exposure, and most obvious forms of environmental damage which has been attributed to ozone. In fact by 1944, new types of foliar injury, "weather fleck", were noticed on vegetation in and around Los Angeles and it was later reported that this injury was caused by smog[44]. In 1952 weather fleck of cigar wrapper tobacco was reported in Connecticut as a disease of unknown cause[45] and in 1959 ozone was identified as the culprit[46]. Activated charcoal filters were first used to protect plants in Los Angeles in 1961[47] and to prevent ozone fleck and premature senescence of tobacco leaves in 1966[48].

In the course of investigating weather fleck in tobacco it was found that certain lines were more sensitive to ozone than others. The cultivar Bel-W3 is extremely sensitive and has become the world's most commonly used bioindicator for ambient ozone[49].

Visible damage to crops has been identified in many other regions of the USA, where ozone is now considered to be the single most important pollutant affecting vegetation. Injury to vegetation caused by ozone has been observed in numerous countries and is the subject of considerable research. Numerous reviews on the toxic have been published[50-57].

A necessary requirement for photosynthesis and respiration is that all terrestrial plants have the capacity to absorb and emit gases. Gas exchange occurs via the stomata - μm-sized pores on leaf surfaces. Ozone and other gaseous pollutants are sorbed onto foliar and other surfaces and move into the plant tissues. Surface deposition may lead to chemical changes in the cuticle and cuticular waxes on the leaf surface, but movement into the interior tissues is the essential prerequisite for most of the biochemical and physiological effects of exposure that have been observed. Ozone uptake is controlled by leaf resistance as well as stomatal density and conductance, and is generally dependent on physical, chemical and biological factors. Sensitivity of plants is modified by any factor influencing stomatal aperture such as light, relative humidity and soil moisture. Genetic and developmental factors, soil fertility of chemicals such as herbicides or fungicides can also influence stomatal aperture.

Once inside a leaf, ozone, or some intermediate like an OH radical, changes the integrity of the cells. If the cells collapse and die, then symptoms occur on the leaf surfaces. In certain broad-leaved plants ozone exposure results in the destruction of the palisade mesophyll cells. Since the affected groups of cells are often minute but distributed over the whole leaf area, the colour of leaves get a bronzing appearance. Each leaf passes through phases of different sensitivity. Generally, they are most sensitive when unfolded and just developed to normal size[58].

A wide array of symptoms on plants occur which depend on the variety of plant, concentration and duration of ozone exposure and presence of other pollutants. Table 7.3 summarises typical ozone injury symptoms for broad leaved plants.

Table 7.3: Crop plants commonly affected by ozone and typical symptoms expressed

Plant	Foliar Symptoms
Bean (*Phaseolus*)	Bronzing and chlorosis
Cucumber (*Cucumis*)	White stipple
Grape (*Vitis*)	Red to black stipple
Morning Glory (*Ipomoea*)	Chlorosis
Onion (*Allium*)	White flecks and tip dieback
Potato (*Solanum*)	Grey fleck and chlorosis
Soybean (*Glycine*)	Red-bronzing and chlorosis
Tobacco (*Nicotiana*)	Metallic to white fleck
Watermelon (*Citrullus*)	Grey fleck

Chronic effects result eventually in changes at the cellular level. This may result in increased leaching of essential nutrients, accompanied or even enhanced by ozone-induced weathering of the cuticle. Photosynthesis may be reduced either directly by damage to the chloroplasts, or indirectly by closure of the stomatal aperture resulting in reduced CO_2 uptake[54,60]. A reduction in the transport of synthesised compounds to roots leads to a decrease in root size and fewer stored reserves. The potential for increased sensitivity to frost, heat and water stress has also been reported[61]. Biochemical perturbations may also result and include the oxidation of sulfhydryl groups, and changes in the content of soluble sugars, starch, phenols, ascorbic acid, amino acids and protein[62,63].

Research has shown that considerable inter- and intra- species variation exists, some of which may be due to the different environmental conditions under which the experiments were performed such as exposure time and ozone concentration. Additionally the classification of plants according to visible injury, taken as a scale of relative sensitivity, is prone to error. Despite this several authors have produced lists of plant species grouped into sensitive, intermediate and less sensitive species. An example of this classification is shown in Table 7.4.

Table 7.4: Sensitivity of selected plants to ozone

Very Sensitive	Sensitive	Less Sensitive
	Crops	
Spinach	Onion	Lettuce
Oat	Beet	Carrot
Rye	Barley	Grape
Bean	Wheat	
Potato	Corn	
Tomato	Alfalfa	
Tobacco		
	Ornamentals	
Petunia	Chrysanthemum	Sultana
	Coleus	Begonia
	Carnation	Geranium
		Gladiolus
		Fuchsia
	Trees	
White Pine	Ponderosa Pine	Scotch Pine
Ash	Apple	Sycamore
	Maple	Birch
	Oak	Walnut
	Poplar	Beech
		Locust
		Yew

(After Prinz, B. (1988). Ozone Effects on Vegetation. Tropospheric Ozone, I.S.A. Isaksen (Ed.), D. Reidel Publishing, Dordrecht, p161-184)

In summary the effects of ozone on plants may include visible leaf injury, reduced plant growth, decreased economic yield, changes in crop quality and alterations in susceptibility to stresses. Not all physiological effects of ozone on plants are reflected in growth of yield reductions. Numerous studies have shown that there are specific combinations of ozone concentrations and exposure duration that plants can experience which will not result in visible injury or reduced yield. Only when these exposure levels are exceeded, and they are highly species dependent, may some form of damage be observed. It has also been pointed out, without any obvious facetiousness, that an increased greenhouse effect caused, in part, by increased ozone levels, will increase the productivity of temperate regions.

Table 7.5: Ozone concentrations averaged over a seven hour day (0900 to 1600 ST) required to produce a 10% yield reduction (after PORG[98]).

Crop	Ozone Concentration (ppbv)
Corn	75-132
Wheat	69-93
Soybean	38-43
Peanut	43-49
Kidney bean	72-86
Cotton	41
Turnip	40-61
Lettuce	53-57
Spinach	41-60

The range of values for each crop are due to a range of different varieties being studied in different seasons. These values were calculated from fitted dose-response models based on 7 h day fumigations.

Research has suggested that ozone substantially reduces the yields of several crops and that the economic effect of these yield reductions may be important[64]. Numerous studies have attempted to assess the economic loss to crop production and a wide range of estimates have been reported. For the US a figure of $ 3 billion annually or about 5-6% of the gross value of farm commodities has been estimated[65]. A 25% reduction in ambient ozone would result in a benefit of approximately $ 1.9 billion dollars[66]. In the Netherlands it has been estimated that ambient ozone causes relatively large crop losses for legumes, potatoes, cut flowers and fodder crops. In 1983 sulphur dioxide, hydrogen fluoride and ozone appeared to reduce crop production by 5% and 70% of the total reduction was caused by ozone[67]. Observations in the US have indicated that of the crop loss due to air pollution, ozone is responsible for 90%, which amounts to 2 to 4% of the total crop production. Many different experimental techniques have been adopted to study the impact of ozone on plant yield. The most comprehensive series of experiments on the relationship between ozone exposure and crop yield began in 1980, when the National Crop Loss Assessment Network (NCLAN) was established[68]. The approach taken was to expose crops to ozone for 7 hours (0900 to 1600 h standard time) per day on every day throughout the growing season at different concentrations, and thus to generate dose-response relationships between the ozone concentration and crop yield. The 7 hour daily period was selected by NCLAN because the parameter was believed to correspond to the period of greatest plant susceptibility to ozone pollution. Towards the end of the program the experimental protocol was redesigned and proportional additions of ozone were applied to crops for 12 hour periods. The results indicate that ozone impacts on a wide range of plant species and the response is different for different crops; some, such as soybean and spinach, show a rapid fall in yield with increasing ozone, whereas others such as wheat and corn do not. The threshold concentrations for a 10% yield reduction range from 40 ppbv for sensitive crops (*e.g.* soybean, turnips) to 75 ppbv and over for the most resistant crops. Table 7.5 and Figure 7.2 summarizes these conclusions. The application of the NCLAN data to agricultural crops in the UK has been questioned. For instance, the period of exposure does not generally coincide with the time of day at which highest ozone levels occur and episodic pattern of ozone occurrence in the UK is not simulated. Current ozone levels in the UK indicate that the yield of some crops could be reduced assuming the NCLAN data is applicable. More research is needed to establish whether the dose-response data developed in the USA are applicable to the crops or climatic conditions in the UK.

Observations from extensive regions in Europe and North America suggest that many forests may be in early stages of ecosystem decline. Research in the US has shown that ozone transported from urban areas can reduce forest productivity, increase the susceptibility of trees to various insect pests and fungal pathogens and render them less resistant to various stresses both biotic and abiotic[69]. Oxidants have been shown from 1956 to damage Ponderosa pine and associated species in the San Bernadino Mountains, California[70]. Injury intensity was found to be consistent with the concentration of oxidant pollutant at the inversion layer juncture with west facing slopes[71]. In the eastern US the most severely affected species are among the conifers and include the eastern white pine and the red spruce[72-74]. Typical symptoms of the affected trees are thin crowns, short needle length, reduced needle retention, needle chlorosis which can occur as mottling, flecking or banding and needle tip necrosis[75-76]. Proof that these injuries are induced by ozone is based

Figure 7.2

Dose-response curves produced from the National Crop Loss Assessment Network. A mean value for the south east of England is also given.

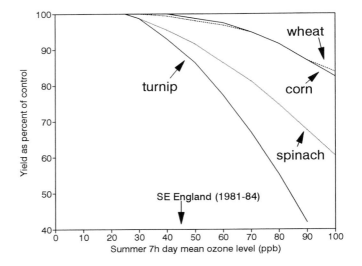

on evidence from injury surveys with concomitant air quality measurements, close-response studies with ozone, and test with clones of individual trees observed to be resistant or susceptible in forests[77-80]. Ozone-induced reductions in growth have been observed under laboratory and greenhouse conditions with concentrations below 100 ppbv[81]. However this effect has not yet been demonstrated under field conditions. No study has shown a loss in productivity of white pine forests.

In Germany and several other central European countries forests have shown 'Neueartige Waldschadenn' (new type of forest damage) since the late 1970s and early 1980s. Damage first appeared on silver fir and then on Norway spruce[82,83]. Similar symptoms have been observed on other tree species, including Scots pine, beech and oak. It has been suggested that ozone may play a role in this forest decline since concentrations in the mountainous areas of southern Germany are comparable to those found in parts of the USA where ozone damage is known to cause visible symptoms on a range of forest species. Additionally, background ozone levels have increased over Germany by 1-2% per annum between 1956 and 1980. Virtually all the evidence indicting ozone as the main culprit is circumstantial. There are indirect mechanisms whereby ozone may induce nutrient leaching from foliage, which in turn affects the sensitivity of trees to chlorophyll photo dissociation, drought and forest stresses. Additionally ozone exposure during the summer, which has been shown to damage the cellular permeability barrier, may predispose conifers to frost injury in the following autumn.

It is now considered that forest decline is not the result of a single phenomenon but is caused by complex interactions between one or more pollutants and other environmental stresses.

7.5   BIOMONITORING

Plant responses, notably visible, characteristic symptoms, have long been used as indications of a number of air and soil pollutants. Bioindicators, therefore, provide a direct method for understanding the risk that pollution presents to the biological components of the affected environment. To perform predictably, the plants should be sensitive to a specific pollutant, genetically uniform, native or adaptable to the region, produce characteristic symptoms, grow indeterminately and respond proportionally to the pollutant exposure[49]. Efforts should also be made to provide uniform soil and water conditions to ensure standard growing conditions. Additionally standard methods for evaluating and quantifying pollutant induced foliar injury must be developed. Bioindicators serve many functions; they provide a cheap and convenient substitute for chemical monitoring and provide an integrated measure of the biological effects of a chemical over a period of time. However, they cannot provide a direct quantitative measure of the amount of a given pollutant in the atmosphere, since the response of any biological organism to a given dose of pollutant is not constant. Substantial progress has been made towards improving our understanding of the variables affecting the performance of indicator species.

Plants have been used as bioindicators for photochemical oxidants - principally ozone and peroxyacl compounds for many years. The organism which has most frequently used to detect ozone in ambient air is the tobacco cultivar Bel-W3[84], although several other plants have been used (Table 7.6). The response of Bel-W3 tobacco is usually the first indication that a country or region has developed an ozone problem. The use of tobacco as an indicator of ozone has been widespread, not only in the US where the first surveys were undertaken. In his detailed review, Heggestad[84] summarises results from 14 different countries as far afield as Australia, Germany, India, Israel, Japan, Mexico and Sweden. The use of Bel-W3 as a biological indicator is expanding: studies are underway in Yugoslavia, Spain and Brazil[84] and recently results have been published from Poland[85]. Two major surveys have been reported in Holland and the United Kingdom.

Table 7.6: Plants that can be used as bioindicators of ambient ozone

Common Name	Latin Name
Bean	Phaseolus Vulgaris
Grape	Vitis Labrusca
Spinach	Spinacea Oleracea
Tobacco	Nicotiana Tabacum
Bel-W3 (Sensitive)	
Bel-C (Intermediate)	
Bel-B (Tolerant)	
White Pine	Pinus Strobus
Grass	Poa Annua
Morning Glory	Ipomoea Purpurea
Trembling Aspen Clones	Populus Tremuloides

As part of the National Monitoring Network for Air Pollution within the Netherlands cultivars of ozone sensitive plants such as bean, spinach, clover and tobacco have been used to indicate the presence of ozone[86-90]. A detailed growing and exposure routine was established and leaf injury was calculated by estimating the percentage of leaf area injured[91]. The results showed there were more adverse effects of ozone in the western half of the country, than in the eastern half. The results show that, at least in geographical terms, a clear cut relationship between ambient ozone and foliar injury does not exist[67]. Varying weather conditions, exposure regime and the presence of other pollutants are likely to have played a role in bringing about the pattern of plant injury. Positive correlations between injury severity and ozone concentrations have, however, been found elsewhere[92].

Nationwide surveys of ozone were carried out in the summers of 1977 and 1978 in the UK. In 1972 tobacco cultivars were exposed at sites to the west of London and Bell and Cox[93] concluded that Bel-W3 was suitable for use as an indicator for ozone under British conditions, although results from a survey in London in 1979 were inconclusive[94]. In 1977 and 1978, Bel-W3 plants were exposed at approximately 50 sites across the UK and the percentage of leaf injury recorded at weekly intervals[95,96]. The geographical distribution of damage, shown in Figure 7.3, was different between the two years, and appeared to be related to sunshine levels. The blank areas in Figure 7.3 represent the areas with the lowest leaf injury/sunshine hours, with increasing density of shading indicating increasing levels of both parameters. The weekly variation in leaf injury was used as a basis to identify possible source regions of photochemical ozone. Results suggested that relatively local

Figure 7.3:

Geographical distribution of leaf injury to Bel W3 indicator plants during May-Sept 1977 and June-August 1987 together with the distribution of sunshine over the same period. The blank areas represent the areas with the lowest leaf injury/sunshine hours. Increasing density of shading indicates increasing level of both parameters. (Reproduced from PORG[98])

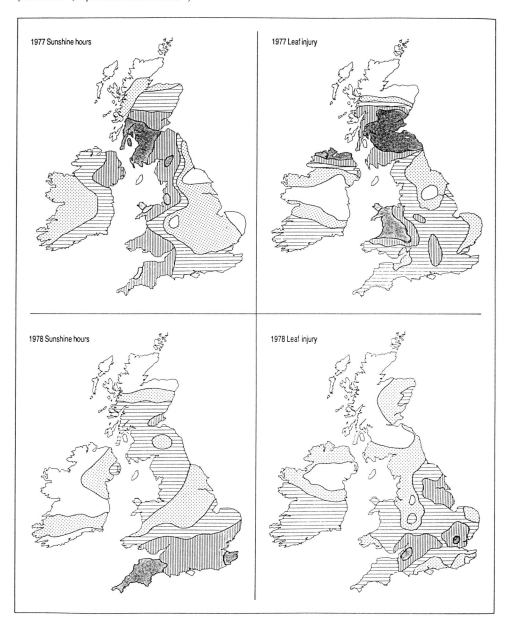

Figure 7.4

Distribution of mean leaf injury index in the summer of 1989 in southern England. Shading as Figure 7.3. (Reproduced from PORG[98])

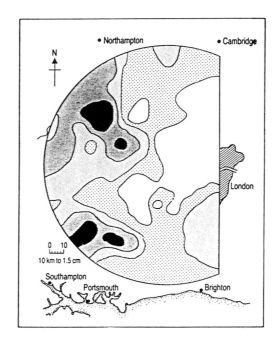

sources (ie distances of 0-150 km) with a population of above 75,000 can have a significant positive influence on the amount of ozone present downwind. The effect of local urban sources can create considerable variation in ozone concentration and hence leaf injury over a relatively small scale. Figure 7.4 shows the distribution in the mean amount of leaf injury to Bel-W3 during the summer of 1980[97]. The results show a very wide range of injury, which varied by a factor of about 5 over the 45 sites used. Winds during this period were predominantly from the west and south-west so highest levels of leaf injury were found in the west of the experimental area, reflecting the importance of urban sources further to the west and south of it.

There is continued use of Bel-W3 in an ozone project in the UK which involved 20,000 or more sites in 1990.

1. Climate Change (1992) The Supplementary Report to the IPCC Scientific Assessment, Cambridge University Press.

2. Mueller, W.J. and Stickney, P.B. (1970) A survey and economic assessment of the effects of air pollution on elastomers, Columbus, OH: Battelle Memorial Institute, NAPCA contract no. CPA-22-69-146.

3. Huevel, H.M, Huisman, R. and Schmidt, J.M (1978) Ozone fading of disperse blue 3 on nylon 6 fibres, In: *The Influence of Physical Fibre Properties*, Text. Res. J., 48, 376-384.

4. Haylock, J.C. and Rush, J.L. (1978) Studies on ozone fading of anthraquinone dyes on nylon fibres, Part II: In-service performance, Text. Res. J., 48, 143-149.

5. Kamath, Y.K., Reutsch, S.B. and Weigmann, H.D. (1983) Micro-spectrophotometric study of ozone fading of disperse dyes in nylon, Text. Res. J., 53, 391-402.

6. Grosjean, D., Whitmore, P., De Moor, C.P. and Cass, G.R. (1987) Fading of alizarin and related artists' pigments by atmospheric ozone reaction products and mechanisms, Environ. Sci. Technol., 21, 635-643.

7. Grosjean, D., Whitmore, P.M., De Moor, C.P., Cass, G.R. and Druzik, J.R. (1988) Ozone fading organic colorants : products and mechanisms of the reaction of ozone with curcumin, Environ. Sci. Technol., 22, 1357-1361.

8. Grosjean, D., Whitmore, P.M., Cass, G.R. and Druzik, J.R. (1988) Ozone fading of natural organic colorants : mechanisms and products of the reaction of ozone with indigos, Environ. Sci. Technol., 22, 292-298.

9. Shaver, C.L., Cass, G.R. and Druzik J.R. (1983) Ozone and the deterioration of works of art, Environ. Sci. Technol., 17, 748-752.

10. Whitmore, P.M. and Cass, G.R. (1987) The ozone fading of traditional natural organic colorants on paper, J. Amer. Inst. Conserv., 26, 45-58.

11. Whitmore, P.M. and Cass, G.R. (1988) The ozone fading of traditional Japanese colorants, Stud. Conserv., 33, 29-40.

12. Swan, A. (1981) Conservation of photographic print collections, Library Trends, 30, 267-296.

13. Shahani, C.J. and Wilson, W.K. (1987) The preservation of libraries and archives, American Scientist, 75, 240-251.

14. Weschler, C.J., Shields, H.C. and Naik, D.V. (1989) Indoor ozone exposures. J. Air Pollut. Control Assoc., 39, 1562-1568.

15. Yocom, J.E. (1982) Indoor-outdoor air quality relationships - a critical review, J. Air Pollut. Control Assoc., 32, 500-520.

16. Brimblecombe, P. (1990) The composition of museum atmospheres, Atmos. Environ., 24B, 1-8.

17. Davies, T.D., Ramer, B., Kaspyzok G. and Delaney, A.C. (1984) Indoor/outdoor ozone concentrations at a contemporary art gallery, J. Air Pollut. Control Assoc., 31, 135-137.

18. Druzik, J.R., Adams, M.S., Tiller, C. and Cass, G.R. (1990) The measurement and model predictions of indoor ozone concentrations in museums, Atmos. Environ., 24A, 1813-1823.

19. Derwent, R.G. (1983) Ozone measurements in an art gallery, Warren Spring Lab. LR 428 (AP).

20. Cass, G.R., Nazaroff, W.W., Tiller, C. and Whitmore, P.M. (1991) Protection of works of art from damage due to atmospheric ozone, Atmos. Environ., 25, 441-451.

21. Hisham, M.W.M. and Grosjean, D. (1991) Air pollution in southern California museums : indoor and outdoor levels of nitrogen dioxide, peroxyacetyl nitrate, nitric acid and chlorinated hydrocarbons. Environ.

343

Sci. Technol., 25, 857-862.

22. Hisham, M.W.M. and Grosjean, D. (1991) Sulfur dioxide, hydrogen sulfide, total reduced sulfur, chlorinated hydrocarbons and photochemical oxidants in southern California museums, Atmos. Environ., 25, 1497-1505.

23. Grosjean, D., Grosjean, E. and Williams, E.L. (1993) Fading of artists colorants by a mixture of photochemical oxidants, Atmos. Environ., 27, 765-722.

24. Bates, D.V. (1989) Ozone myth and reality, Environ. Res., 50, 230-237.

25. Wayne, W.S., Wehrle, P.F. and Carroll, R.E. (1967) Oxidant pollutant and athletic performance. J. Am. Med. Assoc., 199, 901-904.

26. Department of Health. (1991) Advisory Group on the Medical Aspects of Air Pollution Episodes, First Report, Ozone, HMSO.

27. Lippmann, M. (1991) Health effects of tropospheric ozone, Environ. Sci. Technol., 25, 1954-1962.

28. Tilton, B.E. (1989) Health effects of tropospheric ozone, Environ. Sci. Technol., 23, 257-263.

29. Lippmann, M. (1989) Health effects of ozone : a critical review. J. Air Pollut. Control Assoc., 39, 672-695.

30. Koren, H.S., Devlin, R.B., Graham, D.E., Mann, R. and McDonnell, W.F. (1989) The inflammatory response in human lung exposed to ambient levels of ozone. In: *Atmospheric Ozone Research and Its Policy Implications*, T. Schneider, S.D. Lee, G.J.R. Wolters and L.D. Grant (Eds.) Elsevier Science Publishers, Amsterdam.

31. Fouke, J.M., Delemos, R.A. and McFadden, E.R. (1989) Airway response to ultra short-term exposure to ozone. Am. Rev. Respir. Dis., 137, 326-330.

32. McDonnell, W.F., Kehrl, H.R., Abdul-Salaam, S., Ives, P.J., Folinsbee, L.J., Devlin, R.B., O'Neill, J.J. and Horstman, D.H. (1991) Respiratory response of humans exposed to low levels of ozone for 6.6 hours, Arch. Environ. Health, 46, 145-150.

33. Lippmann, M., Lioy, P.J., Leikauf, G., Green, K.B., Baxter, D., Morandi, M. and Pasternak, B.S. (1983) Effects of ozone on the pulmonary function of children, In: *Advances in Environmental Toxicology*, New York: Institute of Environmental Medicine, 423-426.

34. Spektor, D.M., Lippmann, M., Lioy, P.J., Thurston, G.D., Citak, K., James, D.J., Bock, N., Speizer, F.E. and Hayes, C. (1988) Effects of ambient ozone on respiratory function in active normal children. Am. Rev. Respir. Dis., 173, 313-320.

35. Higgins, I.T.T., D'Arcy, J.B., Gibbons, D.I., Avol, E.L. and Gross, K.B. (1990) Effect of exposure to ambient ozone on ventilatory lung function in children, Am. Rev. Respir. Dis., 141, 1136-1146.

36. Kinney, P.L., Ware, J.H. and Spengler, J.D. (1988) A critical evaluation of acute ozone epidemiology results, Arch. Environ. Health, 43, 168-173.

37. Lioy, P.J., Vollmuth, T.A. and Lippmann, M. (1985) Persistence of peak flow decrement in children following ozone exposures exceeding the national ambient air quality standard, J. Air Pollut. Control Assoc., 35, 1068-1071.

38. Bock, N., Lippmann, M., Lioy, P.J., Munoz, A. and Speizer, F.E. (1985) The effects of ozone on the pulmonary function of children, Trans. Air Pollut. Control Assoc., TR-4, 297-308.

39. Kinney, P.L., Ware, J.H., Spengler, J.D., Dockery, D.W., Speizer, F.E. and Ferris, B.G. (1989) Short-term

pulmonary function change in association with ozone levels, Am. Rev. Respir. Dis., 139, 56-61.

40. Spektor, D.M., Lippmann, M., Thurston, G.D., Lioy, P.J., Stecko, J., O'Connor G., Garshick, E., Speizer, F.E., and Hayes, C. (1988) Effects of ambient ozone on respiratory function in healthy adults exercising outdoors, Am. Rev. Respir. Dis., 138, 821-826.

41. Schwartz, J., and Zeger, S. (1990) Passive smoking, air pollution, and acute respiratory symptoms in a diary of student nurses, Am. Rev. Respir. Dis., 141, 62-67.

42. Cody, R.P., Weisel, C.P., Birnbaum, G. and Lioy P.J. (1992) The effect of ozone associated with summertime photochemical smog on the frequency of asthma visits to hospital emergency departments, Environ. Res., 58, 184-194.

43. Kinney, P.L. and Ozkaynak, H. (1991) Associations of daily mortality and air pollution in Los Angeles County, Environ. Res., 54, 99-120.

44. Middleton, J.T., Kendrick, J.B., Jr, and Schwalm, H.W. (1950) Injury to herbaceous plants by smog or air pollution, Plant Disease Reptr., 34, 245-252.

45. Rich, S., Taylor, G.S. and Tomlinson, H. (1969) Crop damaging periods of ambient ozone in Connecticut, Plant Disease Reptr., 53, 969-973.

46. Heggestad, H.E. and Middleton, J.T. (1959) Ozone in high concentrations as cause of tobacco leaf injury, Science, 129, 208-210.

47. Darley, E.F. and Middleton, J.T. (1961) Carbon filter protects plants from damage by air pollution, Florists' Review, 127, 15-16, 43-45.

48. Menser, H.A., Jr, Heggestad, H.E. and Grosso, J.J. (1966) Carbon filter prevents ozone fleck and premature senescence of tobacco leaves, Phytopathology, 56, 466-467.

49. Manning, W.J. and Feder, W.A. (1980) *Biomonitoring Air Pollutants with Plants*, Applied Science Pub., London.

50. Laurence, J.A. and Weinstein, L.H. (1981) Effects of air pollutants on plant productivity, Ann. Rev. Phytopathology, 19, 257-271.

51. Jacobson, J.S. (1982) Ozone and the growth and productivity of agricultural crops, In: *Effects of Gaseous Air Pollution in Agriculture and Horticulture*, M.H. Unsworth and D.P. Ormrod (Eds.), Butterworth, London, 293-304.

52. Taylor, O.C. (1984) Organismal responses of higher plants to atmospheric pollutants: photochemical and other, In: *Air Pollution and Plant Life*, M. Treshow (Ed.), John Wiley and Sons, NY, 215-238.

53. Guderian, R. (Ed.) (1985) *Air Pollution by Photochemical Oxidants: Formation, Transport, Control and Effects on Plants*, Springer-Verlag, Berlin.

54. Prinz, B. and Brandt, C.J. (1985) Effects of air pollution on vegetation, In: *Pollutants and their Ecotoxicological Significance*, H.W. Nurnberg (Ed.), John Wiley and Sons, NY, 67-84.

55. Ashmore, M., Bell, N. and Rutter, J. (1985) The role of ozone in forest damage in West Germany, Ambio, 14, 81-87.

56. Krupa, S.V. and Manning, W.J. (1988) Atmospheric ozone : formation and effects on vegetation, Environ. Pollut., 50, 101-137.

57. Lefohn, A.S. (1992) *Surface Level Ozone Exposures and their Effects on Vegetation*, Lewis Publishers, Michigan.

58. Lacasse, N.L. and Treshow, M. (1976) Diagnosing vegetation injury caused by air pollution, US Environmental Protection Agency Handbook.

59. Amundson, R.G., Kohut, R.J., Schoettle, A.W., Raba, R.M. and Reich, P.B. (1987) Correlative reductions in whole-plant photosynthesis and yield of winter wheat caused by ozone, Phytopathology, 77, 75-79.

60. Reich, P.B. and Amundson, R.G. (1985) Ambient levels of ozone reduce net photosynthesis in tree and crop species, Science, 230, 566-570.

61. Blum, U.T. and Heck, K.W.W. (1980) Effects of acute ozone exposures on snap bean at various stages in its life cycle, Environ. Exp. But., 20, 73-85.

62. Mudd, J.B. and McManus, T.T. (1969) Products of the reaction of peroxyacetyl nitrate with sulfhydryl compounds, Arch. Biochem. Biophys., 132, 237-241.

63. Krause, G.H.M. and Prinz, B. (1989) *Current Knowledge of Ozone on Vegetation/Forest Effects and Emerging Issues*, Elsevier Science Publishers.

64. Heck, W.C., Adams, R.M., Cure, W.W., Heagle, A.S., Heggestad, H.E., Kohat, R.J., Kress, L.W., Rawlings, J.O. and Taylor, O.C. (1983) A reassessment of crop loss from ozone, Environ. Sci. Technol., 17, 572A-581A.

65. Shriner, D.S., Cure, W.W., Heagle, A.S., Heck, W.W., Johnson, D.W., Olson, R.J. and Skelly, J.M. (1982) An analysis of potential agriculture and forestry impacts of long-range transport of air pollutants, Oak Ridge National Laboratory, ORNL Report No. 5910.

66. Adams, R.M., Glyer, J.D., Johnson, S.L. and McCarl, B.A. (1989) A reassessment of the economic effects of ozone on US agriculture, J. Air Pollut. Control Assoc., 39, 960-968.

67. Tonneijck, A.E.G. (1989) Evaluation of ozone effects on vegetation in the Netherlands. In: *Atmospheric Ozone Research and Its Policy Implications*, T. Schneider, S.D. Lee, G.J.R. Wolters and L.D. Grant (Eds.) Elsevier Science Publishers, Amsterdam.

68. Heck, W.W., Taylor, O.C., Adams, R.M., Bingham, G., Miller, J.E., Preston, E.M. and Weinstein, L.H. (1982) Assessment of crop loss from ozone, J. Air Poll. Control Assoc., 32, 353-361.

69. Chevone, B.I. and Linzon, S.N. (1988) Tree decline in North America. Environ. Pollut., 50, 87-99.

70. Miller, P.R. (1983) Ozone effects in the San Bernadino National Forest. In: *Proc. Air Pollution and the Productivity of the Forests*, D.D. Davis, A.A. Miller and L. Dochinger (Eds.), Izaak Walton League, Washington, DC, 161-197.

71. Edinger, J.G., McCutchan, M.H., Miller, P.E., Ryan, W.C., Shroeder, N.J. and Bohan, J.V. (1972) Penetration and duration of oxidant air pollution in the south coast air basin in California, J. Air Pollut. Control Assoc., 22, 882-886.

72. Berry, C.R. (1961) White pine emergence tipburn, a physiogenic disturbance, USDA Forest Service, SE For. Exp. Sta. Paper No. 130.

73. Berry, C.R. and Ripperton, L.A. (1963) Ozone, a possible cause of white pine emergence tipburn, Phytopathology, 53, 552-557.

74. Siccama, T.G., Bliss, M. and Vogelmann, H.W. (1982) Decline of red spruce in the Green Mountains of Vermont, Bull. Torrey Bot. Club., 109, 162-168.

75. Blanchard, R.O., Baas, J. and Van Cotter, H. (1979) Oxidant damage to eastern white pine in New

Hampshire, Plant Dis. Reptr., 63, 177-182.

76. Usher, R.W. and Williams, W.T. (1982) Air pollution toxicity to eastern white pine in Indiana and Wisconsin, Plant Disease, 66, 199-204.

77. Benoit, L.F., Skelly, J.M., Moore, L.D. and Dochinger, L.S. (1982) Radial growth reduction of Pinus Strobus L. correlated with foliar ozone sensitivity as an indicator of ozone-induced losses in eastern forests, Can. J. For. Res., 12, 673-678.

78. Yang, Y.S., Skelly, J.M., Chevone, B.I. and Birch, J.B. (1983) Effects of long-term ozone exposure on photosynthesis and dark respiration of eastern white pine, Environ. Sci. Technol., 17, 371-373.

79. Yang, Y.S., Skelly, J.M. and Chevone, B.I. (1983) Sensitivity of eastern white pine clones to acute doses of ozone, sulfur dioxide, or nitrogen dioxide, Phytopathology, 73, 1234-1237.

80. McLaughlin, S.B. (1985) Effects of air pollution on forests, A critical review. J. Air Pollut. Control Assoc., 35, 512-534.

81. Wang, D., Bormann, F.H. and Kanosky, D.F. (1986) Regional tree growth reductions due to ambient ozone : evidence from field experiments, Environ. Sci. Technol., 20, 1122-1125.

82. Skarby, L. and Sellden, G. (1984) The effects of ozone on crops and forests. Ambio, 13, 68-72.

83. Ashmore, M., Bell, N. and Rutter, J. (1985) The role of ozone in forest damage in West Germany. Ambio, 14, 81-87.

84. Heggestad, H.E. (1991) Origin of Bel-W3, Bel-C and Bel-B tobacco varieties and their use as indicators of ozone. Environ. Pollut., 74, 264-291.

85. Bytnerowicz, A., Manning, W.J., Grosjean., D., Chimielewski, W., Dmuchowski, W., Grodzinska, K. and Godzik, B. (1993) Detecting ozone and demonstrating its phytotoxicity in forested areas of Poland : a pilot study, Environ. Pollut., 80, 301-305.

86. Posthumus, A.C. (1976) The use of higher plants as indicators for air pollution in the Netherlands, In: *Proceedings of the Kuopio Meeting on Plant Damages caused by Air Pollution.* L. Karenlampi (Ed.), 115-120.

87. Posthumus, A.C. (1977) Experimental investigations of the effects of ozone and peroxyacetyl nitrate (PAN) on plants, VDI-Berichte, 270, 153-160.

88. Posthumus, A.C. (1983) Higher plants as indicators and accumulators of gaseous air pollution, Environmental Monitoring and Assessment, 3, 263-272.

89. Posthumus, A.C. and Tonneijck, A.E.G. (1982) Monitoring effects of photo-oxidants on plants, In: *Monitoring of Air Pollutants by Plants Methods and Problems*, L. Steubing and H.J. Jager (Eds.), Junk Publishers, The Hague, 115-120.

90. Tonneijck, A.E.G. and Posthumus, A.C. (1987) Use of indicator plants for biological monitoring of effects of air pollution : the Dutch approach, VDI Berichte, 609, 205-216.

91. Posthumus, A.C. (1982) Morphological symptoms and yield alterations as criteria of evaluation in monitoring the effects of air pollutants with plants, In: *Monitoring of Air Pollutants by Plants Methods and Problems*, L. Steubing and H.J. Jager. W. (Eds.), Junk Publishers, The Hague, 73-77.

92. Jacobson, J.S. and Feder, W.A. (1974) A regional network for environmental monitoring : atmospheric oxidant concentrations and foliar injury with tobacco indicator plants in the eastern United States, Mass. Agric. Exper. Station Bull., 604, 1-31.

93. Bell, J.N.B. and Cox, R.A. (1975) Atmospheric ozone and plant damage in the United Kingdom, Environ. Pollut., 8. 163-170.

94. Ball, D.J. (1981) A study of plants as indicators of photochemical pollution in the London area, The London Naturalist, 60, 27-42.

95. Ashmore, M.R., Bell, J.N.B. and Reily, C.L. (1978) A survey of ozone levels in the British Isles using indicator plants. Nature, 276, 813-815.

96. Ashmore, M.R., Bell, J.N.B. and Reily, C.L. (1980) The distribution of phytotoxic ozone in the British Isles, Environ. Pollut., 1, 195-216.

97. Ashmore, M.R., Dalpra, C., Chaney, J. and Bell, J.N.B. (1982) The distribution and effects on plants of ozone in SE England : a report of field studies during summer 1980, ICCET Series B, No. 1, Imperial College, London.

98. Photochemical Oxidant Review Group (1987) Ozone in the United Kingdom, Harwell Laboratory.

CHAPTER 8

CONTROL STRATEGIES

The need for and existence of legislation to protect the health and welfare of the general public from air pollution is not a modern phenomenon, and laws have been passed since medieval times to reduce atmospheric emissions in order to control certain pollution situations. However despite decades of research and control efforts, photochemical ozone continues to plague many countries, especially in the United States where the problem is now viewed by many as being a near intractable one.

It has been recognized for some time that long range transboundary transport of ozone and/or its precursors is important, and that photochemical air pollution is a problem which requires international action. This is being coordinated by the United Nations Economic Commission for Europe (UNECE) Convention on Long-range Transboundary Air Pollution, whose member states include the European Countries; together with the USA and Canada. Many of these states agreed to a Protocol in Sofia in 1988, binding them to return NO_x emissions to 1987 levels by 1994. More recently the VOC Protocol was agreed in Geneva in 1991 whereby VOC emissions are to be reduced by 30% of 1988 levels[1] by 1999.

8.1   AIR QUALITY CRITERIA

Many countries have formulated ambient air quality standards for ozone. Air quality standards can contain both limit and guide values. Limit values are mandatory, guide values are not. Guidelines, as distinct from standards, have been proposed by the World Health Organization (WHO). Ambient air quality standards are generally set to protect public health, with a margin of safety. Some organizations have adopted maximum observed values for their standards, while others have used the 98th percentile as the criterion. The latter approach allows for a few short-lived pollution episodes per year. The 98th percentile criteria are set at levels that take account of the typical distribution of pollution concentrations and are hence set at lower values than those produced using the maximum value approach.

In the US the 1970 Clean Air Act Amendments required the setting of a National Ambient Air Quality Standards (NAAQS) for major pollutants. Primary standards were intended to protect public health, whilst secondary standards were intended to protect public welfare. The first (NAAQS) for photochemical oxidants were published in 1971 and both standards were set at an hourly concentration of 80 ppbv, not to be exceeded more than once per year. This was amended in 1978 with a change of chemical designation from total oxidants to ozone and a reevaluation of the standard. The primary and secondary standards for ozone are currently 120 ppbv. The standards are attainted "when the expected number of days per calendar year with a maximum hourly average concentration above 120 ppbv (235 $\mu g\ m^{-3}$) is equal to or less than one", on average over a three year period[2]. The State of California has set its own ozone air quality standard at 90 ppbv (1 hour average), not to be exceeded except for exceptional events (e.g. forest fires or stratospheric intrusion) and for events which would be expected to occur less than once every seven years. This standard is extremely stringent and to be able to meet it with a 95% confidence level, an area may need to have an ozone concentration distribution comparable to that for background ozone in remote areas[3].

Table 8.1: WHO guidelines for ozone

Human Health

| 1 hour | 76-100 ppbv |
| 8 hours | 50- 60 ppbv |

Terrestrial Vegetation

1 hour	100 ppbv
24 hours	33 ppbv
Growing season	30 ppbv

The WHO has published guidelines for a range of pollutants including ozone[4]. For ozone the guidelines include a consideration of adverse effect on plants as well as human health. The guidelines are expressed in terms of different averaging times and are given in Table 8.1. The WHO guidelines incorporate safety margins and are not intended to indicate levels at which adverse effects on health would start to develop. Rather, they serve as a trigger to the relevant authorities to consider whether controls need tightening. As is obvious from a comparison of the data in Table 8.1 with that presented in Chapter 6, these guidelines, if realistic suggest that control strategies are required over vast regions of the world.

The European Community adopted air quality standards for SO_2 and smoke in 1980 (EC Directive 80/779/EEC), lead in 1982 (EC Directive 82/884/EEC) and NO_2 in 1985 (EC Directive 85/203/EEC) and the Air Quality Standards Regulation 1989 formally brought

Table 8.2: Thresholds for ozone concentrations in the air[x]

Values are expressed in ppbv.
The original Directive specified values in µg m[-3].
They have been converted to ppbv here using a
factor of 0.5 (assuming a temperature of 20°C)

Health protection threshold

55 ppbv for the mean value over eight hours[xx]

Vegetation protection thresholds

100 ppbv for the mean value over one hour
32.5 ppbv for the mean value over 24 hours

Population information threshold

90 ppbv for the mean value over one hour

Population warning threshold

180 ppbv for the mean value over one hour

(x)   *Concentrations must be measured continuously*
(xx)   *The mean over eight hours is a non-overlapping moving average; it is calculated four times
a day from the eight hourly values between 0 and 09.00, 08.00 and 17.00, 16.00 and 01.00, 12.00
and 21.00.*

into UK legislation the limit and guide values for these species. In September 1992 an EC
Directive (92/72/EEC) on air pollution by ozone was adopted to standardize monitoring and
exchange of information in Europe, and to produce a public information and warning
system when ozone levels exceed certain threshold levels. These values are shown in Table
8.2. Member States must provide the Commission with the maximum, the median and the
98th percentile of the mean values over one hour and eight hours recorded during the year
in each measuring station and also the number, date and duration of periods during which
the thresholds for human protection and vegetation protection are exceeded. It should be
noted that this Directive differs from existing air quality Directives in that the concept of
a legally binding limit value for pollution concentrations has been replaced with trigger
levels above which EC member states are required to issue information or warnings to the
public. In the UK the Department of the Environment now classifies the air quality as very
good, good, poor and very poor in respect to ambient nitrogen dioxide and ozone. Details
are shown in Table 8.3.

Table 8.3: Department of the Environment air pollution classification

Category	Nitrogen Dioxide (ppbv)	Ozone (ppbv)
very good	< 50	< 50
good	50- 99	50- 89
poor	100-299	90-179
very poor	≥300	≥180

Values are based on peak hourly average concentration in a 24 hour period

As discussed in Chapter 7, the results of recent research have substantially expanded our knowledge of exposure-response for many of the various effects of ozone on the respiratory tract. They have also provided a basis for enhanced appreciation of the factors affecting delivered dose, the duration and progression of effects and the variations in responsiveness among individuals. Researchers are now questioning the NAAQS based on average 1 hour peak exposures and whether it offers adequate protection of public health from extended and repeated exposures [4-10]. Human and animal toxicity studies and ambient ozone exposure dynamics have revealed that ozone pollution is impairing people's lung functions when they are exposed to lower concentrations over several hours. These findings support the need to move from a linear to a multi-hour averaging time which adequately protects public health against acute pulmonary effects caused by extended daily exposure. Any new standard should be substantially lower than the current 1 hour standard of 120 ppbv which translates roughly into 100 ppbv over eight hours. Various standards have been suggested including a running eight hour average (rather than the average between two arbitrary times such as 8 a.m. to 5 p.m.) as well as a level of 80 ppbv[11]. There are two important considerations when establishing or revising air quality standards : one is the need to protect public health and welfare; the other is the achievability of the standard itself. A standard that is not achievable results in a waste of national resources and an adverse impact on the economy, with subsequent harm to public health.

8.2   COMPLIANCE WITH AIR QUALITY STANDARDS

Between 1972 and 1985 there was at least one site in the UK where ozone levels exceeded the USA NAAQS. Generally per year it is exceeded fairly regularly in rural, suburban and urban sites in the south east of England. The WHO guidelines are more

Table 8.4: Ozone exceedences of WHO guidelines

Year	1 hour mean ≥76 ppbv		8 hour mean ≥50 ppbv		1 hour mean ≥100 ppbv		Daily mean ≥33 ppbv	Growing season concentration‡
	count	days	count	days	count	days	count	ppbv

Stevenage (suburban)

Year	count	days	count	days	count	days	count	ppbv
76/77*	184	18	37	19	112	15	20	66
77/78	33	6	13	10	5	1	41	26
78/79	23	7	13	9	0	0	17	20
79/80	23	7	7	6	2	1	8	18
80/81	1	1	2	1	0	0	2	11
81/82	0	0	0	0	0	0	0	8
82/83	135	18	39	22	60	13	27	24
83/84	61	17	25	17	14	4	22	24
84/85	120	19	58	41	51	8	53	27
85/86	21	5	16	11	4	2	12	19
86/87~	26	7	17	11	1	1	12	16
87/88	2	1	4	4	0	0	3	14
88/89	0	0	1	1	0	0	4	15
89/90	28	6	14	11	1	1	9	16
90/91	69	13	17	12	23	6	13	19

Central London (urban)

Year	count	days	count	days	count	days	count	ppbv
72/73~	46	15	12	10	13	4	11	19
73/74	146	30	47	33	43	14	28	22
74/75	52	16	26	23	15	4	28	23
75/76	16	7	7	7	0	0	6	15
76/77	80	20	30	22	18	7	28	25
77/78*	8	2	5	5	0	0	8	20
78/79	35	9	15	11	12	3	20	20
79/80~	5	2	4	4	0	0	4	14
80/81	0	0	2	2	0	0	0	12
81/82	21	4	3	3	9	3	3	12
82/83~	1	1	0	0	0	0	0	14
83/84*	1	1	0	0	0	0	0	13
84/85*	0	0	1	1	0	0	1	15
85/86	62	14	31	24	17	5	41	24
86/87~	19	6	9	8	1	1	7	14
87/88	0	0	0	0	0	0	0	6
88/89	1	1	0	0	0	0	0	10
89/90	16	6	9	8	1	1	10	14
90/91*	15	4	5	5	0	0	3	13

Table 8.4(contd.): Ozone exceedences of WHO guidelines

Year	1 hour mean >76 ppbv		8 hour mean >50 ppbv		1 hour mean >100 ppbv		Daily mean >33 ppbv	Growing season concentration‡
	count	days	count	days	count	days	count	ppbv

Sibton (rural)

Year	count	days	count	days	count	days	count	ppbv
73/74*	57	11	32	21	20	4	24	31
74/75*	7	2	7	6	0	0	22	31
75/76*	113	14	78	39	42	7	56	38
76/77*	116	16	46	26	12	5	39	37
77/78~	38	8	51	34	3	2	74	36
78/79*	77	14	67	37	12	5	59	35
79/80*	10	3	13	8	0	0	14	25
80/81	0	0	1	1	0	0	6	19
81/82	207	25	67	34	99	15	70	30
82/83	53	12	35	22	11	4	56	30
83/84~	68	15	48	30	7	2	73	34
84/85	25	5	22	20	6	2	47	27
85/86*	15	3	12	9	0	0	21	24
86/87*	19	6	9	8	1	1	7	-
87/88	7	1	8	5	0	0	37	25
88/89	12	4	25	19	2	1	64	31
89/90	73	12	51	33	13	6	78	34
90/91	73	15	33	22	21	4	56	31

~   75% Data capture
*   50% Data capture

‡   The WHO guideline average concentration not to be exceeded over the growing season is 30 ppbv.

stringent than the NAAQS. Table 8.4 provides a comparison of ozone concentrations at 3 sites in the UK and the WHO guidelines. The 1 hour guideline is exceeded virtually every year at each kind of site.

Despite comprehensive local, state and national regulatory initiatives over the past 20 years, ambient ozone levels in all areas of the United States continue to be a major environmental and health concern. In 1985 the US EPA estimated that 76.4 million people were subject to ozone levels exceeding the standard[12]. By 1987 63 areas did not meet the ozone NAAQS. This figure rose to 101 in 1988 and has since stayed steady with 96 areas in 1989 and 98 in 1990[13,14]. In fact, it has been demonstrated that there are places in the US, especially in the northeast, where at times the NAAQS is not violated, but the workplace permissible exposure limit (PEL) of 100 ppbv for 8 hours is exceeded in outdoor air[7,15]. The analyses indicated that in 25% of the cases where the PEL was violated, the NAAQS was not violated. The Clean Air Act (CAA) requires each state to submit a state implementation plan to provide for attainment of the NAAQS. With the CAA amendments of 1970 and 1977, States were required to develop comprehensive plans and associated control measures that, when implemented, would provide for attainment and maintenance of low ozone concentrations.

A number of reasons have been cited for failure to meet the ozone standard following the 1977 CAA Amendments[16]. These include (1) emission sources growth; (2) incomplete and inadequate emission inventories; (3) underestimates of controls required to meet the ozone standard due to inadequate air quality models and input data; (4) inability to control ozone and its precursors transported from other areas; (5) significant contribution of biogenic VOC sources in some areas; (6) the inability of States to promulgate regulations without EPA support; (7) incomplete and inadequate implementation of existing regulations; and (8) unrealistic deadlines. The CAA has worked to some extent in that ozone concentrations have decreased by 8% from 1982 to 1991[17], although these improvements are not as large as the anticipated reductions in emissions suggested. It has been shown that the health benefits of meeting the NAAQS would be of the order of $2.7 billion just within the South Coast Air Basin of California[18].

8.3   **CONTROL STRATEGIES**

Photochemical ozone can, in principle, be reduced by controlling hydrocarbon and nitrogen oxide emissions. However, the development of an effective strategy for reducing ozone production by controlling anthropogenic emissions of these precursors has proven to be problematic[19]. This is not surprising, in retrospect at least, given the non-linear dependence of ozone production on hydrocarbon and nitrogen oxide emissions (chapter 4). Despite this ineffectiveness, control strategies for ozone have been given extensive consideration throughout the world, particularly in the USA where they have been formulated and enforced over the last 25 years. A significant barrier to progress has been the complexity of the chemistry and transport processes involved.

The EPA's initial efforts to abate ozone formation focused on two fronts: Stationary source emitters and motor vehicles. In particular the control efforts addressed VOC emissions. The CAA Amendments of 1990 reflected a more aggressive emission control philosophy. Like its predecessor, the new law mandates the use of VOC control as the primary approach to ozone reduction but it also calls for NO_x control in addition or in lieu of VOC control in cities for which modelling calculations show this to advantageous [20].

The complex relationships between the precursor species and ozone are often represented by an isopleth diagram. Such isopleths are generally constructed from the results of numerical models (Figure 4.10). Note that the effect on ozone of reducing hydrocarbons and NO_x changes considerably as the concentrations of hydrocarbons and NO_x change. In order to devise an effective strategy for ozone abatement in a given region, it is first necessary to determine where on the ozone isopleth diagram this airshed is located. This is particularly difficult considering the variability in time and space of source and sink terms, and transport. Additionally, the shape of the isopleths generated from model calculations depends upon the formulation of the chemistry and the reactivity of the hydrocarbon mix.

Many different types of photochemical air quality models have been developed [21]. Because of the difficulties in developing and incorporating the complex chemical mechanisms into diffusion-advection models which have been developed to a high level of sophistication for inert pollutants, model development for reactive pollutants has lagged behind that for inert pollutants.

The 1990 CAA Amendment prescribes the use of grid-based photochemical models in developing emissions control plans. Such models were developed in the early 1970s and have undergone continuing evaluation, improvement and refinement [22]. Currently significant deficiencies exist in our knowledge and treatment of the key processes governing oxidant behaviour. For instance, gas phase chemical reactions (of aromatic, carbonyl and longer chain paraffin compounds) are inadequately understood and thus are likely to be inadequately portrayed in extant chemical mechanisms. The use and limitations of these models in regulatory application and policy analysis is now the subject of debate [23]. The Air Resources Board of the California Environmental Protection Agency have produced criteria which they will use to judge the acceptability of a model for control strategy development [24]. These include the soundness of the model's formulation, suitability of the model for the application, validation of the computer code, sensitivity tests and model performance evaluation.

All air quality models require the input of a number of parameters which are currently either poorly defined and/or have large uncertainty. It appears likely that present inventories significantly underestimate the hydrocarbon emissions from mobile sources [25-27].

In the UK, hydrocarbon emission rates are quoted within a range of ±30% but, as was shown in Chapter 3, the current best guess of total hydrocarbon emissions and that for speciated hydrocarbon emissions differ by a factor of 1.6[41]. Additionally areas of uncertainty, of a similar range, exist for the question of speciation. As well as man-made hydrocarbon sources, biogenic emissions have to be treated species by species.

The balance of present evidence is that natural hydrocarbons have no significant part to play in the regional scale photochemical episodes observed over North West Europe[28,29]. However, in certain areas of the USA emissions of biogenic hydrocarbons dominate anthropogenic emissions. In early studies, it was generally concluded that the impact of biogenic hydrocarbons on ozone production was small. Recently this has been shown not to be the case in certain regions. Trainer *et al.*[30] calculated that biogenic hydrocarbons played a dominant role in ozone production over a rural site in Pennsylvania. Chameides *et al.*[31] studied the role of biogenic hydrocarbons in the Atlanta metropolitan area. They found that, even in an urban area, natural emissions can significantly alter ozone levels and their presence can exert a profound influence on the effectiveness of an ozone abatement strategy based on anthropogenic hydrocarbon reductions. This is because the presence of biogenic hydrocarbon emission can result in the region being in the NO_x - limited part of the ozone isopleth diagram (Figure 4.13) rather than in the hydrocarbon-limited part, as would be expected from anthropogenic emissions alone. Similarly Lin *et al.*[32] concluded that biogenic hydrocarbons significantly influence ozone production over rural sites in Ontario, Canada.

The most recent estimate[33] of biogenic emissions for the United States is 28 Tg yr^{-1}. This compares with 20 Tg y^{-1} for anthropogenic hydrocarbon sources. The south eastern and south central portions of the United States account for approximately 43% of the annual US biogenic estimate[33]. It is important to note that the biogenic contribution represents a background hydrocarbon concentration that cannot be removed from the atmosphere by emission control measures. Hence, if, as a result of emission control efforts, anthropogenic emissions are reduced, this background contribution will begin to comprise a larger and more significant fraction of the total hydrocarbon reactivity. It has been predicted that if anthropogenic emissions are totally eliminated, biogenic hydrocarbons will be able to generate ozone concentrations in excess of current guidelines, unless NO_x emissions are controlled[27].

The conclusions concerning ozone control strategies obtained from numerical models depend on how well they accurately reflect what would actually occur in the atmosphere.

357

This is dependent upon the adequacy and completeness of the models themselves as well as the processes represented in them and the simplifications made. The model results do not represent validated predictions of what will happen in the atmosphere in response to changes in emissions. However, they certainly represent our current best guesses.

8.4   **THE VOC PROTOCOL**

Growing concern over photochemical ozone has prompted 21 countries to sign a protocol limiting emissions of VOCs. The Protocol allows for a number of ways for a country to control VOC emissions, given under paragraph 2 (a), (b) and (c) of the Protocol Document.

2 (a)   *It [the country] shall, as soon as possible and as a first step, take effective measures to reduce its national annual emissions of VOCs by at least 30% by the year 1999, using 1988 levels as a basis or any other annual level during the period 1984 to 1990, which it may specify upon signature of or accession to the present Protocol; or*

(b)   *Where its annual emissions contribute to tropospheric ozone concentrations in areas under the jurisdiction of one or more other Parties, and such emissions originate only from areas under its jurisdiction that are specified as TOMAs in annex I [The Lower Fraser Valley, British Columbia; The Windsor-Quebec Corridor, Ontario and Quebec], it shall, as soon as possible and as a first step, take effective measures to:*

(i)   *Reduce its annual emissions of VOCs from the areas so specified by at least 30% by the year 1999, using 1988 levels as a basis or any other annual level during the period 1984-1990, which it may specify upon signature of or accession to the present Protocol; and*

(ii)   *Ensure that its total national annual emissions of VOCs by the year 1999 do not exceed 1988 levels; or*

(c)   *Where its national annual emissions of VOCs were in 1988 lower than 500,000 tonnes and 20 kg/inhabitant and 5 tonnes/km^2, it shall, as soon as possible and as a first step, take effective measures to ensure at least that at the latest by the year 1999 its national annual emissions of VOCs do not exceed 1988 levels.*

On signature of, or accession to, the Protocol, each country should specify which of the above options it wishes to implement, and which base-year it wishes to adopt. Most countries selected 1988 as the base year, although the US chose 1984. The less industrialized nations including Hungary, Bulgaria and the Ukraine committed themselves just to freeze emissions. By April 1992 only 10 out of 21 countries had submitted any form of base-year estimate of VOC and of these only 2 countries has explicitly stated that their VOC emission estimates were for non-methane anthropogenic VOC.

Other obligations of the Protocol are:

No later than two years after the Protocol enters into force,

> - to apply national or international standards to new stationary and mobile sources based on the best available technologies which are economically feasible;

> - to apply national or international measures to products that contain solvents and promote the use of products with low or nil VOC content, including the labelling of products specifying their VOC content;

> - and to foster public participation in VOC emission control programmes through public announcements, encouraging the best use of all modes of transportation and promoting traffic management schemes.

No later than five years after the protocol enters into force, in areas where ozone standards are exceeded or transboundary fluxes originate:

> - to apply best available technologies that are economically feasible to existing stationary sources in the major source categories:

> - and to apply techniques to reduce VOC emissions from petrol distribution and motor vehicle refuelling operations and to reduce the volatility of petrol.

The photochemical ozone creation potential (POCP) approach is used to provide guidance on regional and national control policies for VOCs, taking into account the impact of each VOC species as well as sectoral VOC emissions in episodic ozone formation[34,35]. The POCP is defined as the change in photochemical ozone production due to a change in emission of that particular VOC. POCP values are implicitly dependent on how emission inventories are calculated. This is because POCPs, unlike octane and cetane numbers, are the ratio of two variable parameters - the reactivity of the compound being studied, and the reactivity of the atmosphere into which it is being emitted. Currently there is no consistent method of POCP calculation or information available. Table 8.5 groups VOC species according to their importance in the production of episodic peak ozone concentrations. Importance is expressed on the basis of VOC emission per unit mass. The issue of relative reactivity has already been discussed, from a more chemical standpoint, in Chapter 4. Table 8.6 shows the averaged POCPs for each major source category based on a central POCP

Table 8.5: Classification of VOCs into three groups according to their importance in episodic ozone formation (after ref. 1)

More Important (Group I)

Alkenes	
Aromatics	
Alkanes	> C6 alkanes except 2,3 dimethylpentane
Aldehydes	All aldehydes except benzaldehyde
Biogenics	Isoprene, volatile terpenes

Less Important (Group II)

Alkanes	C3-C5 alkanes and 2,3 dimethylpentane
Ketones	Methyl ethyl ketone and methyl t-butyl ketone
Alcohols	Ethanol
Esters	All esters except methyl acetate

Least Important (Group III)

Alkanes	Methane and ethane
Alkynes	Acetylene
Aromatics	Benzene
Aldehydes	Benzaldehyde
Ketones	Acetone
Alcohols	Methanol
Esters	Methyl acetate
Chlorinated hydrocarbons	Methyl chloroform, Methylene chloride, Trichloroethylene and tetrachloroethylene

estimate for each VOC species in each source category. Many source categories contain a great variety of individual sources and mixtures of hydrocarbons are emitted. Measures to reduce the most reactive VOCs are generally unavailable and reduction measures reduce emissions by mass irrespective of the POCPs.

Vehicles produce significant quantities of hydrocarbons and various legislation is being introduced to control these emissions. More difficult to control are the organic chemicals that evaporate from hundreds of different sources, ranging from large factories to home cleaning fluids. The main sectors and control measures for solvents are listed in Table 8.7. The shortcomings of traditional pollution control methods, such as absorption/incineration, have encourage industry to devise new methods of cleaning the air.

Table 8.6: Sectoral POCPs of the various emission sectors and the percentage by mass of VOCs in each ozone creation class (after ref. 1)

Sector	Sectoral POCP*		Percentage mass in each ozone creation class			
	Canada	United Kingdom	More	Less Important	Least	Unknown
Petrol-engined vehicle exhaust	63	61	76	16	7	1
Diesel vehicle exhaust	60	59	38	19	3	39
Petrol-engined vehicle evaporation	-	51	57	29	2	12
Other transport	63	-	-	-	-	-
Stationary combustion	-	54	34	24	24	18
Solvent usage	42	40	49	26	21	3
Surface coating	48	51	-	-	-	-
Industrial process emissions	45	32	4	41	0	55
Industrial chemicals	70	63	-	-	-	-
Petroleum refining and distribution	54	45	55	42	1	2
Natural gas leakage	-	19	24	8	66	2
Agriculture	-	40	-	-	100	-
Coal mining	-	0	-	-	100	-
Domestic landfill	-	0	-	-	100	-
Dry cleaning	29	-	-	-	-	-
Wood combustion	55	-	-	-	-	-
Slash burn	58	-	-	-	-	-
Food industry	-	37	-	-	-	-

* Relative to ethylene which is given an index of 100

General Dynamics, St. Louis, for instance, use ozone to remove hydrocarbons from the waste stream. Intense uv light breaks down the organic molecules which are then blown over the water and up through a water mist. The water, saturated with reactive oxygen molecules, converts virtually all the organic gases into carbon dioxide and water. Much research is focused on paints and other coatings, which contain VOCs and are among the most polluting products on sale. Paints used in the home tend to use water as the extender,

Table 8.7: VOC-emission Control Measures, Reduction Efficiency and Costs for the Solvent-using Sector (after ref. 1)

Source of emission	Emission control measures	Reduction efficiency	Abatement costs and savings
Industrial surface coating	Conversion to:		
	- powder paints	I	Savings
	- low in/not containing VOCs	I-III	Low costs
	- high solids	I-III	Savings
	Incineration:		
	- thermal	I-II	Medium to high costs
	- catalytic	II-III	Medium costs
	Activated carbon absorption	II-III	Medium costs
Paper surface coating	Incinerator	I-II	Medium costs
	Radiation cure/waterborne inks	I-III	Low costs
Car manufacturing	Conversion to:		
	- powder paints	I	
	- water based systems	I-II	Low costs
	- high solid coating	II	
	Activated carbon absorption	I-II	Low costs
	Incinerated with heat recovery		
	- thermal	I-II	
	- catalytic	I-II	
Commercial painting	Low in/not containing VOCs	I-II	Medium costs
	Low in/not containing VOCS	II-III	Medium costs
Printing	Low-solvent/water-based inks	II-III	Medium costs
	Letterpress: radiation cure	I	Low costs
	Activated carbon adsorption	I-II	High costs
	Absorption		
	Incineration	I-II	
	- thermal		
	-catalytic		
	Biofiltration including buffer filter	I	Medium costs
Metal degreasing	Change-over to systems low in/not containing VOCs	I	
	Closed machines		
	Activated carbon absorption	II	Low to high costs
	Cover, chilled freeboards	III	Low costs
Dry-cleaning	Recovery dryers and good housekeeping (closed cycles)	II-III	Low to medium costs
	Condensation	II	Low costs
	Activated carbon adsorption	II	Low costs
Flat wood panelling	Coatings low in/not containing VOCs	I	Low costs

The process-specific efficiencies are given only qualitatively as follows:

I = > 95%
II = 80-90%
III = < 80%

while industry prefers to use coatings with organic solvents which allow coatings to be applied over a broad range of temperature. One way of reducing emissions is by electrostatic spraying which reduces the amount of paint used. This technique can reduce the volume of evaporating solvent by between 30 and 80%. California intends to make electrostatic painting mandatory for all processes that can use the technique. Solvents can sometimes be dispensed with altogether. One technique is to apply coatings as powders that melt to a smooth finish when heated to between 250 to 300°C. The number of applications of this technique in the US is growing between 10 and 20% per year. Other novel methods include vacuum sealing vehicles in a plastic film that can then be melted permanently into place.

Three direct approaches are proposed for achieving reductions in VOCs within the UK. These are:

(a)   Statutory controls on:
       emissions from stationary sources; emissions from mobile
       sources and the content of materials and products,
(b)   Voluntary action by industry, and
(c)   Public information and participation

Most of the statutory controls result from EEC Directives and the 1990 Environmental Protection Act. The emission limit laid down in the legislation are considerably more stringent than has been the case prior to 1990. The Department of the Environment have investigated the emissions from all the major industries within each of the source categories identified within the VOC Protocol: Solvent use; oil industry; chemical industry; stationary combustion; food industry; iron and steel industry; waste disposal; agriculture; miscellaneous sources and mobile sources. For each specific industry they have estimated the emissions of VOC in 1988, given this figure a confidence limit, identified control standards, speciated total emissions and calculated POCP factor.

The oil industry, as an example, was split into five separate sources: Crude oil production; crude oil distribution; oil refineries; refinery product distribution and 'other' oil industry. Table 8.8 summarises the data reported for oil refineries. It can be seen that VOC emissions from refineries can themselves be divided amongst six distinct sources. The emissions from each separate source are known with reasonable confidence. A reduction of 34% in VOC emissions by 1999 is predicted based on all new plants meeting the new emission limits, existing plants meeting these by 1998, an abatement efficiency of 50%, a turnover rate (annual rate at which existing plants are replaced by new plants) of 10% a^{-1} and a growth rate (annual rate of change in emission assuming no controls) of 2.5% a^{-1}. Possible control techniques are also listed. Similar information is available for another 30 stationary sources of VOCs.

Table 8.8: VOC emissions from oil refineries

Source/process	1988 emission (kt)	data confidence[*]
Fugitive losses	19	C
Waste water	14	C
Point sources	5	C
Cooling water	3	C
Flaring	<1	C
Storage tanks	28	C
Total	69	

Compound class	% wt	Major components involved
Alkanes	>98	ethane, butane, propane, pentane
Aromatics	1	toluene, xylene, benzene
Alkenes	<1	propene

UN ECE groups: 2% group I; 73% group II; 25% group III
POCP Factor: 43

Control techniques
    Preventative maintenance
    Leakproof welded connections
    Improved seal quality
    Mechanical seals, double action seals
    Waste gas reclamation for fuel use
    Floating covers on waste water separators
    Floating roof tanks fitted with secondary rim
     seats for all volatile product storage
    Fixed roof tanks fitted with an integral floating
     roof or be linked to a vapour recovery unit
    Good maintenance procedures to reduce emissions from
     storage tank fittings and seals
    Paint tanks white to reduce heat absorbance
    Vapour recovery during all transfer operations

[*] Data confidence ranges from E (lowest) to A (highest)

Table 8.9: Ways the public can reduce air pollution

1. Try not to use the car

2. Switch off the engine if you are stuck in a traffic jam

3. Do not warm the engine or use choke unnecessarily

4. Keep the car well maintained

5. Do not park in direct sunlight while the engine is hot

6. Avoid overfilling petrol tanks

7. Use water-based or low-solvent paints etc.

8. Put off decorating jobs when poor air quality is forecast

9. Keep the lid on the tin of paint as much as possible

10. Do not spray solvent based paints

11. Limit the use of household products that contain hydrocarbons e.g. furniture polish, nail varnish, shaving cream

12. Do not light bonfires when the air quality is poor

The Department of the Environment has also produced an information leaflet (92EP0015) entitled 'Summertime Smog' and tells people how they can help towards reducing air pollution. Advice is given on action of the road, around the home and in the garden. Some of the suggestions are listed in Table 8.9. While the advice is sound most of it (with the exception of reduced car use) is extremely unlikely to have a major impact on air quality. Public health officials in Los Angeles intend to regulate nearly every source of VOCs. Both volatile fuel used to ignite barbecues (which add up to 2 tons of hydrocarbons a day) and petrol driven lawn movers have been banned. Similarly anti-perspirant sprays are being phased out.

The results of a 30% reduction in annual VOC emissions between 1988 and 1999 over Europe has been carried out by Simpson and Styve[36]. Model simulations for the period April - September 1989 predict a reduction in mean ozone of around 4-8% in north west Europe, with reductions of 1-4% elsewhere; the frequency with which ozone levels exceed 40 ppbv will be reduced by 15-20% in north west Europe, 5-15% in western Europe and less than 5% east of 20° longitude; and ozone levels in excess of 75 ppbv will be reduced by 40-60% in most areas. Despite the many uncertainties in this study, its findings indicate that implementation of the VOC Protocol should have a significant impact on ozone concentrations.

8.5   CONTROLLING EMISSIONS FROM MOTOR VECHICLES

The number of cars in the world is approximately 400 million and this figure is increasing rapidly with a 25% increase between 1981 and 1988 (Figure 8.1). VOC emissions from the transport sector constituted roughly 34% of the total anthropogenic

Figure 8.1

Actual and predicted growth in the global population of vehicles, excluding motorised 2 and 3 wheelers (After Faiz[59]).

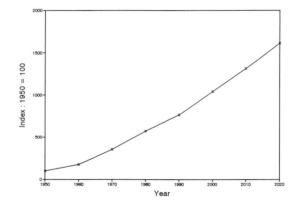

emissions in the US[37] and 41% in the UK[38], both in 1990. Pending the introduction of new engine designs, better fuels and alternative transport systems, the countries of the world are trying to alleviate the adverse effects of road transport by adopting increasingly stringent limits on vehicle exhausts.

The legislative history of vehicle emission controls began in California in 1959 with the adoption of state standards to control exhaust hydrocarbons and carbon monoxide. Federal standards for CO and hydrocarbons began in 1968, and by 1970 the year in which major amendments to the Clean Air Act were enacted, the automobile industry had reduced hydrocarbon emissions by almost 75%. The CAA amendments of 1970 required that

Table 8.10: US light-duty motor vehicle standards (after Calvert et al.[39]).

The emission rates for THC, CO and NO_x are by (or adjusted to the equivalent of) the 1975 Federal Test Procedure. US standards are specified in imperial units, but for consistency, are given here metricated.

Abbreviations: Evap., evaporative HCs

Model year	Federal				California			
	THC (g/km)	CO (g/km)	NO_x (g/km)	Evap (g/test)	THC (g/km)	CO (g/km)	NO_x (g/km)	Evap (g/test)
Pre-control	6.63+	52.5	2.56	47	6.63+	52.5	2.56	47
1966					3.94	31.88	(3.75)	
1968	3.94	31.88	(3.75)#		3.94	31.88		
1970	2.56	21.25			2.56	21.25		6
1971	2.56	21.25			2.56	21.25	2.5	6
1972	1.88	17.5			1.81	21.25	1.82	2
1973	1.88	17.5	1.85		1.81	21.25	1.82	2
1974	1.88	17.5	1.85		1.81	21.25	1.25	2
1975	0.94	9.4	1.94*	2	0.56	5.63	1.25	2
1977	0.94	9.4	1.25	2	0.26	5.63	0.94	2
1978	0.94	9.4	1.25	6*	0.26	5.63	0.94	6*
1980	0.26	4.4	1.25	6	0.24‖	5.63	0.63	2
1981	0.26	2.13	0.63	2	0.24¶	4.38	0.44	2
1983	0.26	2.13	0.63	2	0.24	4.38	0.25^	2
1984	0.26	2.13	0.63	2	0.24	4.38	0.25	2
1985	0.26	2.13	0.63	2	0.24	4.38	0.25	2
1986	0.26	2.13	0.63	2	0.24	4.38	0.25	2
1987	0.26	2.13	0.63	2	0.24	4.38	0.25	2
1989	0.26	2.13	0.63	2	0.24	4.38	0.25	2
1990	0.26	2.13	0.63	2	0.24	4.38	0.25	2
1991	0.26	2.13	0.63	2	0.24	4.38	0.25	2
1992	0.26	2.13	0.63	2	0.24	4.38	0.25	2
1993	0.26	2.13	0.63	2	0.16	2.13	0.25	2

+ Crankcase emissions of 2.56 g/km not included; fully controlled. # Emissions of NO_x (no standard) increased with control of HCs and CO. * Change in test procedure. ‖ Nonmethane HC standard (or 0.26 g/km for total HCs). ¶ Optional HC standard, 0.19 g/km, requires 7-year or 75,000 mile limited recall authority. ^ A 0.7 NO_x optional standard for 1983 and later, but requires limited recall authority for 7 years or 70,000 miles.

emissions from automobiles be reduced by 90% compared to levels already achieved by the 1970 model year. These standards were to be reached by 1975 for CO and hydrocarbons and by 1976 for NO_x. These amendments were unusual in that they were 'technology forcing'. Further reductions have been introduced continuing the downward trend in standards for new car exhaust emissions. These are summarized in Table 8.10. The economic costs of these vehicle emission standards is considerable. It is estimated that the total cost of emission control technology for 1990 light-duty vehicles sold in California is about $698 million[60]. The corresponding cost to consumers per vehicle for emission control ranges from $370 to $2430, with a sales weighted average of $748. The 1990 CAA Amendments require vehicles built in the 1994 model year to emit an average of 0.156g/km of hydrocarbons and 0.25g/km NO_x after 5 years or 50,000 miles. Currently 83% of vehicle VOC emissions come from cars produced before 1981, even through they account for only 40% of the distances travelled. Normal turnover of the fleet is 10% per year, and so by 2000 all the cars on the road in the US will be fitted with technology to control emissions. However, although new production models meet the emission requirements, in actual use, performance is not so good. Measurements of emissions of in use vehicles at 50,000 miles indicated that the standards for CO and hydrocarbons may be exceeded by a factor of $2^{[39]}$. Tampering with control equipment has also been found to be a problem[40].

In June 1991, the European Community adopted Directive 91/441/EEC commonly known as the Consolidated Cars Directive. The new limits are compared with the preceding limits (83/351/EEC also known as ECE Regulation 15.04) in Table 8.11. To meet these new limits almost all new petrol engined cars from 31 December 1992 will be required to use three way catalytic converters. The increasingly stringent limits have been estimated to lead to major reductions in national emissions of NO_x, CO and VOCs from road transport in the UK of 53%, 66% and 68% respectively from 1990 to 2010[41].

In other parts of the world, too, vehicle exhausts are causing growing problems. Several of the industrial countries outside North America and Europe are beginning to introduce stringent emission standards. Japan has for many years applied exhaust standards for cars comparable to forthcoming US rules, and countries such as Australia, Brazil, South Korea, Taiwan and Mexico have set successively lower limit values (Table 8.12).

There are four general approaches to reduce motor vehicle emissions[39,41]: (i) Technological improvements and modifications to the motor vehicle; (ii) the use of fuels that reduce mass emissions of pollutants or result in emissions that are less reactive in the atmosphere; (iii) inspection and maintenance programmes and (iv) transportation control

Table 8.11: EC Directive limits for light duty vehicles

Directive	Reference Weight Min kg	Max kg	Dates Type Approval	Entry into Service	Type Approval CO g/km	HC+NO$_x$ g/km	Partic. g/km	Idle CO%	Conformity of Production CO g/km	HC+NO$_x$ g/km	Partic. g/km	Idle CO%	Comments
83/351/EEC	0	1020	1 Oct 82	1 Oct 85	14.31	4.69		3.5	17.28	5.87		4.5	EC urban cycle
83/351/EEC	1021	1250	1 Oct 82	1 Oct 85	16.54	5.06		3.5	19.74	6.32		4.5	EC urban cycle
83/351/EEC	1251	1470	1 Oct 82	1 Oct 85	18.76	5.43		3.5	22.46	6.79		4.5	EC urban cycle
83/351/EEC	1471	1700	1 Oct 82	1 Oct 85	20.73	5.80		3.5	24.93	7.26		4.5	EC urban cycle
83/351/EEC	1701	1930	1 Oct 82	1 Oct 85	22.95	6.17		3.5	27.64	7.72		4.5	EC urban cycle
83/351/EEC	1931	2150	1 Oct 82	1 Oct 85	24.93	6.4		3.5	29.86	8.17		4.5	EC urban cycle
83/351/EEC	2151	3500	1 Oct 82	1 Oct 85	27.15	6.91		3.5	32.58	8.64		4.5	EC urban cycle
91/441/EEC	0	3500	1 Jul 92	31 Dec 92	2.72	0.97	0.14	3.5	3.16	1.13	0.18	4.5	
93/???/EEC	0	3500	1 Jan 96		2.2	0.5	0.08	Gasoline vehicle					
					1.0	0.7	0.08	IDI Diesel					

369

Table 8.12: Emission limit values : cars

Country	Year	NO$_x$	(g/km) HC	CO	Comments
Sweden Class 3	1993	0.62 0.75(non-urban)	0.25	2.1	Standards must be met for 5 years or 80,000 km - environmental class 3
Class 2	1993	0.25	0.25(methane)	2.1	Entire life: 10 years or 160,000 km
		0.33 (non-urban) 0.37 (entire life)	0.16 (NMHC) 0.19 (entire life)		
Class 1	1993	0.25	0.25	2.1	New requirement: 6.2g/km CO when started from cold
		0.33 0.37	0.078 0.097	2.6 (entire life)	
Japan	1978	0.25	0.25	2.1	
Australia	1986	1.93	0.93	9.3	
Brazil	1990 1992 1997	2.0 1.4 0.6	2.1 1.2 0.3	24 12 2	Applies to manufactures who sell than 2000 cars of any model or total of more than 5000 cars a year
Hong Kong	1992	0.63	0.26	2.10	
Mexico	1990 1991 1993	2.00 1.40 0.62	1.80 0.70 0.25	18.00 7.00 2.11	
South Korea	1987	0.62	0.25	2.11	
Taiwan	1990	0.02	0.25	2.11	

measures. The approach favoured by most automobile manufacturers to achieve emission standards has been the three way catalytic converter[43,44]. Catalyst technology has improved tremendously over the past 15 years and the converter has become something of a milestone in the search for cleaner transport. Recently research has concentrated on reducing the high emissions that occur before the vehicles engine has warmed up. Various car manufacturers are investigating other areas such as lean burn, gas turbine and two stroke engines. The sale of zero emission vehicles (ZEV) in California has been mandated to being in 1998 at the rate of 2% of the annual light-duty vehicle sales and to rise to 10% of these rates by 2003.

The ZEV requirement is tantamount to a mandate for electric vehicles, the principal technology available in the near term to meet the ZEV requirement. The widespread success of electric vehicles depends primarily on the development of advanced batteries that can deliver improved driving range and performance[45,46]. Car manufacturers are keen to point out that the ZEV concept simply diverts emissions from vehicle tailpipes to power station chimneys[47], but the fact remains that point source emissions will be easier to control than widespread emissions from cars.

The 1990 CAA Amendments mandate a clean fuel vehicle and in 1995, the sale of reformulated petrol in the nine worst ozone non-attainment areas. The reformulated fuel should meet the following conditions: Minimum oxygen content (2%), maximum benzene concentration (1% by volume), maximum aromatics content (25% by volume) and reduced vehicular VOC and toxic emissions (15% in 1995 and 25% in 2000). Clean fuel covers a wide range of fuels such as hydrogen, electricity, methanol, propanol or reformulated petrol. These have been discussed in detail a document by the National Research Council[48]. The two changes in fuel properties that appear to have a widespread beneficial effect on emissions from light duty vehicles are reductions in sulphur content and vapour pressure. Reducing sulphur increases catalyst efficiency whereas reduced vapour pressure has a direct effect on the generation of vapour in fuel systems. Evaporative emissions can make up 50% or more of VOC emissions from cars, especially on hot days.

Methanol can be used as motor fuel either on its own or mixed with petrol. The high latent heat of vaporisation of alcohols gives the advantage that charge temperatures before combustion are reduced with a consequential reduction in NO_x formation. As a fuel methanol is not ideal since it is less volatile than petrol, burns with an invisible flame, has half the energy content of petrol and when burnt emits its own pollutant -formaldehyde. To solve these problems manufacturers have commenced an extensive research and development programme. Vehicles fuelled on a mixture of 85% methanol and 15% petrol are the focus of much attention. A number of photochemical modelling studies have been made to estimate the impact of methanol fuelled vehicles on urban ozone[49-52]. Results show that methanol vehicles have a much lower ozone forming potential if the molar non-methane organic compound (as C)/NO_x ratio is equal to 8. At much higher ratios (~20) the ozone forming potential is higher than petrol fuelled vehicles. This is because at high NMHC/NO_x ratios ozone production is NO_x limited and methanol oxidation does not remove NO_x by the formation of organic nitrate, unlike some gasoline fuel components[52]. It appears that the concept of using alternative fuels to control photochemical air pollution is one which is useful close to major urban areas, but will have little effect in downwind areas. Hence the use of alternative fuels, as indeed conventional petrol, must be combined with NO_x control if significant reductions in ozone are to be attained.

Ethanol can be produced from biomass and gives rise to lower emissions of CO_2, ozone forming and toxic substances than petrol. However at present ethanol is almost twice as expensive as petrol in relation to the amount of energy obtained from it. Brazil has by far the largest biofuel programme for motor vehicles in the world[53]. The programme is based on the production of ethanol by fermentation and distillation of sugar cane juice, and is used in its pure form to fuel approximately 4 million cars and vans in the country. The remaining 7 million light vehicles are fuelled with a gasoline and ethanol blend (gasohol) which varies from 10 to 22% of anhydrous ethanol in the mixture.

While options differ widely on the economic and technological advantages and disadvantages of methanol and ethanol. it is generally agreed that formaldehyde and acetaldehyde, the combustion products in the exhaust gases, are much more reactive in polluted atmospheres than their parent alcohols[54]. Additionally, both aldehydes are toxic and possible animal carcinogens. Measurements in Brazil have revealed high concentrations of aldehydes, along with high ratios of acetaldehyde to formaldehyde; about double those observed in major urban areas in other countries[55].

The high energy density and low cost of petrol, extensive production and distribution infrastructure, and availability of effective technologies to control emissions make competition from alternative fuels difficult. In fact Steadman[56] concludes that ten times the improvement in emissions from fuel reformulation could be obtained by identification and repair of 10% of the vehicle fleet. A benefit cost analysis of vehicle emission control alternatives has concluded that VOC can be abated most efficiently by adopting a mixed strategy of reduced vapour pressure control and so-called stage II controls[57]. Vehicle based controls are not only more expensive per gram of hydrocarbon emissions abated than the above combination but they take longer to implement. Stage II controls are equipment installed at petrol stations and onboard controls installed in the motor vehicle. A Stage II control system uses a specially designed petrol dispensing nozzle to collect vapour displaced during refuelling. Such systems have been certified to reduce displaced vapours by at least 95% although inadequate maintenance by station operators can reduce this figure to 64%. Currently, Stage II refuelling controls are in use in most of California, in St. Louis, and in Washington, DC, and are mandated in portions of New Jersey and New York. Stage II onboard refuelling systems route displaced vapours to a carbon absorber on the vehicle.

When the vehicle is in operation, the carbon is regenerated by a reverse flow air sweep through the carbon canister and the vapours are then passed into the engine's air intake and burned. Such onboard systems have been demonstrated to achieve 94% in-use efficiencies[58], although the canisters can become saturated with vapour if the car is left standing in hot weather.

1. Protocol to the 1979 Convention on Long-Range Transboundary Air Pollution Concerning the Control of Volatile Organic Compounds or their Transboundary Fluxes, HMSO, Command Paper 1970.

2. US Environmental Protection Agency (1979) National primary and secondary ambient air quality standards, Fed. Reg., 44, 8202-8233.

3. Chock, D.P. (1991) Issues regarding the ozone air quality standards, J. Air Waste Manage. Assoc., 41, 148-152.

4. World Health Organization (1987) Air quality guidelines for Europe, WHO Regional Publications, No. 23, 7.

5. Lee, S.D., Kaweck, J.M. and Tannahill, G.K. (1985) Evaluation of the scientific basis for ozone/oxidants standards. J. Air Pollut. Control Assoc., 35, 1025-1032.

6. Rombout, P.J.A., Lioy, P.J. and Goldstein, B.D. (1986) Rationale for an eight hour ozone standard. J. Air Pollut. Control Assoc., 36, 913-917.

7. Lioy, P.J. and Dyba, R.V. (1989) Tropospheric ozone : the dynamics of human exposure. Tox. Indust. Health, 5, 493-504.

8. Chock, D.P. (1989) The need for a more robust ozone air quality standard, J. Air Pollut. Control Assoc., 39, 1063-1072.

9. Van Bree, L., Lioy, P.J., Rombout, P.J.A. and Lippmann, M. (1990) A more stringent and longer term standard for tropospheric ozone : emerging new data on health effects and potential exposure, Tox. Applied Pharm., 103, 377-382.

10. Fairley, D. and Blanchard, C.L. (1991) Rethinking the ozone standard. J. Air Waste Manage. Assoc., 41, 928-936.

11. Suri, M. (1988) Tighter ozone standard urged by scientists, Science, 240, 1724-1725.

12. Burke, B. (1987) Smog : its nature and effects. EPA Journal, 13, 9-10.

13. EPA (1991) National Ambient Air Quality and Emissions Trend Report, 1989, EPA Report 450-R-91-003. US EPA, Research Triangle Park.

14. EPA/OAQPS (1991) Ozone and Carbon Monoxide Areas Designated Nonattainment. US EPA, Office of Air Quality Planning and Standards, Research Triangle Park.

15. Berglund, R.L., Dittenhoefer, A.C., Ellis, H.M., Watts, B.J. and Hansen, J.L. (1988) Evaluation of the stringency of alternative forms of a national ambient air quality standard for ozone. In : *The Scientific and Technical Issues Facing Post-1987 Ozone Control Strategies*, G.T. Wolff, J.L. Hanisch and K. Schere (Eds.), Air and Waste Management Association, Pittsburgh, 343-369.

16. OTA (1989) Catching our breath : next steps for reducing urban ozone. Congress of the US Office of Technology Assessment, OTA-0-412.

17. EPA (1991) National Ambient Air Quality and Emissions Trend Report, 1991, EPA Report 450-R-92-001.

18. Hall, J.V., Winer, A.M., Kleinman, M.T., Lurmann, F.W., Brajer, V. and Colome, S.D. (1992) Valuing the health benefits of clean air, Science, 255, 812-817.

19. Lindsay, R.W., Richardson, J.L. and Chameides, W.L. (1989) Ozone trends in Atlanta, Georgia : have emission controls been effective? J. Air Pollut. Control Assoc., 39, 40-43.

20. Helms, G.T., Vitas, J.B. and Nikbakht, P.A. (1993) Regulatory options under the US Clean Air Act: The Federal View. Water Air Soil Pollut., 67, 207-216.

21. Seinfeld, J.H. (1988) Ozone air quality models : a critical review, J. Air Pollut. Control Assoc., 38, 616-645.

22. Russell, A. and Odman, M.T. (1993) Future directions in photochemical air quality modelling. Water Air Soil Pollut., 67, 181-193.

23. Roth, P.M. (1993) Uses and limitations of modelling in analyzing and designing ozone strategies. Water Air Soil Pollut., 67, 195-203.

24. Air Resources Board (1992) Technical Guidance Document: Photochemical Modelling, California Air Resources Board, Technical Support Division.

25. Battye, W. (1993) Trends and uncertainties in volatile organic compound emissions from anthropogenic sources. Water Air Soil Pollut., 67, 47-56.

26. Pierson, W.R., Gertler, A.W. and Bradow, R.L. (1990) Comparison of the SCAQS tunnel study with on-road vehicle emission data. J. Air Pollut. Control Assoc., 40 1495-1504.

27. Chameides, W.L., Fehsenfeld, F., Rodgers, M.O., Cardelino, C., Martinez, J., Parrish, D., Lonneman, W., Lawson, D.R, Rasmussen, R.A., Zimmerman, P., Greenberg, J., Middleton, P. and Wang, T. (1992). Ozone precursor relationships in the ambient atmosphere, J. Geophys. Res., 97, 6037-6055.

28. MacKenzie, A.R., Harrison, R.M., Colbeck, I. and Hewitt, C.N. (1991) The role of biogenic hydrocarbons in the production of ozone in urban plumes in southeast England. Atmos. Environ., 25A, 351-359.

29. Simpson, D. (1992) Long-period modelling of photochemical oxidants in Europe, model calculations for July 1985. Atmos. Environ., 26, 1609-1634.

30. Trainer, M., Buhr, M.P., Curran, C.M., Fehsenfeld, F.C., Hsie, E.Y., Liu, S.C., Norton, R.B., Parrish, D.D. and Williams, E.J. (1991) Observations and modelling of reactive nitrogen photochemistry at a rural site. J. Geophys. Res., 96, 3045-3063.

31. Chameides, W.L., Lindsay, R.W., Richardson, J. and Kiang, C.S. (1988) The role of biogenic hydrocarbons in urban photochemical smog : Atlanta as a case study. Science, 241, 1473-1475.

32. Lin, D., Melo, O.T., Hastie, D.R., Shepson, P.B., Niki, H. and Bottenheim, J.W. (1992) A case study of ozone production in a rural area of central Ontario. Atmos. Environ., 26A, 311-324.

33. Novak, J.H. and Pierce, T.E. (1993) Natural emissions of oxidant precursors, Water Air Soil Pollut., 67, 57-77.

34. Derwent, R.G. and Jenkin M.E. (1991) Hydrocarbons and the long range transport of ozone and PAN across Europe. Atmos. Environ., 25A, 1661-1678.

35. Simpson, D. (1992) Long period modelling of photochemical oxidants in Europe. EMEP MSC-W Note 1/92.

36. Simpson, D. and Styve, H. (1992) The effects of the VOC Protocol on ozone concentrations in Europe.

EMEP MSC-W Note 4/92.

37. EPA (1990) National Air Quality and Emission Trends Report, EPA-450/4-91-023, US Environmental Protection Agency, Research Triangle Park.

38. The UK Environment (1992), HMSO, London.

39. Calvert, J.G., Heywood, J.B., Sawyer, R.F. and Seinfeld, J.H. (1993). Achieving acceptable air quality : some reflections on controlling vehicle emissions, Science, 261, 37-45.

40. Black, F.M. (1988). Motor vehicles as sources of compounds important to tropospheric and stratospheric ozone. EPA 600/0/88/067. US Environmental Protection Agency. Research Triangle Park, NC.

41. Department of the Environment. (1993) Urban Air Quality in the United Kingdom, Department of the Environment.

42. Watkins, L.H. (1991). *Air Pollution From Road Vehicles*. HMSO London.

43. Acres, G.J.K. (1990) Catalyst systems for emission control from motor vehicles. In : *Pollution: Causes, Effects and Control*, R.M. Harrison (Ed.), Royal Society of Chemistry, 221-236.

44. Flagan, R.C. and Seinfeld, J.H. (1988) *Fundamentals of Air Pollution Engineering*, Eaglewood Clifts, N.J.

45. They're New! They're Clean! They're Electric! (1991) EPRI Journal, April/May, 4-15.

46. The Push for Advanced Batteries. (1991) EPRI Journal, April/May, 16-19.

47. Saje, A. (1993). Vehicle emissions, the consumer and industry ; finding the balance. Sci. Total Environ., 134, 353-359.

48. National Research Council (1991). Rethinking the ozone problem in urban and regional air pollution, National Academy Press, Washington, DC.

49. Chang, T.Y. and Rudy, S.J. (1990) Ozone-forming potential of organic emissions from alternative-fuelled vehicles, Atmos. Environ., 24, 2421-2430.

50. Russell, A.G., St. Pierre, D. and Milford, J.B. (1990) Ozone control and methanol fuel use, Science, 247, 201-205.

51. Kaiser, E.W., Siegl, W.O., Heng, Y.I., Anderson, R.W. and Trinker, F.H. (1991) Effect of fuel structure on emissions from a spark-ignited engine, Environ. Sci. Technol., 25, 2005-2011.

52. Dunker, A.M. (1990) Relative reactivity of emissions from methanol fuelled and gasoline fuelled vehicles in forming ozone, Environ. Sci. Technol., 24, 853-862.

53. Boddey, R. (1993) Green energy from sugar cane, Chemistry and Industry, May, 355-358.

54. Finlayson-Pitts, B.J. and Pitts, J.N. (1993) Volatile organic compounds : ozone formation, alternative fuels and toxics, Chemistry and Industry, October, 796-800.

55. Tanner, R.L., Miguel, A.H., Andrade, J.B. de, Gaffney, J.S. and Streit, G.E. (1988) Atmospheric chemistry of aldehydes : enhanced peroxyacetyl nitrate formation from ethanol fuelled vehicular emission, Environ. Sci. Technol., 22, 1026-1034.

56. Steadman, D. (1992) Comments on 'Speciated measurements and calculated reactivities of vehicle exhaust emissions from conventional and reformulated gasolines'. Environ. Sci. Technol., 26, 2036.

57. Lave, L.B., Wecker, W.E., Reis, W.S. and Duncan, A.R. (1990) Controlling emissions from motor vehicles, Environ. Sci. Technol., 24, 1128-1135.

58. Control of air pollution from new motor vehicles and new motor vehicle engines; refuelling emission regulations for gasoline-fuelled light-duty vehicles and trucks and heavy-duty vehicles; notice of proposed rule-making. (1987) Federal Register, 52(160), US Government Printing Office, Washington DC, 31162-31271.

59. Faiz, A. (1993) Automotive emissions in developing countries - relative implications for global warming, acidification and urban air quality, Transpn. Res., 27A, 167-186.

60. Wang, Q., Kling, C. and Sperling, D. (1993) Light-duty vehicle exhaust emission control cost estimates using a part-pricing approach, J. Air Waste Manage. Assoc., 43, 1461-1471.